Design and Manufacturing: An Industry 4.0 Perspective

Design and Manufacturing: An Industry 4.0 Perspective

Guest Editors

Panagiotis Kyratsis
Angelos P. Markopoulos
Henrique de Amorim Almeida
Tatjana Spahiu

MDPI

Basel • Beijing • Wuhan • Barcelona • Belgrade • Novi Sad • Cluj • Manchester

Guest Editors
Panagiotis Kyratsis
Product and Systems
Design Engineering
University of Western
Macedonia
Kila Kozani
Greece

Angelos P. Markopoulos
School of Mechanical
Engineering
National Technical University
of Athens
Athens
Greece

Henrique de Amorim
Almeida
School of Technology
and Management
Polytechnic Institute of Leiria
Leiria
Portugal

Tatjana Spahiu
Textile and Fashion
Department
Polytechnic University
of Tirana
Tirana
Albania

Editorial Office
MDPI AG
Grosspeteranlage 5
4052 Basel, Switzerland

This is a reprint of the Special Issue, published open access by the journal *Machines* (ISSN 2075-1702), freely accessible at: https://www.mdpi.com/journal/machines/special_issues/Design_Manufacturing_Industry_Perspective.

For citation purposes, cite each article independently as indicated on the article page online and as indicated below:

Lastname, A.A.; Lastname, B.B. Article Title. *Journal Name* **Year**, *Volume Number*, Page Range.

ISBN 978-3-7258-6428-7 (Hbk)
ISBN 978-3-7258-6429-4 (PDF)
https://doi.org/10.3390/books978-3-7258-6429-4

Contents

About the Editors

Panagiotis Kyratsis

Panagiotis Kyratsis is Professor at the Department of Product and Systems Design Engineering, University of Western Macedonia, Greece. He is the Director of the Institute of Traditional Architecture and Cultural Heritage, University Research Center "TEMENUS", and the Director of the Computational Design and Digital Fabrication Research Lab (CODE+).

Prof. Panagiotis Kyratsis received his PhD in CAD-based manufacturing process simulations from the Department of Production Engineering and Management, Technical University of Crete, Greece. He holds a diploma in Mechanical Engineering from the Aristotle's University of Thessaloniki, Greece, and he received an M.Sc. in Automotive Product Engineering and an M.Sc. in CAD/CAM from Cranfield University, UK.

He has been involved in several industrial projects and has a great deal of expertise in both the design and the manufacturing aspects of product development. His main research interests include manufacturing/machining, CAD/CAM/CAE systems, computational design, Artificial Intelligence (AI), additive manufacturing (AM), CAD-based applications, product design, reverse engineering, and prototyping. He has published more than 20 books and more than 200 papers in scientific journals and international conferences. He acts as member of the editorial board and reviewer for numerous scientific journals and holds 12 industrial designs and two patents registered within the Greek Patent Office.

Angelos P. Markopoulos

Angelos P. Markopoulos PhD is a Professor at the School of Mechanical Engineering of the National Technical University of Athens (NTUA), Greece. Since 2007, he has worked as a University Professor at several institutes in Greece, and in 2014 he joined the staff of NTUA. He belongs to the editorial boards of several scientific journals and the committees of various international conferences, and he is the PI of many projects funded by the EU. Furthermore, he has fulfilled more than 20 Erasmus+ mobility teaching agreements in various countries in Europe. His research includes topics such as precision and ultraprecision machining processes with special interest in high-speed hard machining, grinding, micromachining and additive manufacturing. Furthermore, he is an expert in manufacturing technology modelling and simulation, including the Finite Element Method, Artificial Intelligence, and Molecular Dynamics. He is the author of more than 200 papers in journals, conferences, and book chapters on the above-mentioned areas and he is the author or editor of 6 books published by Springer and CRC Press. Finally, he was included in the top 2% scientists that are the most influential in their field of expertise, according to the Elsevier and Stanford survey.

Henrique de Amorim Almeida

Henrique de Amorim Almeida is an Associate Professor at the Mechanical Engineering Department of the School for Technology and Management (ESTG) at the Polytechnic Institute of Leiria (IPL). He graduated in Mechanical Engineering at the School for Technology and Management (ESTG). He then went to the Superior Institute of Technology of the Technical University of Lisbon to obtain his Master of Science in Engineering Design. The title of his master's thesis was "Computational Mechanic Design and Optimization of Scaffolds for Tissue Engineering". He then obtained his PhD in Mechanical Engineering from the University of Aveiro with the thesis entitled "Smart design of scaffolds obtained by biofabrication for tissue engineering applications". At ESTG, he lectures on

Antropometrics, Ergonomics and Human Factors, Biomechanical Design, Additive Manufacturing, Biomanufacturing and Mechanical Technology, amongst others. He also participates and has participated in several research projects funded by the following agencies: Portuguese Foundation for Science and Technology, Ciencia Y Tecnologia Para El Desarrollo (CYTED), Portuguese Agency of Inovation (ADI), PRODER, IAPMEI, British Council, European Commission Projects, and the People Marie Curie Actions Program of the International Research Staff Exchange Scheme of the European Commission. He was the Series Editor of the book series "*Springer Tracts in Addititive Manufacturing*" published by Springer Nature. He has also served as a member of the editorial board of journals from Springer, Emerald, and Crimson Publishers and a member of the review editorial board of several journals published by Frontiers. Due to his professional experience in the industry, his technical competences and skills are 3D CAD modelling in several CAD software systems, computer programming, industrial design (reverse engineering and product development) and injection mold design for plastic parts.

Tatjana Spahiu

Tatjana Spahiu is a Lecturer at the Textile and Fashion Department, where she has gained more than 15 years of experience. Her field of expertise is related to the use of digital tools for the garment and footwear industry, including garment and footwear modeling based on computer-aided design. These include the modelling and simulation of garments and footwear and direct manufacturing from digital data (CAD/CAM/CAE), and the engineering and reverse engineering of garments and footwear prototyping. She has been part of different trainings and workshops organized by USAID focused on creating effective professional practice, entrepreneurship, leather, footwear, etc. She has presented and published more than 60 papers in various international scientific conferences and journals. She is also a member and reviewer for scientific journals and conferences. She has been a trainer for several Erasmus+ projects in Romania, Estonia, Bulgaria, Germany, Greece, etc. She has been a member of international and national project teams such as Competitive Enterprise Development in Albania (USAID), Quality and Equity in High Education–Improvement of the CAD/CAM Laboratory at the Department of Textile and Fashion for Garment Design and Production (World Bank, European Bank for Development and Albanian Government), Digi4Wearables Call: ERASMUS-EDU-2022-CB-VET (Capacity building in the field of Vocational Education and Training (VET) (January 2023 December 2025).

machines

Editorial

Design and Manufacturing: An Industry 4.0 Perspective

Panagiotis Kyratsis [1,*], Angelos P. Markopoulos [2], Henrique de Amorim Almeida [3] and Tatjana Spahiu [4]

[1] Department of Product and Systems Design Engineering, University of Western Macedonia, 50100 Kila Kozani, Greece
[2] Section of Manufacturing Technology, School of Mechanical Engineering, National Technical University of Athens, 15780 Athens, Greece; amark@mail.ntua.gr
[3] Escola Superior de Tecnologia e Gestão de Leiria, 2411-901 Leiria, Portugal; henrique.almeida@ipleiria.pt
[4] Faculty of Mechanical Engineering, Polytechnic University of Tirana, 1001 Tirana, Albania; tspahiu@fim.edu.al
* Correspondence: pkyratsis@uowm.gr

Industrial investments include the use of advanced technologies which refine, accelerate, improve quality, and boost stakeholder profitability. As part of the Industry 4.0 strategy, a number of methods and technologies can be applied in order to provide a solid basis for improvements in all aspects of a product's lifecycle, including the following:

- Computer-Aided Design and Manufacturing (CAD/CAM) system integration and Computational Design Engineering (CDE);
- Implementation of Flexible Manufacturing Systems (FMSs) with improved performance based on appropriate parameter optimization, as well as Discrete Event Simulation (DES), computer-based simulations which support FMS implementation, application of digital twin (DT) technology for real-time optimization, and Industry 5.0 transformation.
- Additive Manufacturing (AM) and 3D printing technologies are attracting interest in numerous design and manufacturing engineering areas, including textiles, fashion-related product design, robotic arms, furniture design and manufacturing, etc.
- Business models, the Circular Economy (CE), and Human-Centric Design are also important aspects, along with topology optimization (TO) and finite element analysis (FEA).
- Human–machine interaction using different technologies, i.e., Internet of Things (IoT), Virtual and Augmented Reality (VR and AR), Artificial Intelligence (AI), and Genetic Algorithms (GAs), is also a key consideration.

All of the above factors, among others, offer considerable opportunities for transforming the traditional ways of product design and manufacturing towards more innovative computer-based ways of working. The processes of design and manufacturing have undergone great change, and so to has the work of researchers, engineers, and academics regarding the incorporation of high-end applications into all stages of the product lifecycle (PL), helping to shape the future of industry by creating both new opportunities within specific sectors and new challenging demands [1–3].

A novel conceptual classification based on a multi-objective simulation–optimization (MOSO) method is presented by Jerbi et al., with the aim of designing improved FMSs. Four MOSO alternatives are selected, implementing the Design of Experiments (DoE) principles. Meta model-based approaches use Goal Programming (GP) and Desirability Function (DF). The rest integrate Gray Relational Analysis (GRA) and the VIKOR method. The results prove that GP and the VIKOR method can, in some cases, provide improved performance compared to the DF and GRA methods.

An evaluation of the effect of various 3D printed geometries on textile fabric regarding fabric drapes is presented in Spahiu et al.'s study. The drape coefficient of a created composite structure is measured. The results obtained are compared with those derived from the application of an algorithm developed for determining drape parameters and 3D form representation using digital color images and their image histograms. The measured drape coefficient values are determined with 4% accuracy.

In their study, Konstantinidis et al. identify the digital footprint of Industry 4.0 in the current manufacturing ecosystem, and a systematic literature survey is conducted. The influence of Industry 4.0 on traditional business models, small- and medium-sized enterprises (SMEs), decision-making processes, human–machine interaction, and circularity affairs are investigated. As a result, research gaps and business opportunities, as well as their relevance to Industry 5.0 principles, are identified.

The development of two different applications for designing furniture based on the Computational-based Visual Brand Identity (CbVBI) design methodology is described by Manavis et al. Two alternatives are used—the Application Programming Interface (API) of the Solidworks^{TM} CAD system (VBA event-driven programming language) and the visual programming language of Grasshopper^{TM}, incorporated within Rhinoceros3D^{TM}. Integrating the design process and manufacturing technologies via programming and automating the use of CAD systems are steps that can yield a great deal of design alternatives in a very short time.

Fountas et al., in their research, studied multi-objective optimization of three experimental cases with a laser sintering/melting method using a virus-evolutionary genetic algorithm (GA). The results obtained by the proposed algorithm are statistically comparable to those obtained by the Greywolf (GWO), Multi-verse (MVO), Antlion (ALO), and dragonfly (DA) algorithms. This proves that the virus-evolutionary genetic algorithm is superior to the heuristics that were examined, at least on the basis of evaluating regression models as fitness functions.

Sucuoglu is involved with the design of a structurally optimized mobile robotic system. Topology optimization (TO) and finite element analysis (FEA) were applied to reduce the structural weight. The optimized components were manufactured using Fused Deposition Modeling (FDM) with ABS (Acrylonitrile Butadiene Styrene) as material. A custom power analysis tool was developed for energy optimization. The proposed framework reduces total energy consumption by 5.8%, cuts prototyping time by 56%, and extends mission duration by ~20%.

An advanced autonomous HVAC control system tailored for a chemical fiber factory and emphasizing the human-centric principles and collaborative potential of Industry 5.0 is presented by Balasubramani et al. Central to the system's innovation is the integration of digital twins and physical AI, enhancing real-time monitoring and predictive capabilities. A virtual representation runs in parallel with the physical system, enabling sophisticated simulation and optimization. Digital twins facilitate scenario testing and optimization, while physical AI allows the system to learn from real-time data and simulations.

The review presented by Massaro analyzes the Electronic Digital Twin (EDT) tools characterizing the industrial transformation from Industry 4.0 to Industry 5.0. This research article focuses on the possibility of combining industrial machines' advanced technologies, electronics, and mechatronics with Artificial Intelligence (AI) algorithms. It includes perspectives on the limits and advantages of EDTs.

Funding: This research received no external funding.

Acknowledgments: The Guest Editors of this Special Issue would like to thank the authors for the valuable high-quality work they submitted, the reviewers for their efforts and time spent in order to improve the submissions, and the publisher for their excellent cooperation.

Conflicts of Interest: The authors declare no conflicts of interest.

List of Contributions:

1. Jerbi, A.; Hachicha, W.; Aljuaid, A.M.; Masmoudi, N.K.; Masmoudi, F. Multi-Objective Design Optimization of Flexible Manufacturing Systems Using Design of Simulation Experiments: A Comparative Study. *Machines* **2022**, *10*, 247. https://doi.org/10.3390/machines10040247.
2. Spahiu, T.; Zlatev, Z.; Ibrahimaj, E.; Ilieva, J.; Shehi, E. Drape of Composite Structures Made of Textile and 3D Printed Geometries. *Machines* **2022**, *10*, 587. https://doi.org/10.3390/machines10070587.
3. Konstantinidis, F.K.; Myrillas, N.; Mouroutsos, S.G.; Koulouriotis, D.; Gasteratos, A. Assessment of Industry 4.0 for Modern Manufacturing Ecosystem: A Systematic Survey of Surveys. *Machines* **2022**, *10*, 746. https://doi.org/10.3390/machines10090746.
4. Manavis, A.; Tzotzis, A.; Tsagaris, A.; Kyratsis, P. A Novel Computational-Based Visual Brand Identity (CbVBI) Product Design Methodology. *Machines* **2022**, *10*, 1065. https://doi.org/10.3390/machines10111065.
5. Fountas, N.A.; Kechagias, J.D.; Vaxevanidis, N.M. Optimization of Selective Laser Sintering/Melting Operations by Using a Virus-Evolutionary Genetic Algorithm. *Machines* **2023**, *11*, 95. https://doi.org/10.3390/machines11010095.
6. Sucuoglu, H.S. Development of Topologically Optimized Mobile Robotic System with Machine Learning-Based Energy-Efficient Path Planning Structure. *Machines* **2025**, *13*, 638. https://doi.org/10.3390/machines13080638
7. Balasubramani, M.; Chen, J.; Chang, R.; Shieh, J.-S. Development of a Human-Centric Autonomous Heating, Ventilation, and Air Conditioning Control System Enhanced for Industry 5.0 Chemical Fiber Manufacturing. *Machines* **2025**, *13*, 421. https://doi.org/10.3390/machines13050421.
8. Massaro, A. Electronic Artificial Intelligence and Digital Twins in Industry 5.0: A Systematic Review and Perspectives. *Machines* **2025**, *13*, 755. https://doi.org/10.3390/machines13090755.

References

1. Meka, S.; Dowluru, S.; Dumpala, L. Automatic Feature Recognition Techniques for the Integration of CAD and CAM: A Review. *Smart Sustain. Manuf. Syst.* **2024**, *8*, 83–109. [CrossRef]
2. Sajadieh, S.M.M.; Noh, S.D. From Simulation to Autonomy: Reviews of the Integration of Artificial Intelligence and Digital Twins. *Int. J. Precis. Eng. Manuf.-Green Tech.* **2025**, *12*, 1597–1628. [CrossRef]
3. Li, X.; Zhang, X.; Zhang, Y. Three-dimensional plasticity-based topology optimization with smoothed finite element analysis. *Comput. Mech.* **2024**, *73*, 533–548. [CrossRef]

machines

MDPI

Article

Multi-Objective Design Optimization of Flexible Manufacturing Systems Using Design of Simulation Experiments: A Comparative Study

Abdessalem Jerbi [1], Wafik Hachicha [2,*], Awad M. Aljuaid [2], Neila Khabou Masmoudi [3] and Faouzi Masmoudi [4]

1 OLID Laboratory, Higher Institute of Industrial Management of Sfax (ISGIS), University of Sfax, Sfax 3021, Tunisia; abdessalem.jerbi@isgis.usf.tn
2 Department of Industrial Engineering, College of Engineering, Taif University, P.O. Box 11099, Taif 21944, Saudi Arabia; amjuaid@tu.edu.sa
3 LASEM Laboratory, Department of Mechanical Engineering, National Engineering School of Sfax (ENIS), University of Sfax, Sfax 3038, Tunisia; neila.masmoudikhabou@enis.tn
4 LA2MP Laboratory, Department of Mechanical Engineering, National Engineering School of Sfax (ENIS), University of Sfax, Sfax 3038, Tunisia; faouzi.masmoudi@enis.tn
* Correspondence: wafik.hachicha@isgis.usf.tn; Tel.: +966-53-194-0695

Abstract: One of the basic components of Industry 4.0 is the design of a flexible manufacturing system (FMS), which involves the choice of parameters to optimize its performance. Discrete event simulation (DES) models allow the user to understand the operation of dynamic and stochastic system performance and to support FMS diagnostics and design. In combination with DES models, optimization methods are often used to search for the optimal designs, which, above all, involve more than one objective function to be optimized simultaneously. These methods are called the multi-objective simulation–optimization (MOSO) method. Numerous MOSO methods have been developed in the literature, which spawned many proposed MOSO methods classifications. However, the performance of these methods is not guaranteed because there is an absence of comparative studies. Moreover, previous classifications have been focused on general MOSO methods and rarely related to the specific area of manufacturing design. For this reason, a new conceptual classification of MOSO used in FMS design is proposed. After that, four MOSO methods are selected, according to this classification, and compared through a detailed case study related to the FMS design problem. All of these methods studied are based on Design of Experiments (DoE). Two of them are metamodel-based approaches that integrate Goal Programming (GP) and Desirability Function (DF), respectively. The other two methods are not metamodel-based approaches, which integrate Gray Relational Analysis (GRA) and the VIKOR method, respectively. The comparative results show that the GP and VIKOR methods can result in better optimization than DF and GRA methods. Thus, the use of the simulation metamodel cannot prove its superiority in all situations.

Keywords: flexible manufacturing system; multi-objective simulation-optimization method; discrete event simulation; design of experiments; simulation metamodel; goal programming; desirability function; grey relations analysis; VIKOR method

1. Introduction

The fourth industrial revolution, known as industry 4.0, is considered the upcoming significant technology development as it allows customers to receive their products based on their expectations in terms of product varieties and quantiles [1]. Industry 4.0 can be attributed to its broadening focus on automation, decentralization, system integration, cyber-physical systems, etc. [2]. One of the basic components of Industry 4.0 is the Flexible Manufacturing System (FMS), which is an advanced production system that interconnects machines, workstations, and logistics equipment, the entire manufacturing process being coordinated with the computer. FMS is intended for manufacturing tasks of large typological diversity, for high complexity, for ensuring timely delivery, and for minimal manufacturing costs, while production is unpredictable, organized in small batches, and has frequent changes [3].

Discrete Event Simulation (DES) is a powerful tool for analyzing and optimizing FMS for the purpose of design, modeling, and ongoing performance [4]. A simulation of an entire manufacturing system involves the identification of organization machines, robots, and the layout of the system, also involving multiple processes in the system. In particular, the design of FMS involves the choice of parameters and their value to optimize its performance. DES models allow the user to understand system performance and assist in behavior prediction and to support FMS diagnostics and design. However, DES responds to what-if questions as a tool for system evaluation; by itself, it cannot provide answers to how-if questions [5]. Moreover, DES is essentially a trial-and-error approach and is, therefore, time consuming and does not provide a method for optimization. In fact, many researchers have attempted to combine simulation and optimization procedures to provide a complete design solution with desired properties [6]. The problem of locating the most preferred alternative system design by using experimental evaluations performed using a computer DES is known as the Simulation Optimization (SO) problem.

The main classification criterion for SO approaches is the number of output performance measures. There are two groups of SO methods [7]. The first group is named Single-Objective Simulation Optimization (SOSO) approaches, which are focused on optimizing a single performance measure. The second group, which is studied in this research, covers Multi-Objective Simulation Optimization (MOSO) approaches. MOSO is an area of decision making of multiple criteria that is concerned with mathematical optimization problems that involve more than one objective function to be optimized simultaneously.

In the next section, the classifications of MOSO methods in the literature are presented, and then the MOSO methods used in FMS design are provided. After that, a new conceptual classification of MOSO was applied to FMS design. According to this classification, the aim is to justify the selection of four MOSO methods that are used in the comparative study.

1.1. Literature Overview

1.1.1. General Literature Review of MOSO Methods

MOSO methods are an area of multiple-criterion decision making that optimize multiple performance measures via simulation. In MOSO literature, there are three main classification criteria for organizing these methods.

1. According to the articulation of the preferences of the Decision Maker (DM). This first classification criterion is proposed by Rosen et al. [8]. Four groups of methods are possible and include the following: (1) a priori MOSO methods when the DM expresses their preferences before optimization is conducted; (2) a posteriori MOSO methods (in these methods, the DM selects a solution at the end of the search. Although this approach avoids the disadvantage of the a priori approach by taking into account preference information only at the end of the optimization process, it can lead to extremely high computational costs); (3) a progressive articulation of DM preferences (also named Interactive MOSO Methods) (the progressive approaches repeatedly solicit preference information from the DM to guide the optimization process). These methods enable DM to change his preferences during the optimization process by incorporating knowledge that only becomes available during the search. Interactive methods may be useful when simulation runs are expensive and the DM is readily available to provide input. Finally, the fourth group involves (4) non preference MOSO methods that operate without regard to the preference of DM.
2. According to the research set and variables nature. This second classification criterion is proposed by Hunter et al. [7]. Three groups of methods are possible, including the following: (1) MOSO on finite sets, called Multi-Objective Ranking and Selection (MORS); (2) MOSO with integer-ordered decision variables; and (3) MOSO with continuous decision variables. In the context of integer-ordered and continuous

decision variables, we focus on methods that provably converge to a local efficient set under natural ordering. Furthermore, these methods of the three groups can also be viewed two groups according to the type of the final solution: global solution versus local solution [7]. The MORS methods provide a global solution, in which simulation replications are usually obtained from every point in the finite feasible set, and the estimated solution is the global estimated best. In addition, metaheuristics methods (named also random search) such as simulated annealing, Genetic Algorithms (GA), Tabu Search (TS), etc., also provide global solutions. Metaheuristics methods are efficient because they appropriately control stochastic error. However, the task is more challenging as it results in a number of solutions with different trade-offs among criteria, also known as Pareto optimal or efficient solutions.

3. According to the use or non-use of metamodels. This third classification is proposed implicitly in many research studies such as in Barton and Meckesheimer [9], do Amaral et al. [10], etc. A metamodel or model of the simulation model simplifies the SO in two ways: The metamodel response is deterministic rather than stochastic, and the run times are generally much shorter than the original simulation. The metamodel is used to identify and estimate the relationship between the inputs and outputs of the simulation model, forming a mathematical function that is used to evaluate possible solutions in the optimization process. For example, Hassannayebi et al. [11] highlight that the adoption of metamodel-based SO in industry and service problems has grown due to its potential to reduce the number of simulation rounds necessary in the optimization process. Note that the MOSO methods, which are based on the metamodel, also provide a global solution such as that discussed in the second classification criterion.

1.1.2. FMS Design Literature Review

The study of Diaz et al. [12] presents a MOSO approach for a reconfigurable production lines subject to scalable capacities. The production line produces two product families and is composed of 18 workstations. The authors utilized a Non-Dominated Sorting Genetic Algorithm II (NSGA-II), a variant of GA to address the assignment of the tasks to workstations and buffer allocation for simultaneously maximizing the Throughput Rate (TR) and minimizing total buffer capacity. Červeňanská et al. [13] explored an MOSO of an FMS via a scalar simulation-based optimization method. The authors integrated a simulation with Design of Experiment (DoE) and Weighted Sum and Product multi-objective methods to optimize the total number of products, the Mean Flow Time (MFT), the Machine UTILization (MUTIL), and the average costs per unit of part. The modeled FMS produces two different products with eight workstations using parallel automated work machines.

The paper of Hussain and Ali [14] studied the impact of four design and control factors, control architectures, sequencing flexibility, buffer capacity, and scheduling rule on the performance of an FMS. The studied FMS is composed of six Computer Numerical Control (CNC) machines producing six different types of parts. The system is evaluated on the basis of make-span, average MUTIL, and the average Waiting Time (WT) of parts at the queue using the Taguchi–Grey multi-objective method. Apornak et al. [15] considered a multi-objective optimization of five performance measures in FMS. The authors addressed the optimal set of queues capacity, queues discipline, conveyor and transporter's speed, and operational setup times in an FMS with objectives of minimization of the average WT of raw materials, two average Process Times (PT), as well as the transporter and assembler product outputs. The studied FMS is composed of three work stations producing various kinds of seats for the freight cars. Using DoE, the authors simulated and collected the performance measure of 36 random scenarios. Regression analysis was then used to describe the metamodel of each performance measure. Consequently, the Response Surface Methodology (RSM) was applied to optimize the five objective functions.

Ahmadi et al. [16] proposed two Evolutionary Algorithms (EA): NSGA-II and NRGA are applied and compared to simultaneously combine the improvement of the make-span and stability of the schedule. This stability is evaluated by measuring the deviation of start and completion times of each job between prescheduled and realized schedule. The simulation is used to evaluate the state and condition of the machine breakdowns on a variety of manufacturing systems. Freitag and Hildebrandt [17] used a multi-objective simulation-based optimization to create a control strategy for an FMS by considering earliness and tardiness performance measures. This paper investigates the effect of 10 different attributes, which are the PT, the average PT of all waiting jobs, the Setup Time (ST), the average ST of all waiting jobs, the number of remaining operations, the time in system, the time in queue, the batch family size, the time until operational due date, and the average time until operational due date. The authors used the GA coupled with the simulation to solve the scheduling rule choice problem for a complex FMS.

Ammar et al. [18] investigated the size of the number of workers to be assigned to an FMS as well as the skills that each worker must have in a multi-objective optimization problem. The two objectives considered are minimizing the expected labor cost associated with the manufacturing team and minimizing the expected average task TR. The proposed multi-objective simulation optimization approach is applied to the design of teams of a manufacturing system; using the EA NSGA-II connected to a simulation model developed using Arena. Dengiz et al. [19] implemented a multi-objective optimization method of an FMS based on simulation through DoE, a regression meta-model, and the Goal Programming (GP) method. The authors have modeled and simulated by the ARENA simulation software an FMS with four workstations. Then, they applied the multi-objective optimization method to optimize the TR and MFT in the system by taking into consideration the number of operator, the velocity of material handling, the number of tool, scheduling rules, and the number of pallets as design and control parameters.

Using simulation results, Bouslah et al. [20] developed and solved a mathematical model based on RSM. The main objectives of the authors were to determine the optimal batch size, the optimal hedging level, and the economic sampling plan design, which minimized the average total holding cost, which includes the storage of the Work In Process (WIP) and final inventory stock, the average backlog cost, the average cost of sampling, the average costs of 100% inspection and rectification of the rejected batches, the average cost of transportation, and the average cost of replacement of non-conforming items sold to the consumer. However, the authors did not mention any details on the structure of the simulated manufacturing system. Iç et al. [21] considered a case study of simulation-based multi-objective optimization using the Technique for Order of Preference by Similarity to Ideal Solution (TOPSIS) method hybridized with the Taguchi design technique. The studied production system is an FMS department composed of four CNC machining centers and producing three part types. The authors based their optimization case on the cycle time, TR, and work in queue as performance measures. In addition, they used five factors as decision variables. These factors are the number of cutting tools, the number of operators, the number of pallets, the velocity of transporter robots, and the pallet selection strategy.

The paper of Wang et al. [22] applies an MOSO method to a flexible shop scheduling problem. The two investigated objective functions are the minimum of the maximum PT and the minimum of the maximum machine load. The main considered constraints are the production resources and the technological process. The scheduling model of an FMS is established using simulation software and integrated to NSGA-II EA. In Zhang et al. [23], a hybrid method based on hybrid GA and TS is used to address a multi-objective FMS scheduling problem. Two objectives, which are the make-span and the starting time deviations, are considered to improve schedule efficiency and stability. A case of study of six machines FMS was studied with four different job arrivals rate and six different number of job arrivals.

Azadeh et al. [24] integrated simulations with the GP method and DoE technique to address a multi-objective scheduling problem of an FMS. The proposed method was applied on a real textile shop floor to minimize make-span and tardiness. The authors determined the decision parameters by using the DoE technique by estimating the effects the dyeing machine type, the temperature of the printing, the temperature and the number of center machines, and the scheduling rules through meta-modeling. Then, they used GP to find the optimal values of these decision variables, which are subject to a set of technical and managerial constraints. Um et al. [25] presented the simulation based multi-objective optimization of the design of an FMS with Automated Guided Vehicles (AGVs). Their principal objectives were to minimize congestion and utilization and to maximize TR based on many parameters including the number, velocity, and dispatch rule of AGV, part types, scheduling, and buffer sizes. In this paper, the authors considered a nonlinear programming method combined to evolution strategy. Nonlinear programming was used to determine the design parameters of the system through multi-factorial and regression analyses, and an evolution strategy was used to verify each parameter for simulation-based optimization.

Syberfeldt et al. [26] describe the use of Artificial Neural Networks (ANN) and EA as MOSO methods to the manufacturing cell at Volvo Aero. The two investigated objectives were the maximization of cell utilization and the minimization of overdue components considering the component inter-arrival times and due date as decision criteria. Kuo et al. [27] proposed a practical case of the Grey-based Taguchi method as a MOSO method for a company that provides integrated circuit packaging services. The authors aimed to optimize TR and cycle time performance for ink marking machines to avoid backlog of orders or lost customers, and the TR of the system must be increased. They based their methodology on five three-level control factors, which are the PT, the machine buffer size, the time between adjustment, the ratio of the adjusted PT to original PT, and the mean time between failures.

Oyarbide-Zubillaga et al. [28] focused on the determination of the optimal preventive maintenance frequencies for multi-equipment systems. The authors apply simulation and NSGA-II to the multi-objective optimization problem of preventive maintenance activities to minimize the system's cost and to maximize profit by considering the production speed, the percentage of unavailability of a machine due to corrective maintenance, and the fraction of time before and after the last maintenance as control factors. The system cost was defined as the sum of the preventive and corrective maintenance, the production speed lost, and the quality costs for each of the machines. Profit is the result of selling non-defective products. Park et al. [29] presented a method for determining the design and control parameters of an FMS with multi-objective performance via a fully factorial DoE, regression analysis and trade-off programming. A hypothetical FMS with six workstations was modeled and simulated. The number, speed, and dispatching rules of AGVs, in addition to the number of pallets, the buffer sizes, and the loading, routing scheduling rules, were considered as control parameters. These eight parameters were simultaneously determined by compromising performance measures of TR, delay, MUTIL, and WIP that are formulated using regression analysis.

1.2. The Proposed Conceptual Classification of MOSO for FMS Design

There are many MOSO methods applied for FMS design. According to the previous literature review, it is better to classify them in three main groups: Group A, Group B, and Group C, as detailed in Table 1. This classification is applicable regardless of the articulation of DM's preferences. It should be noted that all of the previous MOSO methods that are applied in the design of FMS are global solutions.

Table 1. The proposed MOSO classification for FMS design.

	The Use of DoE	The Use of Metamodel	Description
Group A	Yes	Yes	First, using designing simulation experiments. Second, applying optimization method on metamodel. A priori DM preferences are generally applied.
Group B	Yes	No	First, using designing simulation experiments. Second, applying multi criteria optimization method on experiments. A priori DM preferences are generally applied.
Group C	No	No	Iterative simulation and optimization using principally metaheuristics for random design research such as simulated annealing, genetic algorithms, etc. Only in this group, the articulation of the preferences of the DM is important.

Table 2 summarizes the methods and techniques used in the MOSO methods used for FMS design. The presence of a cross "X" in a row and column intersection means that the research study stated in row use the method mentioned in column. It shows that all of the previous studies have applied a global solution method. These methods can be classified easily according the proposed classification in three groups (A, B, and C). Group C contains complex optimization techniques using metaheuristics, such as (GA, TS, EA, etc.). The performance of MOSO methods is not guaranteed because there is an absence of comparative studies. None of the previous studies has compared different MOSO methods.

Table 2. MOSO classification for FMS design (since year 2000).

				The Decision Maker's Preferences				The Research Set and Variables Nature			The Use of Meta-Model or Not		The Proposed Classification		
Study	**Method**	**Number of Objectives**	**Number of Factors**	**A Priori**	**A Posteriori**	**Progressive**	**No-Preference**	**Ranking and Selection**	**Integer DV**	**Continuous DV**	**No Metamodel**	**Metamodel**	**Group A**	**Group B**	**Group C**
[12]	NSGA II	2	2				X				X				X
[13]	DoE, Weighted Sum, Weighted Product	3	2	X				X	X			X	X		
[14]	Taguchi design, GRA	2	4		X			X						X	
[15]	DoE, Regression metamodel, RSM	5	8				X		X			X	X		
[16]	NSGA-II, NRGA	2	2		X						X				X
[17]	Genetic Programming	2	10		X						X				X
[18]	NSGA-II	2	2		X						X				X
[19]	DoE, Regression meta-model, GP	2	5	X					X			X	X		
[20]	RSM	8	3				X		X			X	X		
[21]	Taguchi (DoE), TOPSIS	3	5				X	X			X			X	
[22]	NSGA-II	2	2		X						X				X
[23]	GA, TS	2	2				X				X				X

Table 2. *Cont.*

Study	Method	Number of Objectives	Number of Factors	The Decision Maker's Preferences				The Research Set and Variables Nature			The Use of Meta-Model or Not		The Proposed Classification		
				A Priori	A Posteriori	Progressive	No-Preference	Ranking and Selection	Integer DV	Continuous DV	No Metamodel	Metamodel	Group A	Group B	Group C
[24]	DoE, Regression meta-model, GP	2	5	X					X	X		X	X		
[25]	Non-Linear Programming, Evolution Strategy	3	6				X					X			X
[26]	EA, ANN	2	2		X						X				X
[27]	Taguchi, GRA	2	5		X			X				X		X	
[28]	NSGA-II	2	3		X				X		X				X
[29]	DoE, regression metamodel	4	8	X								X	X		

1.3. The Objective of the Case Study

Our main contribution is to fill these gaps in the literature and to conduct a study of several relatively straightforward simulation-based FMS optimization methodologies that cover almost all categories of optimization methods classification. Our study investigates and compares the applicability and performances of the Goal Programming (GP), the Desirability Function (DF) method, the Grey Relational Analysis (GRA), and the VlseKriterijuska Optimizacija I Komoromisno Resenje (VIKOR) method.

All these methods are based on the DoE technique. They must be preceded by a design of experiments to program and sometimes analyze the simulation results. Moreover, these four multi-objective optimization methods have in common the type of preferences of DM; indeed, they are all based on an a priori decision of the DM for the choice of the objectives. On the other hand, the two methods GP and DF use the simulation-based metamodel technique and combine continuous and integer decision variables to solve the multi-objective optimization problem, while the two other methods, GRA and VIKOR, are based on the RS technique and use exclusively integer decision variables. The solutions reached by the GP and DF methods are then global and those reached by the GRA and VIKOR methods are local. In this study, we are interested in four multi-objective optimization methods in the context of FMS. An application on an FMS system will be used as a basis to compare the performances of these methods. It is mainly a matter of comparing the deviations between their results and the expected target values.

2. Materials and Methods

Figure 1 describes in detail the adopted MOSO methodologies applied to FMS. These methodologies are essentially made up of three stages. Each of these stages consists of various steps. In the first stage, the primary step starts with FMS factors levels and performance measures selection and definition. Consequently, DoE is constructed, and the corresponding simulation models are developed using ARENA 14 discrete event simulation software. In the final step of this first stage, simulations are run to collect data for every studied performance measure. These simulation results are then analyzed in the second phase by one of the four adopted multi-objective optimization methods. The steps of this phase are discussed in detail in the following paragraphs. Finally, the optimum factors levels are adopted in the last stage of the multi-objective optimization method.

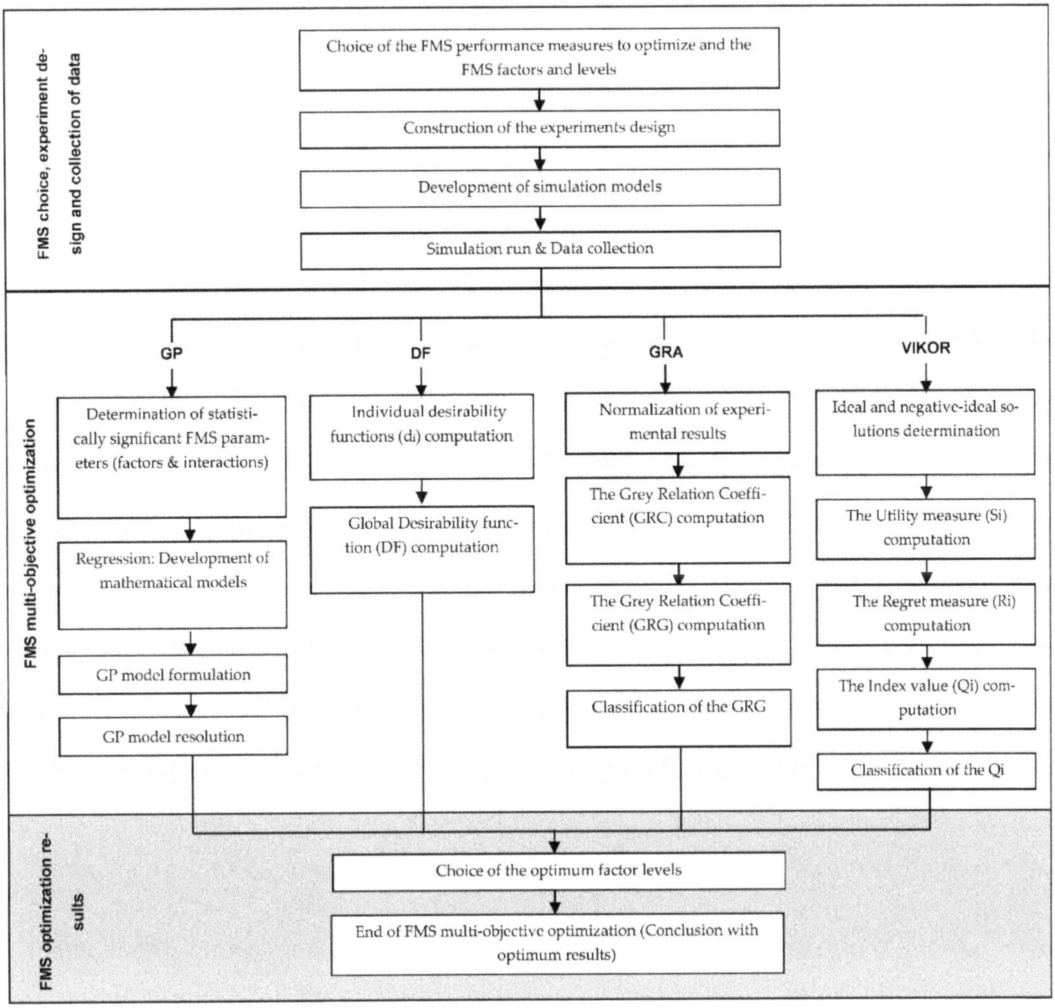

Figure 1. The major steps of the FMS optimization approach with multiple performance objectives.

2.1. Simulation of Possible FMS Designs according to the DoE

2.1.1. The Case Study

The FMS investigated in this study is inspired by Pitchuka et al. [30]. An FMS is a manufacturing system characterized by a certain flexibility that allows the system to react in the case of changes. This flexibility is considered to fall into two categories. The first one, called routing flexibility, generally covers the system's ability to be changed to produce new product types. The second category is called machine flexibility, which consists of the ability to use various machines to perform the same manufacturing operation on a part.

- To capture the FMS flexibility effect on its performance, this research adopts different machine LAYOUT (LAYOUT) for the studied FMS. Indeed, Functional Layout (FL) and Cellular Layout (CL) are the two most used machine layouts in FMSs. In FL, functionally similar machines are grouped into departments, and all machines of every department can perform production operations for any incoming part [31]. However, CL is made up of independent manufacturing cells. Each of these cells is made up of different machine types dedicated to the treatment of similar parts grouped into families. In addition, this work aimed also to measure the effect of part Batch Size (BS), part Inter-Arrival Time (IAT), and scheduling RULEs (RULE) on FMS performances (Table 3). The IAT, defined as the difference between the arrival times to the FMS of two consecutive parts, is generally generated by common probabilistic laws. In addition, parts are grouped into batches to reduce the machine's setup repetitions and transport times between work stations [31]. Furthermore, parts arriving at any work station are made to wait in a queue until the required machine becomes available. Once this required machine is idle, parts must be selected from the waiting queue based on scheduling rules [32–34]. As shown in Table 3, each of the considered FMS factors considered is studied with 2 levels.

Table 3. Studied factors and levels for the FMS case study.

Factor	Levels	
	1	2
LAYOUT	FL	CL
IAT (Minutes/part)	5	25
BS (parts)	5	10
RULE	FCFS	SPT

FL: functional layout; CL: cellular layout; IAT: inter-arrival time; BS: batch size; Rule: dispatching rule; FCFS: first come first served; SPT: short processing time.

- The FMS considered is composed of 8 machines grouped into 3 departments in FL and 2 cells in CL. The two departments "M" and "L" are composed of 3 machines each, while department "M" comprises only 2 machines. This MS is also characterized by two-part families composed of each of 2 part types. Each type of part requires 2 to 5 manufacturing operations (Table 4).

Table 4. Part routing used for the FMS case study.

Part		Functional Layout	Cellular Layout	
Type	**Family**	**Routing Departments**	**Cell**	**Routing Machines**
P1	F2	"L"→ "M"→"D"	C2	"L2"→ "M2"→ "D2"
P2	F1	"L"→"D"→"M"	C1	"L1"→ "D1"→ "M1"
P3	F1	"L"→"M"	C1	"L1"→ "M1"
P4	F2	"L"→"M"→ "D"→ "L"→ "M"	C2	"L2"→ "M2"→ "D2"→ "L3"→ "M3"

- The setup and processing times for each type of part are provided in Table 5. Setup times on every machine can be reduced or cancelled by the setup factor (δ) depending on the similarity of the successive parts family or type. Indeed, if successive parts belong to the same family, the subsequent part setup time must be reduced by a factor of $\delta = 0.5$. On the other hand, if these successive parts have the same type, no machine setup is needed and the subsequent part setup time must be cancelled by a factor of $\delta = 0$. Transfer times in the two layouts follow a statistical uniform law between 10 and 16 min.

Table 5. Parts' setup and processing time used for the FMS case study.

		L	M	D	L	M
P1	ST	T (39, 44, 49)	U (49, 69)	N (65, 15)		
	PT	T (15, 18, 21)	U (9, 13)	N (29, 5)		
P2	ST	T (90, 101, 110)	U (64, 84)	N (107, 35)		
	PT	T (17, 21, 25)	U (20, 28)	N (14, 5)		
P3	ST	T (80, 84, 88)	U (72, 92)			
	PT	T (23, 28, 32)	U (14, 18)			
P4	ST	T (62, 66, 70)	U (50, 58)	N (101, 10)	T (33, 38, 45)	U (78, 98)
	PT	T (18, 20, 22)	U (25, 33)	N (12, 5)	T (20, 23, 26)	U (15, 23)

ST: setup time; PT: processing time; N: normal distribution; U: uniform distribution; T: triangular distribution; all the times are in minutes.

- To characterize the fluidity of parts flow in FMS, different optimization studies used WIP and MFT as major performance measures [31]. WIP has mainly been measured as the number of parts in the system, and MFT is simply obtained by averaging all durations between every part exit times and entry times in FMS. The TR of the production was adopted as the third performance measure. To evaluate TR, it is normal to measure the number of processed parts per unit of time. The maximization of such a measure of performance reflects the best use of material and human resources. To enhance the efficiency of FMS piloting, various optimization studies used the waiting and transfer times (WT and TT) as performance indicators, and they essentially aimed to minimize these two indicators.

2.1.2. The Simulation Model

FMS simulation models were built using Arena 14.0 software. The FL and CL models are composed of three parts: "Parts arriving", "Departments" or "Cells", and "System exit":

- Parts enter to the system through a "Create" module named "Parts Arrival" in which the BS and IAT times are specified. Then, they are grouped into batches by a "Batch" module, named "Arrival Parts Grouping", to assign them their corresponding types through an "Assign" module named "Part Type". Due to the stochastic nature of their PT and ST, these batches are separated into unit products through a "Separate" module called "Parts Separation" to assign them each of their execution times through one of the four "Assign" modules named "Attribute Part i". However, a preliminary step must be performed through a "Decide" module called "Parts Sorting" to direct each type of product to the corresponding "Assign" module. The products then proceed through a "Batch" module named "Parts Grouping" before proceeding through the "Route" module named "Transfer to System" (Figure 2).

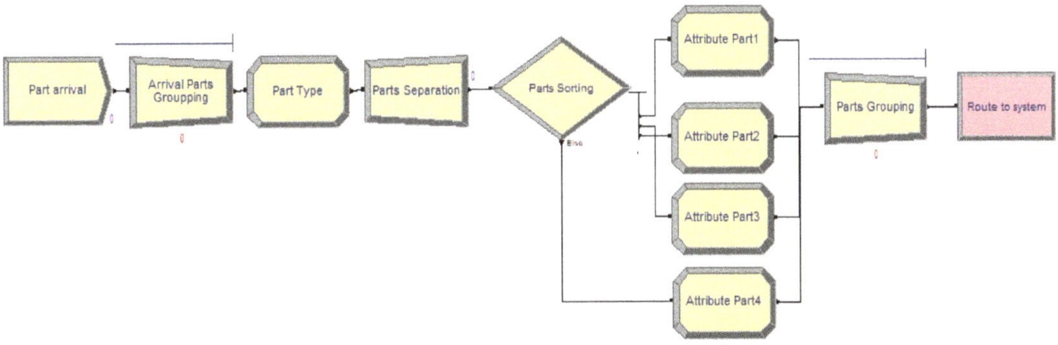

Figure 2. Simulation part model related to parts arriving.

- As soon as one products batch arrives in one of the departments, it is separated into unit products and put on hold in the department queue via the "Hold" module named "Waiting Queue Department i". This queue is governed by a "Queue" module in which the scheduling rule must be specified. Once one of the department machines becomes free, the selected waiting product is released from the "Hold" module. It then passes through a test, represented by the module "Decide" named "Machine Selection", which affects it toward this free machine. The processed products of the machines are grouped again in batches by the "Batch" module named "Grouping of Processed Parts Department i", which succeeds these machines. Finally, each batch of products is transferred to the next step in its production sequence through the module called "Route Department i" (Figure 3).

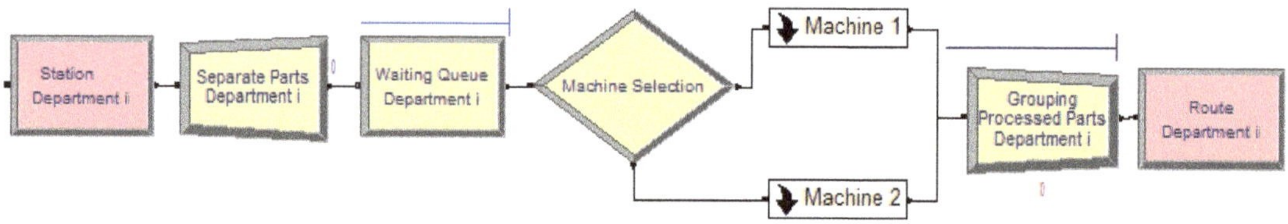

Figure 3. Simulation part model related to department "i".

In the case of the CL simulation model, as soon as a batch of products arrives in one of the cells, it is directed to the first machine in its production sequence. This batch is then separated into unit products by a "Separate" module called "Separation Parts Machine i". These products are then placed on hold in the queue of the machine via a "Hold" module named "Waiting Queue Machine i" until this machine becomes available. The choice of products from the machine queue is made according to the priority rule defined in the "Queue" module corresponding to this "Hold" module. The processed products by one of the machines are grouped into batches via the "Batch" module called "Grouping Parts Machine i". This batch is transferred to the next machine in its production sequence via the "Intracellular Route Cell i" module. By using this module, the transfer is performed in the cells, and the transfer time in this case is equal to zero. Each product with a completed production sequence must be evacuated to the system's output section. Hence, the "Cell i Output Route" module is used with a non-zero transfer time (Figure 4).

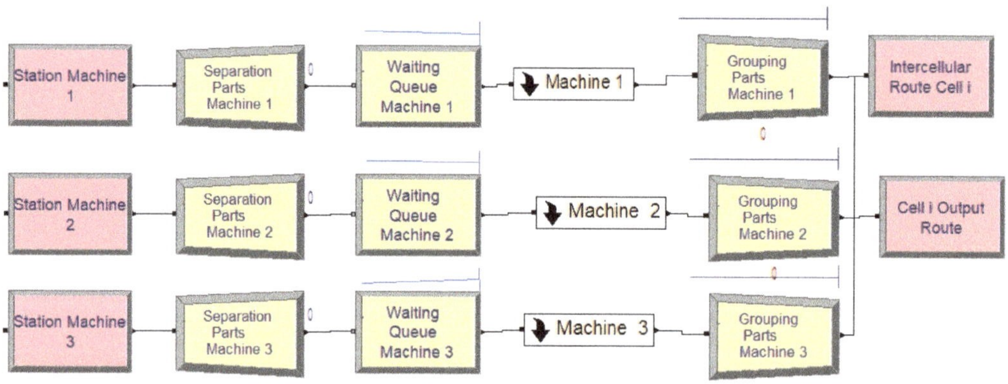

Figure 4. Simulation part model related to cell "i".

- In the two FL and CL simulation models, the machines are modeled by "Process" modules. In these modules, the transformation times are defined, which are function of PT and ST weighted by factor δ. Thus, Transformation time = PT + δxST. The value of factor δ depends on the similarity of the types of products entering and leaving the machine. In fact, a module called "Selection Delta Value Machine Selection i" applies a test on all incoming products to the machine to look for the value of this factor. For this, it compares two variables named "Part Type" and "Part Family" defined in the two "Assign" modules, named "Part Type in Machine i" and "Part Type Out Machine i". If the two variables "Part Type" are identical, the module "Delta value machine selection i" directs the incoming product to the module "Assign" named "Delta Equal 0 Machine i" corresponding to the value of factor δ = 0. If the two variables "Part Type" are different but the two variables "Part Family" are identical, the module "Delta Value Machine i" directs the incoming product to the module "Assign" named "Delta Equal 0.5 Machine i" corresponding to the value of factor δ = 0.5. Otherwise, module "Selection Delta Value Selection Machine i" directs the incoming product to the module "Assign" named "Delta Equal 1 Machine i", corresponding to the value of factor δ = 1 (Figure 5).

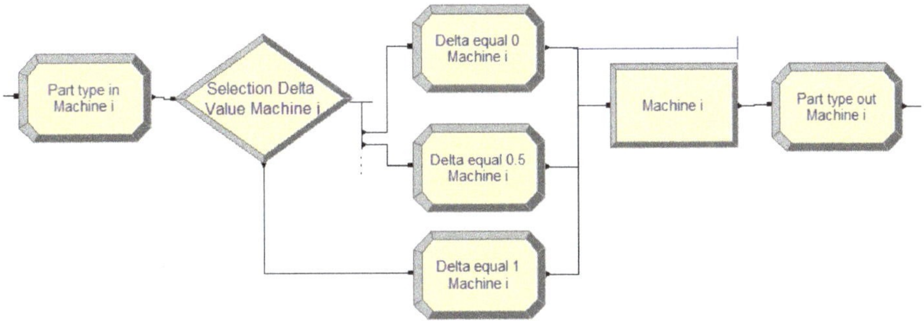

Figure 5. Simulation part model related to machine "i".

- The leaving products batch proceeds through an "Assig" module, called "Output Performance Measures", for computing and updating all variables defined as performance measures. The acquired data are then stored in an Excel file using a "Readwrite" module for further treatment and analysis. Finally, the batches of products are evacuated from the simulation model via the "Dispose" module named "System Exit" (Figure 6).

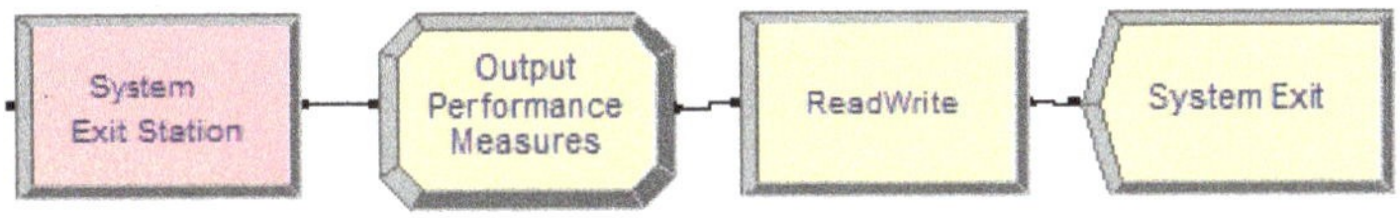

Figure 6. Simulation part model related to system exit.

2.2. Design of Experiments (DoE)

In this phase, we determine the number of distinct model settings to be run and the specific values of the factors for each of these simulation runs. There are many strategies for selecting the number of runs and the factor settings for each run include the following: random designs, combinatorial designs, sequential designs, factorial designs, etc.

Factorial designs are based on a grid, with each factor tested in combination with every level of every other factor. Factorial designs are attractive for three reasons: (i) The number of levels that are required for each factor is one greater than the highest-order power of that variable in the model, and the resulting design permits the estimation of coefficients for all cross-product terms; (ii) they are probably the most commonly used class of designs; and (iii) the resulting set of run conditions are easy to visualize graphically for as many as nine factors [35].

The case study is about an FMS design with four factors, and each factor has two levels, as mentioned in the Table 3. Therefore, a 2^4 full factorial design was used to collect simulation results.

2.3. Multi-Objective Optimization Methods

2.3.1. The GP Method

The GP is an optimization technique to solve problems with variety of objectives, which are generally incommensurable and often conflict each other in a decision-making horizon. The standard version of GP was first introduced by Charnes and Coper [31]. The GP model is based on an objective function formulated to find the most satisfactory solution that minimizes the total sum of positive and negative deviations from the level of attainment of the objectives levels (goals) set by the decision maker. This objective function is subject to physical and operating constraints of the system. The first type of constraints represents operating physical limits of the studied system. As for the second constraints, they are generally described by mathematical connections between the FMS factors and interactions and the performance measures to optimize. Hence, the principal purpose of the second stage of two first steps of the DoE-GP hybridization method is to build mathematical connections between FMS factors and responses. Statistical analyses are applied on the obtained simulation results to identify significant factors and interactions, and the relationships between the identified significant factors and interactions and the performance measures are translated to mathematical models by using the regression technique. In the third step of this stage, the GP model is developed setting the performance measures as goals and including other FMS constraints. Finally, this model is resolved using resolved using LINGO 18.0 software. The aim of this GP model is to find the most suitable levels of FMS factors that lower the total deviation of each performance measure from their respective target levels obtained in DoE.

The GP model takes the following form.

$$\text{Minimise } Z = \sum_{i=1}^{p} \delta_i^+ + \delta_i^-, \tag{1}$$

Moreover, it is subject to the following:

$$\sum_{j=1}^{n} a_{ij}x_j - \delta_i^+ + \delta_i^- = g_i \ (i = 1 \ldots p), \tag{2}$$

$$\rho x \leq C \text{ (the operating physical constrain of the system)} \tag{3}$$

$$x_j \geq 0 \ (j = 1 \ldots n), \tag{4}$$

$$\delta_i^+ \text{and} \delta_i^- \geq 0 \ (i = 1 \ldots p), \tag{5}$$

where the following is the case:

1. g_i: The goal set for the ith objective for (i =1 ... p) (the objectives here are the performance measures);
2. x_j: The jth decision variable for (j = 1 ... n) (the decision variables here are the significant FMS factors and interactions);
3. a_{ij}: The technological parameters (these parameters are the coefficients of the developed mathematical models relating the performance measures to significant FMS factors and interactions);
4. ρ: The matrix of coefficients related to the physical FMS constraints;
5. C: The vector of available physical FMS resources;
6. δ_i^+, δ_i^- : The positive and negative deviations from the goals values.

2.3.2. The DF Method

The DF method is based on two steps. The first defines a desirability function by assigning values to responses that reflect their desirability. This involves transforming each value of the performance measure 'j' of experiment 'i', y_{ij}, into a partial dimensionless desirability function d_i, where $0 \leq d_i \leq 1$. This function includes the choices of the decision maker when constructing the optimization procedure.

A one-sided desirability transformation arises when the goal is to maximize or minimize the response, and two values A and B must be specified as the lower and upper limits. Equations (6) and (7) present the one-sided transformation equations that will be used for minimization and maximization goals, respectively.

$$d_i = \begin{cases} 1 & A \geq y_{ij} \\ \left(\frac{y_{ij}-A}{B-A}\right)^{\omega_j} & A \leq y_{ij} \leq B \\ 0 & y_{ij} \geq B \end{cases}, \quad (6)$$

$$d_i = \begin{cases} 1 & y_{ij} \geq B \\ \left(\frac{y_{ij}-A}{B-A}\right)^{\omega_j} & A \leq y_{ij} \leq B \\ 0 & A \geq y_{ij} \end{cases}, \quad (7)$$

The parameter ω_j can be described as a power value or weight allocated according to the researcher subjective impression about the role of the response in the total desirability of the product.

A value of ω_j equal to 1 implies that a linear desirability function is applied. If the value of ω_j is less than 1, the obtained desirability function means that performance does not have to be close to the lower or upper limit, depending on the optimization goal, to have a higher desirability value. In contrast, if the value ω_j is greater than 1, the desirability function implying that the performance has to be closest to the lower or upper limit, depending on the optimization goal, to have a higher desirability value.

To simultaneously optimize multiple performance sets, the individual desirability is combined using a geometric mean in the composite desirability.

$$DF = \left(\prod_{i=1}^{n} d_i\right)^{\frac{1}{n}}, \quad (8)$$

A value of DF different from zero implies that all performances are in a desirable range simultaneously. In addition, a value of DF close to 1 means that the combination of the different criteria is globally optimal and the performances values are near the target values.

2.3.3. The GRA Method

Units of performance measurement are often different, so the influence of some of them may be neglected. This can also happen if some performance measures have a very wide range compared to others. In addition, if the expected optimization goals are contradictory, this will result in incorrect results in the analysis [36]. It is, therefore, necessary to normalize all performance values for each experiment in the first step of the multi-objective GRA-based optimization method's second stage.

In the developed DoE, for each of the "m" simulation experiments, "n" performance measures are measured. The ith experiment trial can be expressed as $Y_i = (y_{i1}, y_{i2}, \ldots, y_{ij}, \ldots, y_{in})$. Here, y_{ij} is the value of the performance measure "j" of experiment "i". The term Y_i can be translated into the comparability sequence $X_i = (x_{i1}, x_{i2}, \ldots, x_{ij}, \ldots, x_{in})$

using one of Equations (9) and (10), which are, respectively, used for larger-the-better and smaller-the-better objective values:

$$x_{ij} = \frac{y_{ij} - \underline{y_j}}{\overline{y_j} - \underline{y_j}} i = 1, 2, \ldots, m j = 1, 2, \ldots, n, \tag{9}$$

$$x_{ij} = \frac{\overline{y_j} - y_{ij}}{\overline{y_j} - \underline{y_j}} i = 1, 2, \ldots, m j = 1, 2, \ldots, n, \tag{10}$$

where the following is the case.

$$\overline{y_j} = \text{Max}\left\{y_{ij}, \ i = 1, 2, \ldots, m\right\}, \tag{11}$$

$$\underline{y_j} = \text{Min}\left\{y_{ij}, \ i = 1, 2, \ldots, m\right\}, \tag{12}$$

After the normalization procedure, all x_{ij} values, relative to the performance measures, will be scaled in [0, 1]. The Grey Relational Coefficient (GRC) is then computed to determine how close x_{ij} is to $x_{0j} = \text{Max}\{x_{ij}, i = 1, 2, \ldots, m\}$. The larger the grey relational coefficient, the closer x_{ij} and x_{0j} are. The grey relational coefficient can be calculated in the second step by the following:

$$\gamma\left(x_{0j}, x_{ij}\right) = \frac{\Delta_{min} + \zeta\Delta_{max}}{\Delta_{ij} + \zeta\Delta_{max}} i = 1, 2, \ldots, m j = 1, 2, \ldots, n, \tag{13}$$

where the following is the case.

$$\Delta_{ij} = \left|x_{0j} - x_{ij}\right|, \tag{14}$$

$$\Delta_{min} = \text{Min}\left\{\Delta_{ij}, \ i = 1, 2, \ldots, m; j = 1, 2, \ldots, n\right\}, \tag{15}$$

$$\Delta_{max} = \text{Max}\left\{\Delta_{ij}, \ i = 1, 2, \ldots, m; j = 1, 2, \ldots, n\right\}, \tag{16}$$

Note that ζ is the distinguishing coefficient, $\zeta \in [0, 1]$. The purpose of this coefficient is to expand or compress the range of the grey relational coefficient; usually, it is set equal to 0.5.

Once the entire GRC is computed, the Grey Relational Grade (GRG) is calculated in the third step based on the comparability and the reference sequence $X_i = (x_{i1}, x_{i2}, \ldots, x_{ij}, \ldots, x_{in})$ and $X_0 = (x_{01}, x_{02}, \ldots, x_{0j}, \ldots, x_{0n})$ using the following:

$$\Gamma(X_0, X_i) = \sum_{j=1}^{n} \omega_j \gamma\left(x_{0j}, x_{ij}\right); i = 1, 2, \ldots, m, \tag{17}$$

where ω_j is the weight for the jth response, chosen by the decision makers. Of course, the sum of ω_j is equal to 1.

In the final step of the GRA method, the GRG values are ranked in decreasing order. The optimal trial corresponds to the GRG maximum value.

2.3.4. The VIKOR Method

As in the case of the GRA method, which is based on GRG ranking, the VIKOR method is based on the computation of the VIKOR index and its ranking. In the first step of the VIKOR method, the ideal solution (A^*) and the negative-ideal solution (A^-) are to be determinate. A^* and A^- represent, respectively, the maximum and minimum performance measure values of every experimental trial, and they are described as follows.

$$A^* = \text{Max}\left\{y_{ij}, \ i = 1, 2, \ldots, m\right\} = \left\{y_1^*, y_2^*, \ldots, y_j^*, \ldots, y_n^*\right\}, \tag{18}$$

$$A^- = \text{Min}\left\{y_{ij}, \ i = 1, 2, \ldots, m\right\} = \left\{y_1^-, y_2^-, \ldots, y_j^-, \ldots, y_n^-\right\}, \tag{19}$$

In the two following steps of the VIKOR method application the utility and the regret measures for the ith experimental trial, S_i and R_i respectively, are computed as follows:

$$S_i = \sum_{j=1}^{n} \omega_j \left(y_j^* - y_{ij}\right) / \left(y_j^* - y_j^-\right), \tag{20}$$

$$R_i = \text{Max}_i \left[\omega_j \left(y_j^* - y_{ij}\right) / \left(y_j^* - y_j^-\right)\right], \tag{21}$$

where ω_j is the weight for the jth response, chosen by the decision makers. Of course, the sum of ω_j is equal to 1.

In the fourth step, the VIKOR index of the ith experimental trial is computed as follows:

$$Q_i = v\left[\frac{S_i - S^-}{S^* - S^-}\right] + (1 - v)\left[\frac{R_i - R^-}{R^* - R^-}\right], \tag{22}$$

where the following is the case.

$$S^- = \text{Min}S_i, \tag{23}$$

$$S^* = \text{Max}S_i, \tag{24}$$

$$R^- = \text{Min}R_i, \tag{25}$$

$$R^* = \text{Max}R_i, \tag{26}$$

Note that v is the weight of the maximum group's utility. It is usually set to 0.5.

In the final step of the VIKOR method application, the VIKOR index values are ranked in decreasing order, and the optimal trials correspond to the maximum value.

3. Results

3.1. Simulation Results

The case study is about an FMS design with four factors, each factor has two levels, as mentioned in Table 3. Therefore, a 2^4 full factorial design was used to collect simulation results. Each of the 16 simulation experiments was replicated 10 times. Simulation results show that a warm-up period of 10,000 min is needed, and models can then be run for 90,000 min. All final simulation results are provided in Appendix A. MFT simulation results are stated in Table A1, WIP simulation results are in Table A2, TR simulation results are in Table A3, WT simulation results are in Table A4, and TT simulation results are in Table A5.

3.2. Multi-Objective Optimization Methods

3.2.1. The GP Method

The use of GP as an MOSO method contains mainly four phases. The first phase is about the selection of the significant coefficient of the metamodel using Student's *t*-test. The second phase provides the final metamodel of each performance measure. The third and fourth phases concern the application of GP optimization.

1. Determination of statistically significant FMS parameters: The main effects of the studied factors and interactions were analyzed in $\alpha = 0.05$ of significance levels using the MINITAB statistical package (Table 6). Significant factors and interactions ($p \leq 0.05$) are shown in bold.

Table 6. Estimated coefficients of the simulation metamodel.

	MFT			WIP			TR			WT			TT		
Term	**Coef**	**T**	**P**	**Coef**	**T**	**P**	**Coef**	**T**	**P**	**Coef**	**T**	**P**	**Coef**	**T**	**P**
Constant	2034.8	80.78	**0.000**	51.541	85.46	**0.000**	34.7833	647.60	**0.000**	12,804	51.20	**0.000**	243.865	1955.65	**0.000**
BS	−883.4	−35.07	**0.000**	−32.921	−54.59	**0.000**	−8.6261	−160.60	**0.000**	−2499	−9.99	**0.000**	81.199	651.16	**0.000**
IAT	−1452.7	−57.67	**0.000**	−45.593	−75.60	**0.000**	−13.4205	−249.86	**0.000**	−9222	−36.87	**0.000**	−0.175	−1.40	0.163
RULE	−977.3	−38.80	**0.000**	2.231	3.70	**0.000**	0.1465	2.73	**0.007**	−4889	−19.55	**0.000**	0.243	1.95	0.054
LAYOUT	−1237.8	−49.14	**0.000**	−39.896	−66.15	**0.000**	5.1784	96.41	**0.000**	−7200	−28.79	**0.000**	−48.646	−390.11	**0.000**
BS*IAT	967.3	38.40	**0.000**	33.700	55.88	**0.000**	1.5064	28.05	**0.000**	4323	17.29	**0.000**	0.015	0.12	0.903
BS*RULE	828.1	32.87	**0.000**	−3.119	−5.17	**0.000**	−0.0592	−1.10	0.273	3444	13.77	**0.000**	0.065	0.52	0.603
BS*LAYOUT	1032.5	40.99	**0.000**	34.680	57.50	**0.000**	−4.7818	−89.03	**0.000**	5362	21.44	**0.000**	−15.899	−127.50	**0.000**
IAT*RULE	977.6	38.81	**0.000**	−2.155	−3.57	**0.000**	−0.1495	−2.78	**0.006**	4887	19.54	**0.000**	−0.111	−0.89	0.376
IAT*LAYOUT	1258.3	49.95	**0.000**	40.047	66.40	**0.000**	−5.1661	−96.18	**0.000**	7831	31.31	**0.000**	0.026	0.21	0.836
RULE*LAYOUT	954	37.87	**0.000**	−2.313	−3.83	**0.000**	−0.1110	−2.07	**0.041**	4762	19.04	**0.000**	−0.191	−1.53	0.128
BS*IAT*RULE	−829	−32.91	**0.000**	3.065	5.08	**0.000**	0.0650	1.21	0.228	−3447	−13.78	**0.000**	0.024	0.19	0.850
BS*IAT*LAYOUT	−941.1	−37.36	**0.000**	−33.724	−55.92	**0.000**	4.7214	87.90	**0.000**	−4624	−18.49	**0.000**	−0.066	−0.53	0.600
BS*RULE*LAYOUT	−808.6	−32.10	**0.000**	3.316	5.50	**0.000**	0.0375	0.70	0.486	−3366	−13.46	**0.000**	0.070	0.56	0.577
IAT*RULE*LAYOUT	−954.0	−37.87	**0.000**	2.269	3.76	**0.000**	0.1120	2.08	**0.039**	−4766	−19.06	**0.000**	0.027	0.21	0.831
BS*IAT*RULE*LAYOUT	809.1	32.12	**0.000**	−3.273	−5.43	**0.000**	−0.0413	−0.77	0.444	3359	13.43	**0.000**	−0.126	−1.01	0.314

A*B: interaction between factor A and factor B (For example, BS*IAT means interaction between factor BS and factor IAT.

2. Development of mathematical models: Based on the simulation results, mathematical models were computed. Regression Equations (27)–(31) with identified significant factors been derived for MFT, WIP, TT, and WT.

$$\begin{aligned}\text{MFT} = {} & 144.633 - 13633 \times \text{BS} - 5742.6 \times \text{IAT} - 68028 \times \text{RULE} - 71627 \times \text{LAYOUT} + \\ & 542.4 \times \text{BS} \times \text{IAT} + 6511 \times \text{BS} \times \text{RULE} + 6809 \times \text{BS} \times \text{LAYOUT} + 2721.8 \times \text{IAT} \times \text{RULE} \\ & + 2845.2 \times \text{IAT} \times \text{LAYOUT} + 33807 \times \text{RULE} \times \text{LAYOUT} - 260.52 \times \text{BS} \times \text{IAT} \times \text{RULE} - \\ & 269.48 \times \text{BS} \times \text{IAT} \times \text{LAYOUT} - 3235.7 \times \text{BS} \times \text{RULE} \times \text{LAYOUT} - 1352.5 \times \text{IAT} \times \\ & \text{RULE} \times \text{LAYOUT} + 129.46 \times \text{BS} \times \text{IAT} \times \text{RULE} \times \text{LAYOUT} \\ & \textit{With } R^2 = 99.41\% \textit{ and } R^2(\textit{adj}) = 99.34\%, \end{aligned} \quad (27)$$

$$\begin{aligned}\text{WIP} = {} & 1078.3 - 96.84 \times \text{BS} - 42.75 \times \text{IAT} + 239.6 \times \text{RULE} - 529.2 \times \text{LAYOUT} + 3.849 \times \\ & \text{BS} \times \text{IAT} - 25.91 \times \text{BS} \times \text{RULE} + 48.47 \times \text{BS} \times \text{LAYOUT} - 9.52 \times \text{IAT} \times \text{RULE} + 20.99 \\ & \times \text{IAT} \times \text{LAYOUT} - 121.6 \times \text{RULE} \times \text{LAYOUT} + 1.031 \times \text{BS} \times \text{IAT} \times \text{RULE} - 1.912 \times \\ & \text{BS} \times \text{IAT} \times \text{LAYOUT} + 13.16 \times \text{BS} \times \text{RULE} \times \text{LAYOUT} + 4.835 \times \text{IAT} \times \text{RULE} \times \\ & \text{LAYOUT} - 0.5237 \times \text{BS} \times \text{IAT} \times \text{RULE} \times \text{LAYOUT} \\ & \textit{With } R^2 = 99.47\% \textit{ and } R^2(\textit{adj}) = 99.42\%, \end{aligned} \quad (28)$$

$$\begin{aligned}\text{TP} = {} & -61.61 + 9.883 \times \text{BS} + 4.1508 \times \text{IAT} + 2.415 \times \text{RULE} + 98.713 \times \text{LAYOUT} - 0.50632 \\ & \times \text{BS} \times \text{IAT} - 9.4912 \times \text{BS} \times \text{LAYOUT} - 0.0971 \times \text{IAT} \times \text{RULE} - 3.9333 \times \text{IAT} \times \\ & \text{LAYOUT} - 1.116 \times \text{RULE} \times \text{LAYOUT} + 0.37771 \times \text{BS} \times \text{IAT} \times \text{LAYOUT} + 0.0448 \times \text{IAT} \\ & \times \text{RULE} \times \text{LAYOUT} \\ & \textit{With } R^2 = 99.88\% \textit{ and } R^2(\textit{adj}) = 99.87\%, \end{aligned} \quad (29)$$

$$\begin{aligned}\text{TT} = {} & 3.12 + 51.558 \times \text{BS} - 1.898 \times \text{LAYOUT} - 12.7192 \times \text{BS} \times \text{LAYOUT} \\ & \textit{With } R^2 = 99.97\% \textit{ and } R^2(\textit{adj}) = 99.97\%, \end{aligned} \quad (30)$$

$$\begin{aligned}\text{WT} = {} & 669{,}420 - 58944 \times \text{BS} - 26657 \times \text{IAT} - 298872 \times \text{RULE} - 334431 \times \text{LAYOUT} + \\ & 2351 \times \text{BS} \times \text{IAT} + 27062 \times \text{BS} \times \text{RULE} + 30009 \times \text{BS} \times \text{LAYOUT} + 11951 \times \text{IAT} \times \text{RULE} \\ & + 13247 \times \text{IAT} \times \text{LAYOUT} + 148498 \times \text{RULE} \times \text{LAYOUT} - 1081.9 \times \text{BS} \times \text{IAT} \times \text{RULE} \\ & - 1176.1 \times \text{BS} \times \text{IAT} \times \text{LAYOUT} - 13447 \times \text{BS} \times \text{RULE} \times \text{LAYOUT} - 5937 \times \text{IAT} \times \\ & \text{RULE} \times \text{LAYOUT} + 537.4 \times \text{BS} \times \text{IAT} \times \text{RULE} \times \text{LAYOUT} \\ & \textit{With } R^2 = 97.86\% \textit{ and } R^2(\textit{adj}) = 97.64\%, \end{aligned} \quad (31)$$

Every constant in each of these equations corresponds to the average responses for each performance measure, and the coefficients assigned to the factors and interactions correspond to their respective effects.

3. GP model formulation and resolution: We propose a GP model in which the selected performance measures are considered. The optimal configuration of decision variables minimizes the sum of penalties (dj). The parameter dj are deviations from the desired levels of the goals that are subject to series constraints. With the regression equations presented previously, the above-mentioned goal programming model can be stated as shown in Equations (32)–(40):

$$\text{Min Z} = d^+_{\text{MFT}} + d^+_{\text{WIP}} + d^-_{\text{TR}} + d^+_{\text{WT}} + d^+_{\text{TT}}, \quad (32)$$

which are subject to the following.

$$144633 - 13633 \times BS - 5742.6 \times IAT - 68028 \times RULE - 71627 \times LAYOUT + 542.4 \times BS \times IAT + 6511 \times BS \times RULE + 6809 \times BS \times LAYOUT + 2721.8 \times IAT \times RULE + 2845.2 \times IAT \times LAYOUT + 33807 \times RULE \times LAYOUT - 260.52 \times BS \times IAT \times RULE - 269.48 \times BS \times IAT \times LAYOUT - 3235.7 \times BS \times RULE \times LAYOUT - 1352.5 \times IAT \times RULE \times LAYOUT + 129.46 \times BS \times IAT \times RULE \times LAYOUT + d^-_{MFT} - d^+_{MFT} = G_{MFT}, \quad (33)$$

$$1078.3 - 96.84 \times BS - 42.75 \times IAT + 239.6 \times RULE - 529.2 \times LAYOUT + 3.849 \times BS \times IAT - 25.91 \times BS \times RULE + 48.47 \times BS \times LAYOUT - 9.52 \times IAT \times RULE + 20.99 \times IAT \times LAYOUT - 121.6 \times RULE \times LAYOUT + 1.031 \times BS \times IAT \times RULE - 1.912 \times BS \times IAT \times LAYOUT + 13.16 \times BS \times RULE \times LAYOUT + 4.835 \times IAT \times RULE \times LAYOUT - 0.5237 \times BS \times IAT \times RULE \times LAYOUT + d^-_{WIP} - d^+_{WIP} = G_{WIP}, \quad (34)$$

$$-61.61 + 9.883 \times BS + 4.1508 \times IAT + 2.415 \times RULE + 98.713 \times LAYOUT - 0.50632 \times BS \times IAT - 9.4912 \times BS \times LAYOUT - 0.0971 \times IAT \times RULE - 3.9333 \times IAT \times LAYOUT - 1.116 \times RULE \times LAYOUT + 0.37771 \times BS \times IAT \times LAYOUT + 0.0448 \times IAT \times RULE \times LAYOUT + d^-_{TR} - d^+_{TR} = G_{TR}, \quad (35)$$

$$3.12 + 51.558 \times BS - 1.898 \times LAYOUT - 12.7192 \times BS \times LAYOUT + d^-_{TT} - d^+_{TT} = G_{TT}, \quad (36)$$

$$669420 - 58944 \times BS - 26657 \times IAT - 298872 \times RULE - 334431 \times LAYOUT + 2351 \times BS \times IAT + 27062 \times BS \times RULE + 30009 \times BS \times LAYOUT + 11951 \times IAT \times RULE + 13247 \times IAT \times LAYOUT + 148498 \times RULE \times LAYOUT - 1081.9 \times BS \times IAT \times RULE - 1176.1 \times BS \times IAT \times LAYOUT - 13447 \times BS \times RULE \times LAYOUT - 5937 \times IAT \times RULE \times LAYOUT + 537.4 \times BS \times IAT \times RULE \times LAYOUT + d^-_{WT} - d^+_{WT} = G_{WT}, \quad (37)$$

LAYOUT and RULE are binary (1 or 2), (38)

$$5 \leq IAT \leq 25, \quad (39)$$

$$5 \leq BS \leq 10, \quad (40)$$

The objective G_{MFT}, G_{WIP}, G_{TR}, G_{TT}, and G_{WT} goal values were fixed basing on the experimental design results.

4. The GP model was solved using the mathematical software LINGO 18.0. The best value of the objective function was found to be equal to 136.99 and was obtained for the following levels of the studied factors: LAYOUT = CL, RULE = FCFS, IAT = 25, and BS = 5.

3.2.2. The DF Method

Applying Equations (6) and (7) for the studied performance measures, the individual desirability functions 'd' are very close to 1.0, as shown in Figure 7. Furthermore, Figure 7 illustrates the effect of each factor (columns) on the FMSs' performance measures and the desirability of the composite (rows). The red vertical lines and the corresponding numbers in red indicate the levels of optimal factors. The blue horizontal lines and the corresponding numbers in blue represent the values of the performance measures corresponding to the levels of optimal factors. Each of the performance measures is accompanied by the corresponding desirability function values 'd_i'. In addition, the first row provides the value of the composite desirability 'DF', as presented in Equation (8), corresponding to the levels of the optimal factors. The obtained DF is equal to 0.984, which represents an ideal case of optimization. To obtain this desirability, the factors' levels must be set to the values shown below the global solution in Figure 7. That is, BS = 5, IAT = 5, LAYOUT = CL, and RULE = SPT.

3.2.3. The GRA Method

Based on Equations (9)–(17), the simulation results were normalized, and the GRC and GRG were calculated (Table 7). Once GRG was ranked, it appears that the optimum performance measures were obtained for the factor levels LAYOUT = CL, RULE = SPT, IAT = 5, and BS = 5. The row in bold in Table 7 indicate the optimal solution obtained using the GRA method, which has a rank equal to 1.

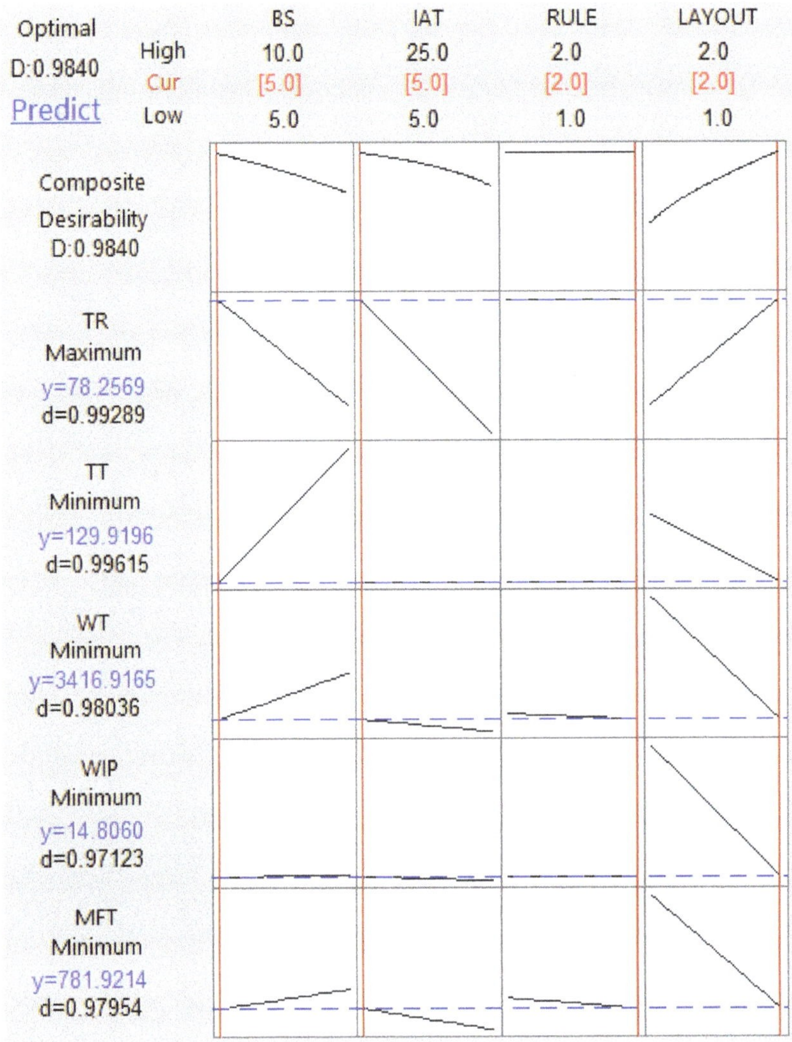

Figure 7. Desirability optimization results.

3.2.4. The VIKOR Method

Based on Equations (18)–(26), the utility and the regret measures as well as the VIKOR index were computed (Table 8). Once the VIKOR index was ranked, it appears that the optimum performance measures were obtained for factor levels LAYOUT = CL, RULE = SPT, IAT = 25, and BS = 5. The row in bold in Table 8 indicates the optimal solution obtained using VIKOR method, which has a rank equal to 1.

Table 7. GRA results.

Exp	Normalization					GRC					GRG	Rank
	MFT	WIP	TR	TT	WT	MFT	WIP	TR	TT	WT		
1	0.000	0.132	0.368	0.750	0.000	0.333	0.365	0.442	0.666	0.333	0.428	16
2	0.860	0.874	0.354	0.003	0.715	0.781	0.799	0.436	0.334	0.637	0.597	14
3	0.991	0.996	0.222	0.750	0.997	0.983	0.991	0.391	0.666	0.995	0.805	5
4	0.992	0.996	0.001	0.004	0.972	0.985	0.992	0.334	0.334	0.947	0.718	11
5	0.864	0.000	0.390	0.744	0.773	0.786	0.333	0.451	0.661	0.688	0.584	15
6	0.931	0.897	0.364	0.000	0.848	0.878	0.829	0.440	0.333	0.767	0.650	13
7	0.991	0.994	0.222	0.748	0.997	0.983	0.989	0.391	0.665	0.995	0.805	6
8	0.992	0.996	0.002	0.001	0.972	0.985	0.992	0.334	0.334	0.947	0.718	12
9	0.968	0.965	0.996	0.999	0.970	0.940	0.934	0.993	0.997	0.943	0.961	2
10	0.958	0.956	0.385	0.498	0.899	0.922	0.919	0.448	0.499	0.832	0.724	10
11	1.000	1.000	0.224	1.000	1.000	1.000	1.000	0.392	0.999	1.000	0.878	4
12	0.979	0.989	0.000	0.499	0.940	0.959	0.979	0.333	0.499	0.892	0.733	8
13	**0.978**	**0.968**	**1.000**	**0.999**	**0.979**	**0.959**	**0.940**	**1.000**	**0.999**	**0.960**	**0.972**	**1**
14	0.959	0.955	0.386	0.496	0.901	0.924	0.917	0.449	0.498	0.835	0.725	9
15	1.000	1.000	0.224	1.000	1.000	1.000	0.999	0.392	1.000	1.000	0.878	3
16	0.979	0.989	0.000	0.499	0.940	0.959	0.979	0.333	0.499	0.893	0.733	7

Table 8. VIKOR results.

Exp.	The Utility Measure (S_i)	The Regret Measure (R_i)	The VIKOR Index (Q_i)	Rank
1	0.305	0.150	0.000	16
2	0.646	0.175	0.493	14
3	0.947	0.200	0.959	3
4	0.715	0.199	0.787	7
5	0.603	0.173	0.443	15
6	0.691	0.186	0.639	13
7	0.946	0.200	0.958	4
8	0.593	0.199	0.699	12
9	0.903	0.200	0.928	6
10	0.818	0.192	0.785	9
11	1.000	0.200	1.000	2
12	0.681	0.198	0.749	10
13	0.908	0.200	0.932	5
14	0.817	0.192	0.786	8
15	**1.000**	**0.200**	**1.000**	**1**
16	0.681	0.198	0.749	11

4. Discussion

The application of the four optimization methods in the context of FMS shows good results for four of the five performances in the case of the two methods GP and VIKOR, and only two performance measures for DF and GRA methods (Table 9).

The results show that the MFT, WIP, TT, and WT performance measures met their targets for GP and VIKOR methods. Indeed, they all show relatively minor deviations from their target values. Only the deviation of RT reaches, respectively, −63.53% and −81.87% in the case of these two methods. On the other hand, in the case of DF and GRA methods, only the optimal values of TT and TR were close to their corresponding targets, while the deviations between the achieved values and the objective values in the case of WIP, WT, and MFT can reach +242.529%, +107.377%, and +83.293% respectively. Hence, the optimization results can be considered satisfactory in the case of GP and VIKOR methods. Meanwhile, it was not the case for the optimization results of the DF and GRA methods.

Table 9. MOSO results.

Performance Measure		MFT	WIP	TR	TT	WT
Goal value		426.596	4.323	78.303	129.757	1647.683
GP	Optimized value	433.000	4.425	28.557	129.922	1704.500
	Deviation value	+6.404	+0.102	−49.746	+0.165	+56.817
	Deviation in %	+1.501%	+2.359%	−63.530%	+0.127%	+3.448%
DF	Optimized value	781.921	14.806	78.257	129.971	3416.917
	Deviation value	+355.325	+10.483	−0.046	+0.214	+1769.234
	Deviation in %	+83.293%	+242.494%	−0.059%	+0.165%	+107.377%
GRA	Optimized value	781.921	14.806	78.303	129.917	3416.917
	Deviation value	+355.325	+10.484	0	+0.160	+1769.234
	Deviation in %	+83.293%	+242.529%	0%	+0.124%	+107.377%
VIKOR	Optimized value	428.164	4.409	14.195	129.757	1655.257
	Deviation value	+1.568	+0.086	−64.108	0	+7.574
	Deviation in %	+0.368%	+1.982%	−81.872%	0%	+0.460%

The two methods GP and DF require a higher level of analysis effort than the two methods of GRA and VIKOR. Indeed, in addition to the modeling and development of the simulation models, which is a common point tọ the four compared optimization methods, as well as the planning of experiments with the DoE method, the two methods GP and DF require relatively higher levels of expertise in the use of the analysis software MINITAB and LINGO. On the opposite side, the two methods GRA and VIKOR only need the development of the equations on Excel, which is within the reach of the majority of DMs. This has an impact on the applicability of the MOSO method.

Table 10 summarizes the performance of the four MSOSO methods being compared in this study. Signs '+' and "−" are assigned to the optimization methods based on their achieved optimization results and their applicability. A '+' is assigned to each method resulting in a good optimization result, which is expressed by reasonable or small deviations. On the other hand, a "−" is assigned to each method that leads to an optimization result characterized by high deviations. For applicability, a "−" is assigned to each method that requires a high level of analysis and expertise. In the opposite case, a '+' is assigned to this optimization method. These methods are then classified according to assigned signs. Any method obtaining two "+" signs will be considered the most efficient. On the other hand, if it obtains two "−" signs, it will be considered as the most mediocre one. In the case where the optimization method obtains both signs "+" and "−", the classification gives priority to the obtained optimization result. The best method is VIKOR, which belongs to group B in the proposed classification. It is followed by the GP method, from group A, since it reaches good optimization results, although it requires a considerable analysis effort. The GRA method, from group B, comes in third rank and the DF method, from the group A, closes the classification at the last rank. This classification shows that the use of optimization methods based on a metamodel does not always produce the best results.

Table 10. MOSO performances.

MOSO	Group	Optimization Result	Applicability	Rank
GP	A	+	−	2
DF	A	−	−	4
GRA	B	−	+	3
VIKOR	B	+	+	1

5. Conclusions

Various MOSO methods have been presented, developed, and used in the literature. These methods have been the subject of numerous classifications. However, the performance of these methods is not guaranteed due to the lack of comparative studies. Moreover, these classifications have been very diverse and are rarely related to the specific domain of manufacturing systems.

The objective of this research is two-fold. First, we proposed a new conceptual classification of MOSO methods applied to the context of MFS design. Second, four MOSO methods are selected according to this classification and compared through a case study related to the FMS design problem inspired by the literature. This comparison is based on the quality of the optimal solutions obtained by these methods as well as the degree of difficulty of their applicability through the necessary analysis effort and the degree of expertise of the user of these methods. All these studied methods are based on DoE. Two of them are metamodel-based approaches that incorporate the GP and the DF, respectively. The other two methods are not metamodel-based approaches and incorporate GRA and VIKOR, respectively. The comparative results show that the VIKOR method can result in a better optimization than GP, GRA, and DF methods in that order. It is clear, thus, that the use of MOSOs based on meta-models does not produce the best solution in all situations.

This research compares four MOSO methods applied in the context of FMS design. Some future research perspectives should be addressed:

- In this study, four MOSO methods are compared. Two methods belong to group A of the proposed new classification, while the other two belong to group B. The extension of the current comparison to other MOSO methods belonging to group C is the first objective of our interesting perspectives.
- The studied MOSO methods have been applied on a model of an FMS inspired from the literature. This model has six machines grouped in two cells in the CL and three departments in FL. In addition, this FMS processes only four products grouped into two families. Extending the comparison performed in this study to real and more complex FMSs to evaluate the reliability of MOSO methods is the second objective of our interesting perspectives.
- The experimental design developed in this comparison study and which is the basis for the simulation results used in the analysis and generation of optimization solutions is based on four factors: IAT, BS, RULE, and LAYOUT. These four factors are explored on the basis of two levels each. This number of factors and levels remains relatively limited and generates a limited number of experiments. The comparison of MOSO Methods in Manufacturing Systems characterized by a large number of factors and levels is the third objective of our interesting perspectives.
- The application of the compared MOSO methods proceeds through different steps to generate optimization solutions. These steps usually require the intervention of a user to transfer the results from one step to another. The integration of these analysis and optimization steps into the simulation software, as in the case of the OptQuest tool in several simulation tools, would be a very interesting perspective.

Author Contributions: Conceptualization, A.J. and W.H.; methodology, A.J. and W.H.; software, A.J.; validation, A.J., N.K.M., A.M.A. and F.M.; formal analysis, W.H. and A.M.A.; investigation, A.J. and N.K.M.; resources, A.J. and N.K.M.; data curation, A.J., A.M.A. and F.M.; writing—original draft preparation, A.J. and F.M.; writing—review and editing, A.J., A.M.A., N.K.M. and W.H.; visualization, A.J. and W.H.; supervision, W.H. and F.M.; project administration, A.J., A.M.A., W.H. and F.M.; funding acquisition, A.M.A. All authors have read and agreed to the published version of the manuscript.

Funding: This research was supported and funded by Taif University Researchers Supporting Project number (TURSP-2020/229), Taif University, Taif, Saudi Arabia.

Institutional Review Board Statement: Not applicable.

Informed Consent Statement: Not applicable.

Data Availability Statement: Data are contained within the article.

Acknowledgments: This research was supported by Taif University Researchers Supporting Project number (TURSP-2020/229), Taif University, Taif, Saudi Arabia. First, the authors are grateful for the financial support. Second, the authors would like to thank the editors and the three anonymous referees for their valuable and constructive comments on the first draft of this manuscript.

Conflicts of Interest: The authors declare no conflict of interest.

Appendix A

Table A1. MFT simulation results according to DoE.

EXP	Factor Levels				Replication									
	LAYOUT	RULE	IAT	BS	1	2	3	4	5	6	7	8	9	10
1	1	1	5	5	16,334.1	16,645.9	16,612.2	16,866.2	17,446.5	16,523.3	16,778.2	16,601.8	17,529.3	18,119.7
2	1	1	5	10	2312.8	1636.8	2435.7	3132.7	2374.0	3222.1	3128.3	1968.7	4831.2	2432.0
3	1	1	25	5	597.0	549.5	559.6	528.5	614.4	540.6	583.7	577.4	552.6	569.8
4	1	1	25	10	538.0	565.0	560.4	577.7	559.5	581.1	546.9	538.7	538.2	547.6
5	1	2	5	5	2574.7	3339.8	2014.4	2735.9	2348.5	2484.6	2698.9	1959.8	2784.8	3760.3
6	1	2	5	10	1881.7	1261.2	1463.6	1614.9	1166.6	1919.4	1276.1	1449.3	1662.2	2019.3
7	1	2	25	5	547.0	570.4	567.4	564.8	552.6	559.2	599.0	565.1	606.4	577.2
8	1	2	25	10	546.7	544.0	545.1	561.6	544.9	565.4	543.0	558.3	559.2	562.0
9	2	1	5	5	1460.9	1149.0	1314.0	711.9	825.9	699.7	848.5	846.1	832.2	859.4
10	2	1	5	10	1087.7	1052.0	1305.7	993.0	1214.9	1155.9	1038.4	1132.8	1153.9	1085.4
11	2	1	25	5	434.5	431.8	423.1	422.7	419.7	427.9	423.9	423.9	424.8	433.5
12	2	1	25	10	773.0	769.2	796.1	783.9	771.2	768.1	772.8	769.6	791.4	783.7
13	2	2	5	5	690.2	717.1	752.2	721.7	1260.2	741.1	721.4	748.1	690.2	777.0
14	2	2	5	10	1159.5	1053.5	1133.9	1121.7	1080.1	1187.4	1071.3	1085.1	1080.9	1094.4
15	2	2	25	5	438.0	427.5	421.1	422.6	428.0	425.6	423.8	426.3	436.3	432.3
16	2	2	25	10	773.0	769.0	796.1	783.9	771.0	768.1	772.8	769.6	791.4	783.4

LAYOUT: 1 = FL, 2 = C; RULE: 1 = FCFS, 2 = SPT.

Table A2. WIP simulation results according to DoE.

EXP	Factor Levels				Replication									
	LAYOUT	RULE	IAT	BS	1	2	3	4	5	6	7	8	9	10
1	1	1	5	5	276.4	285.9	287.2	288.6	296.7	286.0	286.4	288.0	299.9	308.5
2	1	1	5	10	38.5	27.3	41.1	52.3	39.8	53.7	52.1	33.1	80.0	40.6
3	1	1	25	5	6.1	5.6	5.7	5.4	6.2	5.5	5.9	5.9	5.6	5.8
4	1	1	25	10	5.4	5.7	5.6	5.8	5.6	5.9	5.5	5.4	5.4	5.5
5	1	2	5	5	325.0	345.3	302.0	349.1	318.0	322.5	332.9	329.0	371.0	343.7
6	1	2	5	10	58.2	23.5	38.3	29.4	22.4	65.8	22.7	35.2	48.4	38.5
7	1	2	25	5	5.9	6.2	6.2	6.1	6.0	6.0	6.5	6.1	6.6	6.3
8	1	2	25	10	5.6	5.6	5.6	5.7	5.6	5.8	5.5	5.7	5.7	5.7
9	2	1	5	5	24.4	19.2	22.0	12.0	13.9	11.7	14.2	14.2	14.0	14.4
10	2	1	5	10	18.2	17.6	21.9	16.6	20.3	19.3	17.4	18.9	19.3	18.1
11	2	1	25	5	4.4	4.4	4.3	4.3	4.3	4.3	4.3	4.3	4.3	4.4
12	2	1	25	10	7.8	7.7	8.0	7.9	7.7	7.7	7.8	7.7	8.0	7.9
13	2	2	5	5	14.4	13.1	14.2	13.2	24.0	13.6	16.3	13.7	12.5	13.0
14	2	2	5	10	20.1	18.1	19.7	19.4	18.6	20.6	19.1	18.7	18.7	18.8
15	2	2	25	5	4.5	4.4	4.3	4.3	4.4	4.4	4.4	4.4	4.5	4.5
16	2	2	25	10	7.8	7.8	8.0	7.9	7.8	7.8	7.8	7.8	8.0	7.9

LAYOUT: 1 = FL, 2 = C; RULE: 1 = FCFS, 2 = SPT.

Table A3. TR simulation results according to DoE.

EXP	Factor Levels				Replication									
	LAYOUT	RULE	IAT	BS	1	2	3	4	5	6	7	8	9	10
1	1	1	5	5	39.54	38.37	37.72	38.23	37.29	37.99	38.23	37.76	36.74	35.8
2	1	1	5	10	37.84	38.59	36.66	36.33	38.07	35.98	35.96	37.22	34.69	37.81
3	1	1	25	5	28.52	28.46	28.31	28.48	28.4	28.36	28.42	28.37	28.46	28.45
4	1	1	25	10	14.31	14.31	14.31	14.26	14.31	14.26	14.31	14.28	14.24	14.27
5	1	2	5	5	39.7	39.05	41.83	37.92	40.49	40.27	39.34	39.88	35.48	38.14
6	1	2	5	10	36.3	38.67	37.49	38.27	38.46	35.59	39.18	37.21	36.57	37.71
7	1	2	25	5	28.52	28.48	28.35	28.36	28.41	28.45	28.37	28.41	28.33	28.28
8	1	2	25	10	14.27	14.31	14.27	14.3	14.27	14.34	14.31	14.35	14.27	14.27
9	2	1	5	5	76.95	77.61	77.27	78.35	78.27	78.71	78.5	78.53	78.23	78.24
10	2	1	5	10	38.79	38.88	38.56	39.05	38.91	38.88	38.89	38.91	38.87	39.12
11	2	1	25	5	28.54	28.5	28.58	28.54	28.58	28.5	28.62	28.54	28.58	28.62
12	2	1	25	10	14.19	14.19	14.15	14.23	14.22	14.16	14.18	14.23	14.18	14.22
13	2	2	5	5	78.5	78.52	78.24	78.47	77.5	78.6	77.39	78.63	78.72	78.46
14	2	2	5	10	38.92	39.06	38.73	39.03	38.92	38.97	38.77	39.03	38.93	39.05
15	2	2	25	5	28.53	28.5	28.58	28.5	28.54	28.55	28.62	28.54	28.57	28.58
16	2	2	25	10	14.19	14.19	14.15	14.23	14.22	14.16	14.18	14.23	14.18	14.22

LAYOUT: 1 = FL, 2 = C; RULE: 1 = FCFS, 2 = SPT.

Table A4. WT simulation results according to DoE.

EXP	Factor Levels				Replication									
	LAYOUT	RULE	IAT	BS	1	2	3	4	5	6	7	8	9	10
1	1	1	5	5	82,836.8	85,752.2	86,140.3	86,578.0	89,250.2	84,996.3	85,347.6	85,214.9	89,138.4	92,596.2
2	1	1	5	10	21,801.7	15,089.4	22,715.8	29,844.8	22,789.5	30,417.2	29,423.0	18,018.8	46,237.5	23,032.6
3	1	1	25	5	1927.4	1838.2	1837.4	1681.1	2073.2	1759.2	1939.3	1923.3	1779.0	1920.5
4	1	1	25	10	3938.8	4089.8	4115.0	4115.8	4030.9	4087.6	3949.3	3991.3	4038.8	3955.6
5	1	2	5	5	26,779.8	25,064.3	14,124.1	24,226.4	13,939.3	19,553.4	21,059.2	12,639.7	14,090.3	37,979.1
6	1	2	5	10	17,439.0	11,361.6	13,119.9	15,247.0	10,391.8	19,065.1	11,499.0	13,196.1	15,287.0	19,011.7
7	1	2	25	5	1776.7	1854.7	1843.8	1829.6	1754.0	1829.4	1942.5	1837.2	2073.1	1883.9
8	1	2	25	10	4019.1	4010.4	4017.5	4086.7	4034.9	4129.1	3973.3	4076.8	4030.6	4063.8
9	2	1	5	5	6628.1	5230.5	5937.7	3069.7	3627.5	3008.1	3746.4	3740.8	3663.6	3781.3
10	2	1	5	10	10,025.4	9574.9	12,052.1	8981.6	11,220.2	10,425.2	9458.4	10,387.8	10,438.4	9835.0
11	2	1	25	5	1664.4	1655.0	1637.7	1637.6	1634.2	1676.7	1646.6	1630.7	1640.7	1653.1
12	2	1	25	10	6773.5	6798.3	6731.1	6853.6	6657.2	6739.5	6990.0	6786.5	6747.4	6830.7
13	2	2	5	5	3367.2	3093.6	3233.1	3102.4	5799.1	3205.1	3084.3	3216.5	3006.0	3062.0
14	2	2	5	10	10,596.6	9474.9	10,320.7	10,257.9	9831.7	10,804.9	9753.2	9931.9	9828.1	9990.0
15	2	2	25	5	1671.3	1631.5	1636.3	1632.5	1646.6	1654.3	1646.3	1651.4	1693.2	1689.2
16	2	2	25	10	6772.9	6796.7	6730.7	6853.6	6654.9	6739.5	6707.0	6786.5	6747.4	6827.4

LAYOUT: 1 = FL, 2 = C; RULE: 1 = FCFS, 2 = SPT.

Table A5. TT simulation results according to DoE.

EXP	Factor Levels				Replication									
	LAYOUT	RULE	IAT	BS	1	2	3	4	5	6	7	8	9	10
1	1	1	5	5	194.0	194.6	195.2	195.2	196.3	194.5	195.0	195.2	195.6	194.2
2	1	1	5	10	389.9	389.1	390.2	389.2	387.4	389.8	389.4	389.2	390.6	388.4
3	1	1	25	5	194.7	195.4	195.0	194.7	195.7	195.5	195.4	194.9	194.7	194.1
4	1	1	25	10	388.3	389.2	388.9	390.1	389.0	389.7	390.0	389.6	386.9	388.8
5	1	2	5	5	195.7	195.2	194.6	197.1	193.3	193.9	195.2	210.8	194.0	194.4
6	1	2	5	10	392.3	387.7	390.0	388.1	390.3	392.1	389.9	390.8	390.7	389.5

Table A5. *Cont.*

EXP	Factor Levels				Replication									
	LAYOUT	RULE	IAT	BS	1	2	3	4	5	6	7	8	9	10
7	1	2	25	5	195.5	194.8	195.1	195.9	195.2	195.4	194.9	195.3	195.3	195.5
8	1	2	25	10	391.7	388.7	390.9	389.4	389.7	390.2	388.3	387.6	392.7	390.1
9	2	1	5	5	130.0	130.1	130.3	130.4	129.9	129.9	129.7	130.7	129.7	130.3
10	2	1	5	10	262.1	259.5	261.0	260.6	260.6	259.9	258.8	261.0	260.5	259.6
11	2	1	25	5	130.2	129.0	130.2	130.5	129.4	130.1	129.2	130.3	129.6	130.4
12	2	1	25	10	258.4	258.8	261.2	259.6	261.2	260.4	261.0	259.3	262.8	260.4
13	2	2	5	5	129.4	129.7	129.8	129.9	129.8	130.4	130.1	129.9	130.3	129.9
14	2	2	5	10	262.6	260.9	259.8	261.4	260.5	261.7	260.1	260.8	262.4	260.7
15	2	2	25	5	130.0	128.9	130.0	130.2	130.0	129.9	129.2	129.9	129.4	130.2
16	2	2	25	10	258.4	258.8	261.2	259.6	261.2	260.4	261.0	259.3	262.8	260.4

LAYOUT: 1 = FL, 2 = C; RULE: 1 = FCFS, 2 = SPT.

References

1. Salah, B.; Alsamhan, A.M.; Khan, S.; Ruzayqat, M. Designing and Developing a Smart Yogurt Filling Machine in the Industry 4.0 Era. *Machines* **2021**, *9*, 300. [CrossRef]
2. Mian, S.H.; Salah, B.; Ameen, W.; Moiduddin, K.; Alkhalefah, H. Adapting Universities for Sustainability Education in Industry 4.0: Channel of Challenges and Opportunities. *Sustainability* **2020**, *12*, 6100. [CrossRef]
3. Florescu, A.; Barabas, S.A. Modeling and Simulation of a Flexible Manufacturing System—A Basic Component of Industry 4.0. *Appl. Sci.* **2020**, *10*, 8300. [CrossRef]
4. Saren, S.K.; Tiberiu, V. Review of flexible manufacturing system based on modeling and simulation. *Fascicle Manag. Technol. Eng.* **2016**, *24*, 113–118. [CrossRef]
5. Ammeri, A.; Hachicha, W.; Chabchoub, H.; Masmoudi, F. A comprehensive literature review of mono-objective simulation optimization methods. *Adv. Prod. Eng. Manag.* **2011**, *6*, 291–302.
6. Hachicha, W. A Simulation Metamodeling based Neural Networks for Lot-sizing problem in MTO sector. *Int. J. Simul. Model.* **2011**, *10*, 191–203. [CrossRef]
7. Hunter, S.R.; Applegate, E.A.; Arora, V.; Chong, B.; Cooper, K.; Rincón-Guevara, O.; Vivas-Valencia, C. An Introduction to Multiobjective Simulation Optimization. *ACM Trans. Modeling Comput. Simul.* **2019**, *29*, 1–36. [CrossRef]
8. Rosen, L.; Harmonosky, C.M.; Traband, M.T. Optimization of systems with multiple performance measures via simulation: Survey and recommendations. *Comput. Ind. Eng.* **2008**, *54*, 327–339. [CrossRef]
9. Barton, R.R.; Meckesheimer, M. Chapter 18 metamodel-based simulation optimization Handb. *Oper. Res. Manag. Sci.* **2006**, *13*, 535–574.
10. Amaral, J.V.S.; Montevechi, J.A.B.; Miranda, R.C.; de Sousa Junior, W.T. Metamodel-based simulation optimization: A systematic literature review. *Simul. Model. Pract. Theory* **2022**, *114*, 102403. [CrossRef]
11. Hassannayebi, E.; Boroun, M.; Jordehi, S.A.; Kor, H. Train schedule optimization in a high-speed railway system using a hybrid simulation and meta-model approach. *Comput. Ind. Eng.* **2019**, *138*, 106110. [CrossRef]
12. Diaz, C.A.B.; Aslam, T.; Ng, A.H.C. Optimizing Reconfigurable Manufacturing Systems for Fluctuating Production Volumes: A Simulation-Based Multi-Objective Approach. *IEEE Access* **2021**, *9*, 144195–144210. [CrossRef]
13. Červeňanská, Z.; Kotianová, J.; Važan, P.; Juhásová, B.; Juhás, M. Multi-Objective Optimization of Production Objectives Based on Surrogate Model. *Appl. Sci.* **2020**, *10*, 7870. [CrossRef]
14. Hussain, M.S.; Ali, M.A. Multi-agent Based Dynamic Scheduling of Flexible Manufacturing Systems. *Glob. J. Flex. Syst. Manag.* **2019**, *20*, 267–290. [CrossRef]
15. Apornak, A.; Raissi, S.; Javadi, M.; Tourzani, N.A.; Kazem, A. A simulation-based multi-objective optimisation approach in flexible manufacturing system planning. *Int. J. Ind. Syst. Eng.* **2018**, *29*, 494. [CrossRef]
16. Ahmadi, E.; Zandieh, M.; Farrokh, M.; Emami, S.M. A multi objective optimization approach for flexible job shop scheduling problem under random machine breakdown by evolutionary algorithms. *Comput. Oper. Res.* **2016**, *73*, 56–66. [CrossRef]
17. Freitag, M.; Hildebrandt, T. Automatic design of scheduling rules for complex manufacturing systems by multi-objective simulation-based optimization. *CIRP Ann.* **2016**, *65*, 433–436. [CrossRef]
18. Ammar, A.; Pierreval, H.; Elkosantini, S. A multiobjective simulation optimization approach to define teams of workers in stochastic production systems. In Proceedings of the 2015 International Conference on Industrial Engineering and Systems Management (IESM), Seville, Spain, 21–23 October 2015; pp. 977–986. [CrossRef]
19. Dengiz, B.; Tansel, Y.; Coskun, S.; Dagsalı, N.; Aksoy, D.; Çizmeci, G. A new design of FMS with multiple objectives using goal programming. In Proceedings of the International Conference on Modeling and Applied Simulation, Vienna, Austria, 19–21 September 2012; pp. 42–46.

20. Bouslah, B.; Gharbi, A.; Pellerin, R. Joint production and quality control of unreliable batch manufacturing systems with rectifying inspection. *Int. J. Prod. Res.* **2014**, *52*, 4103–4117. [CrossRef]
21. Iç, Y.T.; Dengiz, B.; Dengiz, O.; Cizmeci, G. Topsis based Taguchi method for multi-response simulation optimization of flexible manufacturing system. In Proceedings of the Winter Simulation Conference, Savannah, GA, USA, 7–10 December 2014; pp. 2147–2155.
22. Wang, G.C.; Li, C.P.; Cui, H.Y. Simulation Optimization of Multi-Objective Flexible Job Shop Scheduling. *Appl. Mech. Mater.* **2013**, *365–366*, 602–605. [CrossRef]
23. Zhang, L.P.; Gao, L.; Li, X.Y. A hybrid genetic algorithm and tabu search for a multi-objective dynamic job shop scheduling problem. *Int. J. Prod. Res.* **2013**, *51*, 3516–3531. [CrossRef]
24. Azadeh, A.; Ghaderi, S.F.; Dehghanbaghi, M.; Dabbaghi, A. Integration of simulation, design of experiment and goal programming for minimization of makespan and tardiness. *Int. J. Adv. Manuf. Technol.* **2010**, *46*, 431–444. [CrossRef]
25. Um, I.; Cheon, H.; Lee, H. The simulation design and analysis of a Flexible Manufacturing System with Automated Guided Vehicle System. *J. Manuf. Syst.* **2009**, *28*, 115–122. [CrossRef]
26. Syberfeldt, A.; Ng, A.; John, R.; Moore, P. Multi-objective evolutionary simulation-optimisation of a real-world manufacturing problem. *Robot. Comput.-Integr. Manuf.* **2009**, *25*, 926–931. [CrossRef]
27. Kuo, Y.; Yang, T.; Huang, G.W. The Use of a Grey-Based Taguchi Method for Optimizing Multi-Response Simulation Problems. *Eng. Optim.* **2008**, *40*, 517–528. [CrossRef]
28. Oyarbide-Zubillaga, A.; Goti, A.; Sanchez, A. Preventive maintenance optimisation of multi-equipment manufacturing systems by combining discrete event simulation and multi-objective evolutionary algorithms. *Prod. Plan. Control* **2008**, *19*, 342–355. [CrossRef]
29. Park, T.; Lee, H.; Lee, H. FMS design model with multiple objectives using compromise programming. *Int. J. Prod. Res.* **2001**, *39*, 3513–3528. [CrossRef]
30. Pitchuka, L.N.; Adil, G.K.; Ananthakumar, U. Effect of the conversion of the functional layout to a cellular layout on the queue time performance: Some new insights. *Int. J. Adv. Manuf. Technol.* **2006**, *31*, 594–601. [CrossRef]
31. Jerbi, A.; Chtourou, H.; Maalej, A.Y. Comparing functional and cellular layouts using simulation and Taguchi method. *J. Manuf. Technol. Manag.* **2010**, *21*, 529–538. [CrossRef]
32. Jerbi, A.; Chtourou, H.; Maalej, A.Y. Comparing Functional and Cellular Layouts: Simulation models. *Int. J. Simul. Model.* **2009**, *8*, 215–224. [CrossRef]
33. Chan, F.T.S. Impact of operation flexibility and dispatching rules on the performance of a flexible manufacturing system. *Int. J. Adv. Manuf. Technol.* **2004**, *24*, 447–459. [CrossRef]
34. Dominic, P.D.D.; Kaliyamoorthy, S.; Saravana, K.M. Efficient dispatching rules for dynamic job shop scheduling. *Int. J. Adv. Manuf. Technol.* **2004**, *24*, 70–75. [CrossRef]
35. Barton, R.R. Designing simulation experiments. In Proceedings of the 2013 Winter Simulation Conference, Washington, DC, USA, 8–11 December 2013; Pasupathy, R., Kim, S.-H., Tolk, A., Hill, R., Kuhl, M.E., Eds.; IEEE: Piscataway, NJ, USA, 2013; pp. 342–353.
36. Huang, J.T.; Liao, Y.S. Optimization of machining parameters of wire-EDM based on grey relational and statistical analyses. *Int. J. Prod. Res.* **2003**, *41*, 1707–1720. [CrossRef]

machines

MDPI

Communication

Drape of Composite Structures Made of Textile and 3D Printed Geometries

Tatjana Spahiu [1,*], Zlatin Zlatev [2], Elita Ibrahimaj [1], Julieta Ilieva [2] and Ermira Shehi [1]

[1] Faculty of Mechanical Engineering, Polytechnic University of Tirana, 1019 Tirana, Albania; elitaibrahimaj96@gmail.com (E.I.); eshehi@fim.edu.al (E.S.)

[2] Faculty of Technics and Technologies, Trakia University, 8602 Yambol, Bulgaria; zlatin.zlatev@trakia-uni.bg (Z.Z.); zhulieta.ilieva@trakia-uni.bg (J.I.)

* Correspondence: tspahiu@fim.edu.al

Abstract: Applications of 3D printing in the fashion industry have continued to attract interest from academia and industry in order to improve and add functionalities to products. Among these applications, an interesting one is 3D printing on textile fabric. Composite structures created by 3D printing and textile fabric change a drape by improving or worsening its appearance. The scope of this work is to evaluate the effect of various 3D printed geometries on textile fabric regarding fabric drapes. The drape coefficient of the created composite structure is evaluated using a drape tester built according to EN ISO 9073-9. The results taken are compared with an algorithm developed for determining drape parameters and 3D form representation using color digital images and their image histograms. The measured values of the drape coefficient are close, with a minimal difference, up to 4%. The 3D printed patterns show a significant effect on the drape coefficient of textile fabrics by depicting another way to modify fabric drapes and create complex shapes by using less material. This can be seen as an advantage in the fashion industry where complex geometries can be added to textile fabrics, while changing fabric drape and product personalization and adding functionalities for garments and technical textiles.

Keywords: 3D geometries; textile fabric; 3D printer; algorithm; drape parameters

1. Introduction

Continuing development of technology has shown its impact in different areas of production. The fashion industry has adapted these technologies to revolutionize it brands. However, one of the main requests is fulfilling or meeting consumers' preferences, especially the preferences regarding product personalization. Digitalization in the fashion industry speeds up product customization through different technologies, as part of the fourth industrial revolution. Industry 4.0, even known as smart manufacturing because of its main pillars such as 3D modeling, simulation, additive manufacturing, and digital data processing, has optimized the development of a wide range of products. Moreover, I4.0 is a synonym for sustainable production. As a sensitive topic, it has gained the attention of various researchers who highlight that there is a link between Industry 4.0 and sustainability, where I4.0 is a key factor in speeding up sustainable development [1].

The fashion industry is more exposed to the issue of sustainability due to its negative contribution through fast fashion, overconsumption, technological process, etc. Due to this fact, there are initiatives from companies that try to make an impact or give their contribution to sustainable ways of production. Improvements are seen in developed countries which are more likely to adopt Industry 4.0, but its implementation is still in its infancy [2]. In addition, researchers have found that retail companies are the first to use pillars of Industry 4.0 compared to textile companies. However, findings reveal a wide spread in fashion management [3]. Among the pillars of I4.0, 3D printing is a well-known technology for various applications in developing fashion products. Having a wide range of applications is driven by improvements in technology and a wide range of materials available. Of increasing interest are the applications for fashion products in order to improve or add functionalities. Objects are created by laying down on the printing support successive layers of materials extruded from the heated nozzle. Combinations of printing materials with textile fabric can be a great solution to add functionalities to garments or technical textiles. The composite structure created by these combinations may present adhesion problems that are linked with textile properties and 3D printing parameters. Various researchers have studied the influence of these parameters in order to find the optimum value or parameters that result in good adhesion [4,5].

According to Denker et al. [6], 3D printers are widely used devices because they make prototyping easy, but they come with limitations such as the type of material used and the relatively long printing time. The authors point out that the use of fabric as the main material on which to print overcomes some of the limitations of 3D printers. This expands the application of 3D printers in general.

The technology for 3D printing on fabrics was developed by Peng et al. [7]. The authors propose a technology in which the textile fabric is cut with a laser and applied layer by layer until a three-dimensional object is obtained. The method is suitable for the development of touch-sensitive sensing devices applied directly to the textile fabric. The disadvantage of this technology is that it requires complex equipment and long preparation of the 3D CAD model before its physical production.

Effective results with 3D printing on fabrics were reported by Pei et al. [8]. The authors managed to combine polymer with textile fabrics. According to the authors, the advantage of combining the two materials is that the textile fabric is a breathable and deformable material, and the polymer can strengthen its structure. As a result of this study, PLA exhibited the best properties compared to other polymer threads. Tested on eight types of fabrics, the material has good adhesion, low distortion, satisfactory bending strength, and high print quality. Grimmelsmann et al. [4] explored the possibilities of the presented 3D printing technologies as a tool for adding sensors and actuators to textile fabrics and creating smart products. The authors point out that it is necessary for the electronic devices attached by 3D printing to retain the characteristics for the entire planned service life of the textile fabrics. In addition, more research is needed on the combination of textile fabrics and polymeric materials that will ensure the efficient use of electronic devices for smart textiles.

Another interesting application is embroideries created by 3D printing on textile fabric. Three-dimensional printing in the form of embroidery on textile fabrics is proposed by Goudswaard et al. [5]. The authors apply their development as non-retaining buttons for smart textiles. The authors point out that more research is needed to find methods for maintaining flexibility and a design approach to making smart textiles using 3D printing techniques.

One of the important characteristics of fabrics is their drape. A number of methods have been developed to determine it. Methods for obtaining, processing, and analyzing images are suitable for such purposes as obtaining fabric drapes. They are affordable, and they do not require complex equipment and complex computational procedures. They are suitable for application directly on the production line when assessing the condition and quality of textile fabrics with 3D prints on them.

Modifications of the existing method for automated evaluation of drapery characteristics are proposed by Sanad et al. [9]. The authors found that image processing techniques were more effective for weaving drapes in nonwovens than in woolen fabrics.

Capdevila et al. [10], determine the drape characteristics of woolen fabrics with an accuracy of 76%. Using classifiers, the authors reported errors of up to 4% when using data from color digital images to determine the drape of woolen textile fabrics.

A modified method for determining the drape characteristics of digital images is proposed by Ragab et al. [11] by achieving a maximum accuracy of 90% compared with the standard method for determining the drape characteristics.

It has been found that the direct use of data on textile fibers embedded in textile fabrics is not a sufficient source of informative data when predicting the characteristics of draperies. In another work proposed by Pan et al. [12], the accuracy achieved is only 55–84% when using data on the characteristics of textile fibers to predict the drape of textile fabrics on their color digital images.

The characteristics of textile fabrics were used as input data to predict the drape characteristics. Using digital image data, Hussain et al. [13] achieved an accuracy of 76–82% through a modified draping coefficient and degree of bending of the textile fabric.

From the review of the available literature sources, it can be summarized that it is necessary to look for more effective methods for the analysis of images of draperies. The aim is to make them suitable for use in practice. One such method is image histogram analysis [14]. The advantage of histogram analysis is that it is an affordable image analysis method that does not require complex computational procedures and does not require a long time to process the data obtained. In this method, a digital image is divided into multiple segments, sets of pixels, also known as regions of interest in the image. This type of analysis is not sufficiently applied in the analysis of fabric drapes, on their color digital images.

An important practical application of the drape images, which has become necessary in recent years, is their use for three-dimensional visualization and simulation in the analysis of textile fabrics. According to Petrak et al. [15], an important parameter for adjustment in the 3D simulation of garments is the density of the used grid of the polygonal model. By optimally choosing this density, the resulting three-dimensional model will be simple enough and will carry enough information about the condition of the drapery.

Comparative analysis of actually measured drape characteristics and results of three-dimensional simulation was performed by Rudolf et al. [16]. The authors point out that there is no complete correspondence between the obtained 3D model and the real drape. According to them, it is necessary to look for more methods that will lead to sufficient accuracy in the three-dimensional drape simulation. One such solution is the precise determination and importation of characteristics of the fabrics in the simulation software, as indicated by Han et al. [17].

Suitable for determining the fabric drape are methods developed using 3D scanning and stereo video cameras. The disadvantage of these methods is that they are time-consuming and require the use of complex equipment for operation and maintenance, at high cost.

Methods for a more accurate 3D drape suitable for use with CAD systems have been proposed. Efendioglu et al. [18] indicate that 3D scanning, laser scanning, and stereo video cameras are applicable for this purpose. Common disadvantages of these methods are that they require the use of complex technical equipment, require a long time to obtain the three-dimensional shape of the scanned fabric drape, and involve a high cost of technical means used for this purpose.

Ju et al. [19] offer options, using different methods of draping the fabric and three-dimensional scanning. Such draping methods are hanging drape, Cusick's drapemeter, and modified Cusick's drapemeter.

Improving the 3D simulations of drapery requires prior knowledge of the composition and properties of the fabric, as well as accurate determination of the drape characteristics of their digital images. When looking for a connection between the type of fibers used in the textile fabric and the drape characteristics, it is necessary to look for appropriate calculation procedures because the direct use of data on the types of fibers in the fabric does not provide sufficient accuracy.

From the review of the available literature sources related to obtaining a three-dimensional shape of the fabric drape, it can be concluded that the existing approaches cannot provide sufficient accuracy of the obtained results. There are also insufficiently defined criteria for an objective assessment of the resulting drape from different types of fabrics.

The Cusick's drape method is suitable for creating three-dimensional shapes of fabric drapes on their two-dimensional visual images. Creating an algorithm through which it is possible to reproduce 3D drapery is suitable. The main research on the automation of this method aims to determine the drape parameters on 2D images. Insufficient research has been found in the available literature related to the reproduction of a three-dimensional shape from two-dimensional images obtained by this method. Reproduction of 3D shapes of fabric drapes obtained from 2D images is possible because the conditions under which the images were obtained, the location of the object in space, and the physical dimensions of the object are known in advance.

The aim of the present work is to review the possibilities, limitations, and evaluation of fabric drape with 3D printed geometries through comparative analyses and application of methods for image processing and analysis.

The main contributions that can be defined are as follows:

1. A comparative analysis was performed on the use of histograms of images as a criterion for assessing changes in textile fabrics with and without printed 3D elements on them.
2. An algorithm is proposed that realizes the basic principles for obtaining a 3D shape of fabric drape based on its 2D images. The algorithm is supplemented with basic functions to estimate the changes in the fabric drape after 3D printing geometries.
3. A comparative analysis is performed between the results obtained with the proposed algorithm and a standard drape tester. It was found that the proposed algorithm can be used to determine the characteristics of a fabric drape with printed 3D geometries with sufficient accuracy.

The article is organized in the following order: After the literature review and definition of the main problems and limitations related to the analysis of fabric drapes with 3D printed geometries, color digital images describe the materials and methods used to solve some of these problems. Textile fabrics, 3D printing materials, and methodology for image analysis of histograms are described. The main criteria for evaluating the changes on fabric drapes after printing 3D geometries are presented. The results of comparative analyses related to the determination of the drape coefficient by a standard method and by image analysis of histograms are reported. Finally, a discussion is presented in which the results obtained are compared with those of the available literature, and the content of the present work is summarized as conclusions.

2. Materials and Methods

2.1. Preparation of 3D Printing on Textile Fabric

Three-dimensional printing on textile fabric is realized using an Orcabot XXL (Prodim International B. V., Helmond, The Netherlands) fused deposition modeling (FDM) printer, which is equipped with the slicing software Orcabot Repetier Host V1.0.6 (Prodim International B. V., Helmond, The Netherlands).

Table 1 presents the main 3D printing parameters used for 3D printing for poly (lactic acid) (PLA).

Table 1. The 3D printing parameters.

3D Printing Parameters	PLA
Nozzle diameter	0.4 mm
Nozzle temperature	210 °C
Printing bed temperature	60 °C
Layer thickness	0.2 mm

The textile fabric chosen for 3D printing was 100% linen. Linen is a flax-based textile and is naturally hydrophilic like other textiles produced from natural fibers. The main fabric parameters presented are presented in Table 2. This property applied as a hydrophilic finish acts as a penetration and adhesion promoter [20]. Moreover, tests conducted on various textile fabrics confirm that adhesion is higher even in pure 100% linen textile [21].

Table 2. Fabric parameters.

Weave	Thickness	Areal Weight	Warp Density (1/cm)	Weft Density (1/cm)
Plain	0.2 mm	84 g/m^2	25 threads	22 threads

Various geometries such as triangles, pyramids, hexagons, and circles were 3D printed on textile fabrics and arranged in different ways on the circular sample. The selection of 3D printed geometries is arbitrary. This is necessary to give a broad overview of different possibilities of how fabric drapes are influenced by 3D printed patterns. It is well known that the use of plastic materials as 3D printed geometries in our case will increase the mass of these composite structures by influencing the drape coefficient. However, through these tests, we want to present a way to change a fabric drape with a low value of fabric masses due to 3D printed structures with very low heights. Of course, to make a comparison of the effects of printed geometries on fabric drape with almost unchanged masses of textiles, fabric drape is first measured without printing on it and then measured with 3D printed geometries. Tests are conducted with diverse 3D printed patterns, using a drape tester built according to EN ISO 9073-9 [22]. Tinker CAD from Autodesk (Autodesk, Inc., San Francisco, CA, USA) was used for modeling, and 3D models were exported as *.stl files by following their preparation for the 3D printing process using the "OrcabotRepetier-Host V1.0.6" slicing software.

2.2. *Image Histogram Analysis*

The sample image was transferred to the PC through a mobile phone. Photographs of the partly shaded areas were taken and processed using Matlab 2017b software (The Mathworks Inc., Natick, MA, USA). A sample graph is shown in Figure 1. The image histogram might be used for comparing two histograms from images that have different numbers of pixels. In addition, the range of change of these pixels is important too. The process of obtaining the image histogram of Sample 5 (hexagonal geometry arranged as "Y") is presented.

Figure 2 shows the principle of obtaining a fabric drape histogram. The original RGB image is converted to HSV. The data of the converted image are in "double" format. Of the three components of HSV, the V-component represents the regions of interest in the image. These regions are: support disk, drape and paper ring. These regions are indicated on the histogram of the V-component, which changes in the interval [0, 1]. The thresholds for the separation of the three object areas in the histogram, denoted as T_1 and T_2, respectively, are also shown.

One of the main tasks in histogram analysis is to determine the threshold at which object areas differ on the histogram. Figure 3 shows the basic principle corresponding to the histograms used in this work with three object areas.

The main error ΔT, which occurs when determining the threshold between the object areas in the histogram, is due to their overlap.

A basic algorithm for determining the global threshold was used to determine the value of "T" [23]. Implemented in Matlab environment, the algorithm has the following form:

```
count = 0;
T = mean2(f); %Mean of matrix elements
ready = false;
while (-ready)
end
```

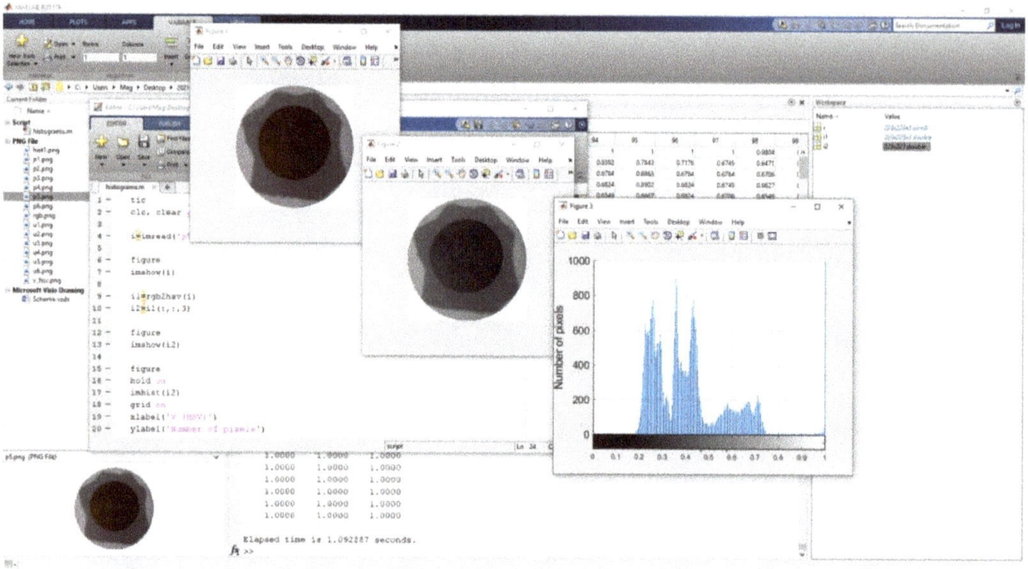

Figure 1. Image processing in Matlab environment.

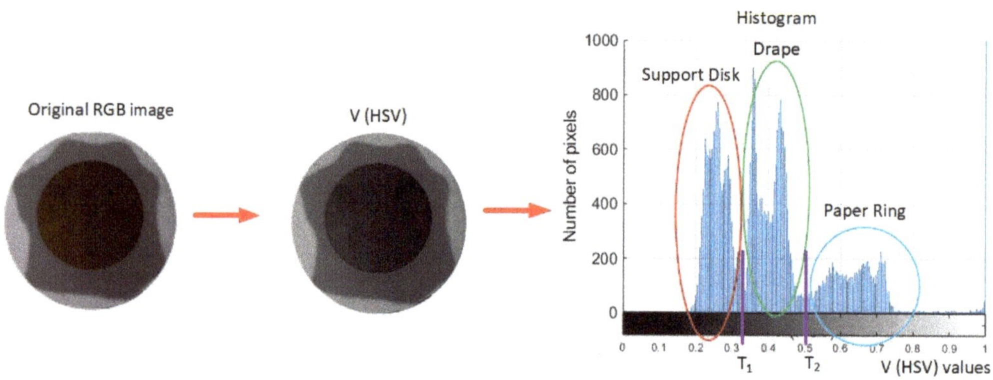

Figure 2. Obtaining histogram of drape image.

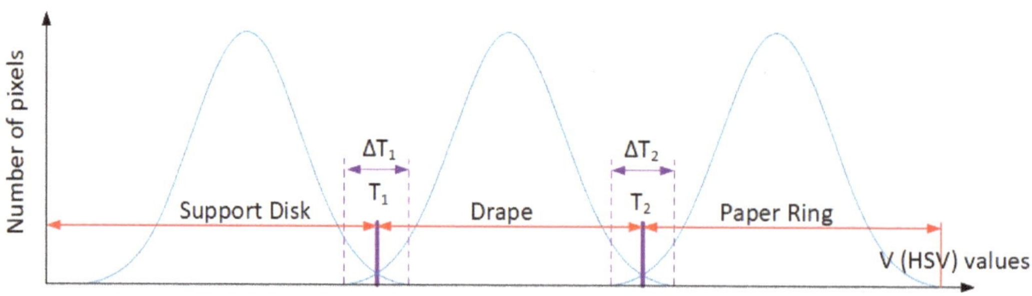

Figure 3. Determination of threshold values.

The presented text is colored is according to the formatting in the Matlab environment. Since the values of the V (HSV) component change in the interval [0, 1], it is not necessary to normalize the calculated value of the threshold T. The conditions for obtaining drape images are relatively the same. The lighting of the shot scene is homogeneous. As a result of the calculations with this algorithm, the error ΔT is less than 10%, as there is a relatively good separability of object areas. This makes the analysis acceptable for practical purposes. The histogram analysis is also used to determine the drape coefficient dependences.

2.3. Main Calculations for Development of an Algorithm for Determining Drape Parameters and 3D Form Simulation

Part of this work was the comparison of data generated by the use of an algorithm for determining drape parameters and 3D form representation by color digital images. In the last stage of algorithm implementation, the mathematical formulas presented by Sanad et al. [9] and Alikhanov et al. [14] are used. Using these equations, the drape coefficient and the coefficients of form, eccentricity, and minimum rectangle are calculated.

The drape coefficient (DC) is a coefficient for determining the drape ability of the fabric.

$$DC = \frac{A_{fs} - A_{sd}}{A_o - A_{sd}} \times 100,\ \% \quad (1)$$

Where A_{sd} is the area of the supporting disk (f_s—fabric shadow), A_o is the area of the circle around the drape, and A_{fs} is the drape area.

The coefficient of the form (K) represents the relationship between the drape contours and dimensions.

$$K = \frac{{P_{fs}}^2}{A_{fs}} \quad (2)$$

where A_{fs} is the drape area and P_{fs} is the perimeter of the drape.

The eccentricity (E) shows how the drape contour differs from an ideal circle.

$$E = \frac{D_{fs}}{d_{fs}} \quad (3)$$

where d_{fs} is the small diameter of the drape and D_{fs} is the large diameter of the drape.

The coefficient of the minimum rectangle (AR) shows how close the drape contour is to the minimum rectangle that describes it.

$$A_{mr} = d_{fs}D_{fs}; \quad AR = \frac{A_{fs}}{A_{mr}} \quad (4)$$

where A_{mr} is the area of the minimum rectangle described around the drape and A_{fs} is the drape area.

The ANOVA method was applied to assess the statistical difference between groups of samples, using the Statistica 12 software (TIBCO Software Inc., Palo Alto, CA, USA).

The proposed algorithm was compared with the standard drape test method. For this purpose, the correlation method was used. Using this method, the strength of the relationship (connection) between two measurement methods is checked. The distribution of the data was checked using the following methods: Shapiro–Wilks test, Kolmogorov–Smirnov test, and Lilliefors test. As a criterion for evaluation, a correlation coefficient R was used. At $R < 0.3$, there is very weak connection between the data; at $0.3 < R < 0.5$, the connection is weak; at $0.5 < R < 0.7$, the connection is moderate; and for $R > 0.7$, the connection is strong.

All data were processed at a level of significance $\alpha = 0.05$.

The next part of the algorithm was used to visualize the 3D shape of the fabric drape based on data obtained from the digital image, according to the standard ISO 9073-9; Figure 4 shows how to determine the drape parameters in order to restore its 3D shape. The method corresponds to that presented in Efendioglu et al. [18]. The most important drape parameters that participate in determining its three-dimensional contour are as follows:

radius of the work-piece (ε); support disk radius (ξ); the distance from each point of the outer contour of the drape to the support disk (λ); the height of the draped fabric at each reading point of the drape (d). Hence, the inclined distance (ε-ξ) for each point of the drape contour is determined.

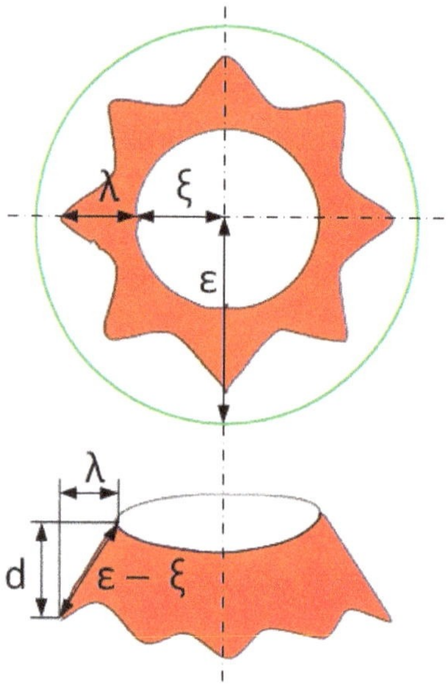

Figure 4. Determining the 3D shape of the drape.

3. Results

3.1. Obtaining Drapes with 3D Prints

One of the samples with 3D printed objects, still placed on the printing bed, is depicted in Figure 5. Due to the sufficiently large printing bed, it was possible to print the whole pattern in one printing session.

Figure 5. The 3D printing geometry on textile fabric—general view.

Drape tests of pure textile fabric and fabric with 3D printed geometries show differences in textile fabric given by the shaded areas. The types of single object geometry, their numbers, and their positions on the sample modify the drape of the textile fabric. Figure 6 presents textile samples with 3D printed geometries and the drape profiles of textile samples before and after 3D printing different geometries on them.

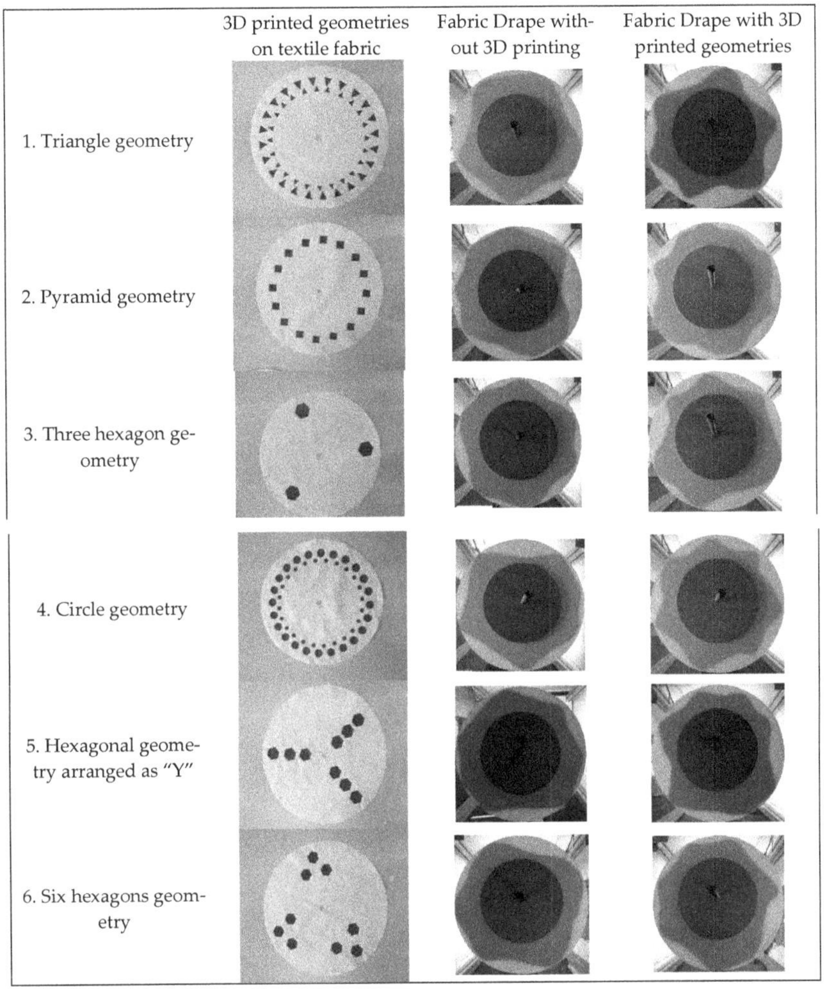

Figure 6. Drape of textile fabrics before (left images) and after 3D printing different geometries on them (right images).

3.2. Results from Histogram Analysis

The histograms taken for each sample before and after 3D printing geometries are presented in a graph. Figure 7 shows an example of histograms of the same fabrics, one without prints and the other with 3D prints. This figure is only an example to show what the histogram of one drape looks like. In this case, drapes with the number 5 hexagonal geometry are arranged as a "Y". It can be seen that in the pure fabric, the histogram becomes more compact and the areas of interest are closer to each other. After the implementation of 3D prints, the regions of interest are wider, but the threshold between different regions in the histogram can be defined with lower error rates, because there is less overlap, between different regions on the histogram. The same tendency was found for other fabrics with and without 3D printing.

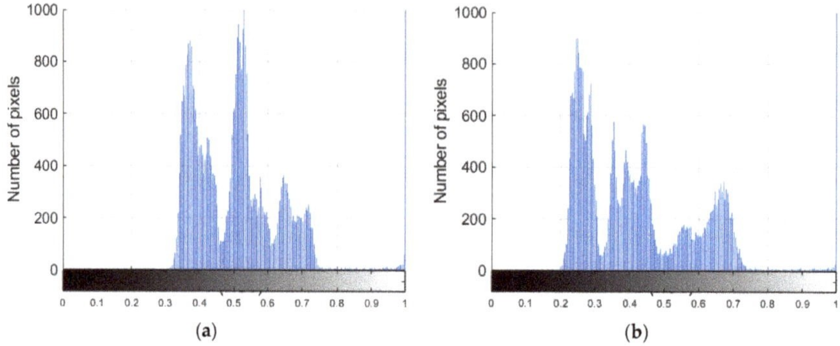

Figure 7. An example of image histograms of drapes. (**a**) Fabric drape without 3D printing; (**b**) fabric drape with 3D printed geometries.

Figure 8 shows the results of determining the maximum number of pixels of the region in the histogram in which the drapery is. This figure represents summarized data for all of the analyzed fabrics: with and without 3D printed geometries with their mean and standard deviation of the number of pixels and range of V (HSV) component for every region of interest in the drape images. It can be seen that the maximum number of pixels is higher for textiles without printing than for those with 3D prints. Conversely, in non-printed textile fabrics, the range of the V-component on which the histograms are made is wider than that in the draperies of 3D printed textile fabrics. In both types of textile fabrics, the highest number of pixels in the histogram of the drapery is observed in hexagonal geometries arranged as "Y" and six-hexagon geometries, and the lowest is observed in those with three-hexagon geometries and circle geometries. The widest ranges of the V-component with a drapery region are observed in the last two mentioned variants, both in the printed and in the unprinted textile fabrics.

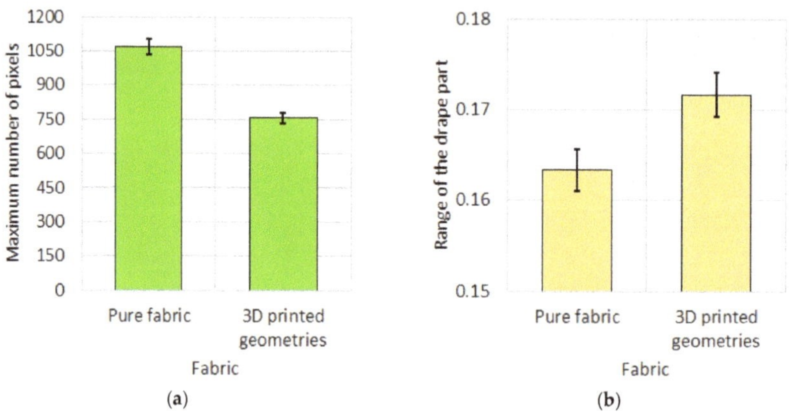

Figure 8. Amendment of the 3D prints on the image histograms of drapes. (**a**) Maximum number of pixels of the drape region in the histogram; (**b**) dimension of the range of V (HSV) component of drape region in the histogram. The values are significantly different at $p < 0.05$.

The presented results from the analysis of image histograms correspond to the changes in the values measured with a standard drape meter. Drape coefficients tested for pure textile fabric and fabric with 3D printed geometries are depicted in Table 3 below. The results show that there is a slight reduction of up to 4% in the drape coefficient after 3D printing is applied to the fabric. The biggest differences are in fabrics with circles and three-hexagon geometries. Pyramidal elements have the least influence on the draping of textile fabrics.

Table 3. The 3D printing drape parameters (DC, %).

Fabric Type / Geometries	Pure Fabric, %	3D Printed Geometries, %	Standard Deviation, %
Triangle geometries	40	37	2
Pyramid geometries	41	41	0
Three-hexagon geometries	45	40	4
Circle geometries	40	35	4
Hexagonal geometries arranged as "Y"	43	40	2
Six-hexagon geometries	43	42	1

Figure 9 shows the histograms of drape coefficients. The drape coefficient of the textile with 3D printed geometries is lower compared with the pure textile fabric, thus showing a positive effect of 3D printing on the drape coefficient. However, when analyzing the shape and number of nodes, it can be seen that the last one is less influenced by the imprinted 3D patterns [24]. All the 3D printed patterns show a significant effect on the fabric drape. This effect is explained by the number of 3D printed patterns, their positioning on the specimen, and their shape and size, as their height is the same for all geometries.

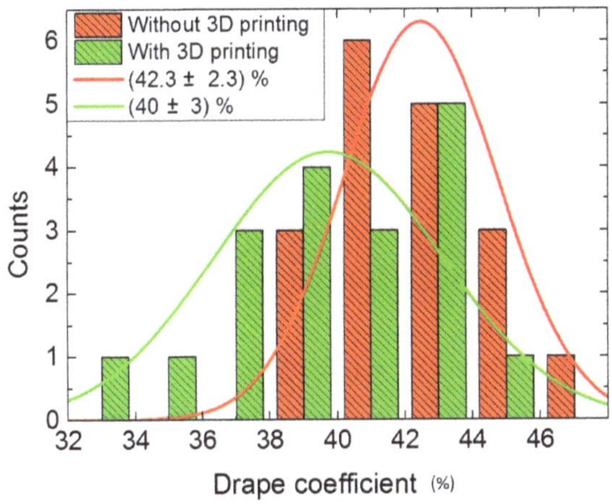

Figure 9. Drape coefficients without and with 3D printed geometries. The values are significantly different at $p < 0.05$.

3.3. An Algorithm for Determining Drape Parameters and 3D Form Representation by Color Digital Images

The proposed algorithm for determining drape parameters and 3D form representation by color digital images has steps that include not only the 3D representation of the drape, but also some calculations that describe the changes in the fabric when the drape is formed.

Each RGB image of fabric drape without or with 3D printed geometries is converted to HSV, and only the V-component is used. Contour noise is filtered out, and the contour area is compacted. The contour of the object areas is determined and drawn, including the supporting disk, drape contour, and the paper ring. The histogram of the image determines the object areas of the support disc, the drape contour, and the background. Moreover, the fold number is determined, which is an indicator of the flexibility of the textile fabric. For this purpose, a radius-vector function is used. In the last stage of algorithm implementation, mathematical formulas are used to calculate the drape parameters.

Figure 10 shows the results of the work from the proposed algorithms for determining the drape characteristics and simulating their 3D appearance through visualization. It can be seen that the obtained three-dimensional models are visually similar to those of digital images.

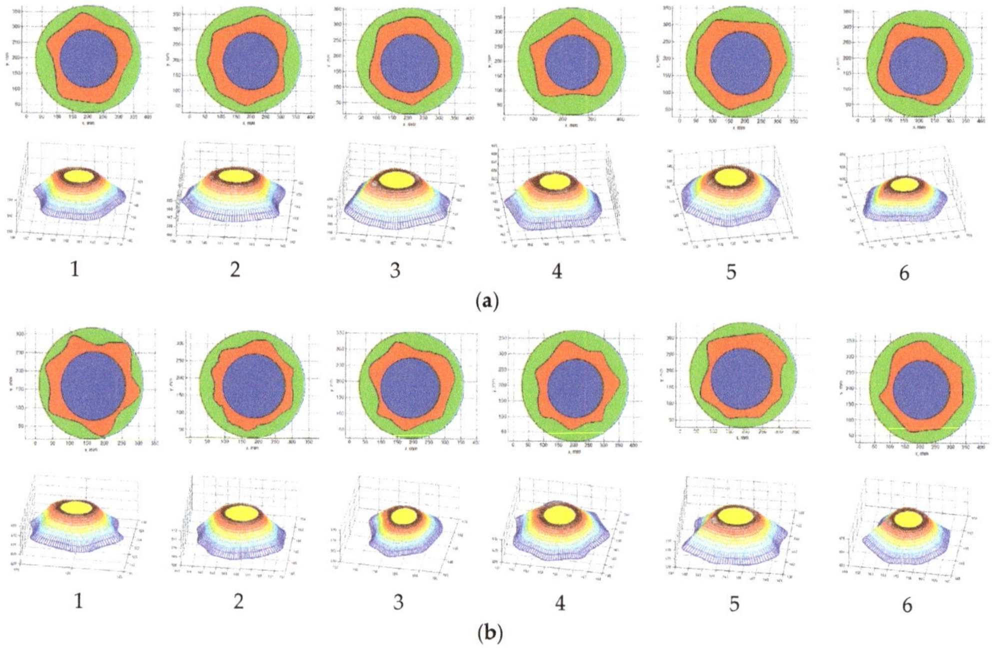

Figure 10. The 3D representation of the drape. (**a**) Without printing; (**b**) with 3D printed geometries. Description: Red, Drape; Green, Paper disk; Blue, Supporting disk. The color changes of 3D variants are according to the height of the drape.

Table 4 shows the results from the drape analysis with the proposed algorithm. For textiles with 3D prints, the number of folds (N) increases. There is an increase in the draping coefficient (DC) in the samples with 3D printed elements. For samples with 3D elements, the distance coefficient (RD) increases, while for fabrics without prints, it remains close to constant values. The same tendency is shown by the roundness factor (C) and the minimum rectangle factor (AR). The form factor (K) decreases for fabrics with 3D printing, while for those without print, it remains close and constant. Printing 3D elements does not affect the eccentricity (E). It has the same values for both types of textiles.

Figure 11 shows the results for the number of peaks. The highest number of peaks is observed in Sample 2 with 3D prints. The lowest number is in samples with circle geometry and six-hexagon geometry. When observing the average values of the number of peaks, it is clear that those with 3D prints have a higher number of peaks compared to the pure fabric.

Figure 12 shows changes in RD, according to the type of fabric with and without 3D prints. It can be seen that the application of 3D prints reduces the flexibility of the fabric. It is lowest in samples with tightly spaced prints. The fabric with hexagonal geometry arranged as "Y" has higher values of RD, and the fabrics with triangle geometry prints have lower values of RD.

Table 4. Characteristics of draperies determined by the proposed algorithm.

Sample	Pure Fabric (Without 3D Printing)						Fabric with 3D Printing Geometries					
Number	1	2	3	4	5	6	1	2	3	4	5	6
N	6	6	5	4	5	4	6	8	7	7	7	6
DC, %	47.46	50.06	47.14	38.67	54.40	45.75	56.07	46.02	44.16	40.21	49.27	44.75
RD	188.73	221.12	173.71	151.61	180.30	149.71	42.99	85.09	92.09	135.00	201.79	178.65
C	0.68	0.67	0.72	0.67	0.72	0.72	0.61	0.68	0.72	0.67	0.68	0.68
K	18.50	18.75	17.45	18.82	17.45	17.49	20.62	18.38	17.55	18.89	18.36	18.35
E	1.08	1.12	1.05	1.09	1.07	1.07	1.04	1.05	1.03	1.02	1.08	1.09
AR	0.57	0.56	0.59	0.57	0.59	0.60	0.56	0.64	0.62	0.58	0.59	0.57

N, number of peaks; DC, drape coefficient; RD, drape distance ratio; C, circularity; K, coefficient of form; E, eccentricity; AR, coefficient of minimum rectangle.

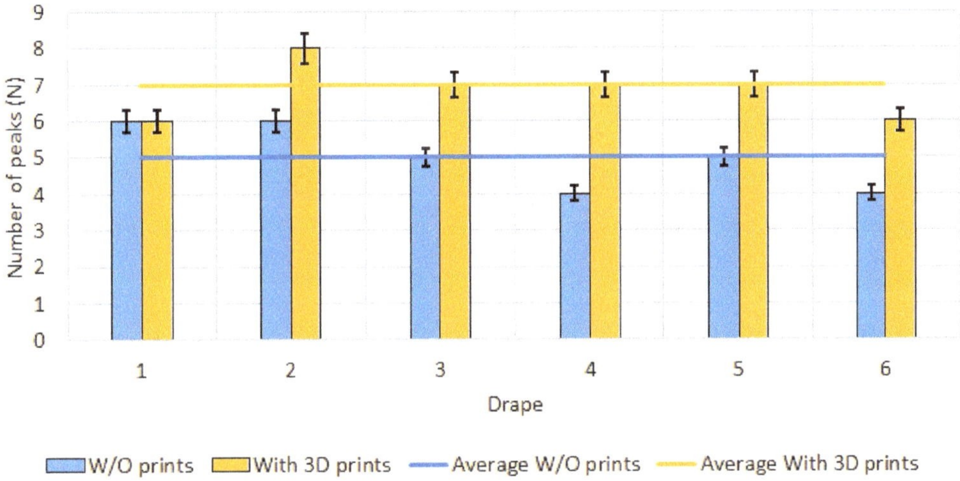

Figure 11. Comparison between obtained number of peaks for drapes with and without 3D printing (N). The values are significantly different at $p < 0.05$.

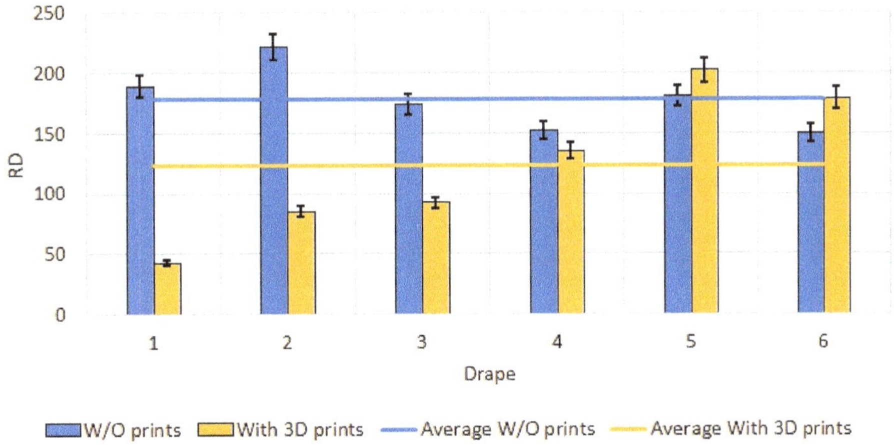

Figure 12. Changes in RD. The values are significantly different at $p < 0.05$.

Figure 13 shows changes in the coefficient of form (K), according to the type of textile with and without 3D prints. Adding 3D printed elements increases the values of the aspect ratio. The highest values of this coefficient are observed in samples with triangle geometry and circle geometry, which have more closely spaced elements compared to the other samples. The lowest values are exhibited by the sample with three-hexagon geometry. That geometry does not affect the coefficient of form, in comparison with other geometries.

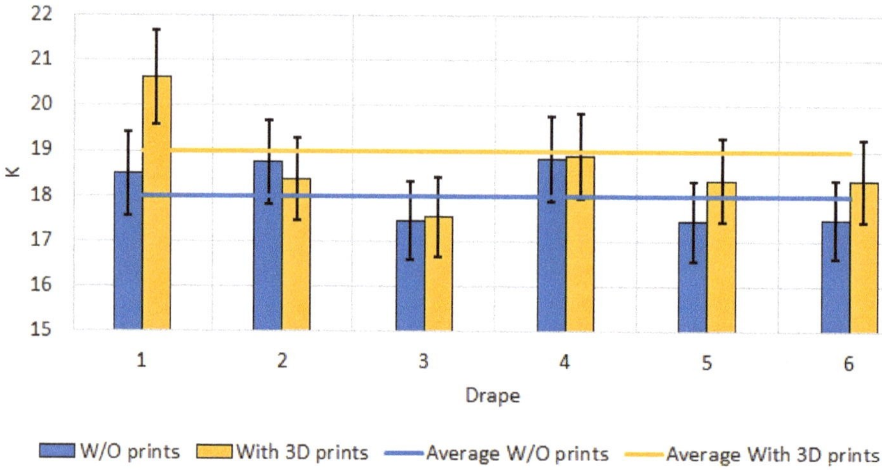

Figure 13. Changes in coefficient of form K. The values are significantly different at $p < 0.05$.

Table 5 shows the results of a comparative analysis of the drape coefficient determined with a laboratory drape meter and the algorithm proposed here. It can be seen that the measured values are close, with minimal difference. This can be proved by the obtained strong correlation coefficient (R = 0.7).

Table 5. Comparative analysis of drape coefficient (DC, %) measurement methods.

Number	Structure of the Sample	Fabric without 3D Printed Geometries DC, %		Fabric with 3D Printed Geometries DC, %	
		DM	A	DM	A
1	Triangle geometries	40.33	47.46	37.37	56.07
2	Pyramid geometries	41.93	50.06	41.67	46.02
3	Three-hexagon geometries	45.39	47.14	40.21	44.16
4	Circle geometries	40.04	38.67	35.96	40.21
5	Hexagonal geometries arranged as "Y"	43.73	54.40	40.38	49.27
6	Six-hexagon geometries	43.36	45.75	42.71	44.75

DC, drape coefficient; DM, drape meter; A, algorithm.

4. Discussion

In the present work, the results indicated in the available literature are supplemented, updated, and partially improved. Three-dimensional printing technologies have been proven to be an effective method for improving the appearance and properties of textile fabrics. This complements the summaries of Denker et al. [6], according to which the use of fabric as the main material on which to print overcomes some of the limitations of 3D printers.

In this paper, a clear and objective assessment of the properties of the obtained textile fabrics with 3D printing is proposed. The results of Capdevila et al. [10] depict that an accuracy of 76% was achieved in the analysis of draperies of woolen fabrics. In the present work, a close accuracy of 70% has been achieved in the analysis of textile linen fabrics. Compared to the standard method, the errors are of the order of 4%. In addition, in the present work, the data for fabrics are enriched with 3D elements printed on them.

The results obtained here complement those of Hussain et al. [13]. By using histograms of images, an accuracy of over 80% was achieved compared to the standard method for determining drape characteristics, close to that of the presented authors. An accuracy higher than this was achieved by Ragab et al. [11], who reported a maximum accuracy of 90%.

Sanad et al. [9] found that image processing techniques are more effective for weaving draperies in nonwovens than in woolen fabrics. This is complemented in the present work by methods for analyzing histograms of images, as well as an algorithm for obtaining three-dimensional shapes from two-dimensional images, with added features to assess the basic drape characteristics.

The methods proposed here for obtaining the 3D shape of drapes from their 2D images are suitable for use with 3D CAD systems. This complements the findings of Efendioglu et al. [18], who showed that 3D scanning, laser scanning, and stereo video cameras are applicable for this purpose. The use of color digital images removes the limitations of these methods, which require the use of complex technical equipment and a long time to obtain the three-dimensional shape of the scanned fabric drape. The use of digital images does not require complex and expensive equipment or complex computational procedures for processing the data obtained.

The results of the present work fulfill one of the recommendations of Rudolf et al. [16], which indicates that it is necessary to look for more methods that will lead to sufficient accuracy in the three-dimensional simulation of fabric drape. The algorithm proposed in this paper for obtaining a 3D shape of the fabric drape implements the recommendation.

The following research will be related to removing the limitations of the algorithm proposed here. More research is needed to assess the impact of factors influencing textile drape. This will improve the accuracy of the laziness obtained by removing the limitations of the algorithm proposed here.

5. Conclusions

The drape is an important property for garment appearance and is considered by pattern makers for garment design and production. Effects of design or seams are studied by different authors to modify garment drapes. With the application of innovative technologies such as 3D printing, we have tested the effect of 3D printing various geometries according to number, shape, and positioning or arrangement on linen fabric. Drape coefficients measured on pure fabric and fabric with 3D printed patterns depict another way to modify fabric drape and create complex shapes by using less material compared to the traditional manufacturing methods, which is seen as an advantage in the fashion industry.

A comparative analysis of the use of image histograms as a criterion for assessing changes in textile fabrics with and without printed 3D elements on them was conducted. It was found that in the textile fabrics without printed 3D elements, the region of interest in the histogram has a smaller range and more pixels compared to the textile fabrics on which 3D elements are printed. The results were confirmed by the measurement with a standard drape tester. In this analysis, higher values of the draping coefficient were obtained for fabrics without 3D elements than for those without such elements. The difference in the drape coefficient is up to 4% for both types of fabrics.

An algorithm is proposed that implements the basic principles for obtaining a 3D shape of a drape based on its 2D images. The algorithm is supplemented with basic functions for estimating the changes in fabric drape after printing 3D elements on it. Compared to the standard drape tester, the proposed algorithm offers 70% accuracy in determining the drape characteristics of fabrics with 3D printed shapes on them.

From the comparative analysis of the results obtained in the present work with those indicated in the available literature, it was found that the data from the available literature have been updated, improved, and supplemented. Software and hardware are offered, through which the characteristics of fabric drape with and without 3D printed elements can be determined in an accessible and objective way. These instruments can be used for these purposes, ensuring sufficient accuracy of measurements.

The achieved accuracy is sufficient for the application of the algorithm for practical purposes, in determining the drape characteristics and their 3D reproduction from their two-dimensional color digital images.

The methods and technical means proposed in the present work can be used in the analysis of fabrics from which to produce garments, curtains, upholstery, etc.

Author Contributions: Conceptualization, T.S. and Z.Z.; methodology, T.S.; algorithm, Z.Z.; validation, E.I., E.S. and J.I.; formal analysis, E.S.; investigation, Z.Z., J.I., E.S. and E.I.; resources, T.S., E.S., E.I. and Z.Z.; data curation, T.S.; writing—original draft preparation, T.S. and Z.Z.; writing—review and editing, T.S.; visualization, Z.Z. and T.S.; supervision, T.S.; project administration, T.S.; funding acquisition, Z.Z. and J.I. All authors have read and agreed to the published version of the manuscript.

Funding: This research received no external funding.

Institutional Review Board Statement: Not applicable.

Informed Consent Statement: Not applicable.

Data Availability Statement: The data will be available on request.

Acknowledgments: The authors would like to thank Andrea Ehrmann for the support during the work in the framework of Students Erasmus + mobility between the Polytechnic University of Tirana and Fachhochschule Bielefeld University of Applied Sciences in Germany.

Conflicts of Interest: The authors declare no conflict of interest.

References

1. Beltrami, M.; Orzes, G.; Sarkis, J.; Sartor, M. Industry 4.0 and sustainability: Towards conceptualization and theory. *J. Clean. Prod.* **2021**, *312*, 12773. [CrossRef]
2. Ahmad, S.; Miskon, S.; Alabdan, R.; Tlili, I. Towards Sustainable Textile and Apparel Industry: Exploring the Role of Business Intelligence Systems in the Era of Industry 4.0. *Sustainability* **2020**, *12*, 2632. [CrossRef]
3. Madsen, D. The Emergence and Rise of Industry 4.0 Viewed through the Lens of Management Fashion Theory. *Adm. Sci.* **2019**, *9*, 71. [CrossRef]
4. Grimmelsmann, N.; Martens, Y.; Schäl, P.; Meissner, H.; Ehrmann, A. Mechanical and electrical contacting of electronic components on textiles by 3D printing. *Procedia Technol.* **2016**, *26*, 66–71. [CrossRef]
5. Goudswaard, M.; Abraham, A.; Goveia da Rocha, B.; Andersen, K.; Liang, R.-H. FabriClick: Interweaving pushbuttons into fabrics using 3D printing and digital embroidery. In Proceedings of the Crafting Textiles, DIS '20, Eindhoven, The Netherlands, 6–10 July 2020; pp. 379–393.
6. Denker, M.; Diec, D.; Saglam, M. Possibilities of Fabric Embeds in 3D Printing. 2021. Available online: https://merlindenker.de/files/Fabric_Embeds_in_3D_Printing.pdf (accessed on 23 June 2022).
7. Peng, H.; Mankoff, J.; Hudson, S.; McCann, J. A Layered Fabric 3D Printer for Soft Interactive Objects. In Proceedings of the 33rd Annual ACM Conference on Human Factors in Computing Systems (CHI '15), Seoul, Korea, 18 April 2015; Association for Computing Machinery: New York, NY, USA, 2015; pp. 1789–1798. [CrossRef]
8. Pei, E.; Shen, J.; Watling, J. Direct 3D printing of polymers onto textiles: Experimental studies and applications. *Rapid Prototyp. J.* **2015**, *21*, 556–571. [CrossRef]
9. Sanad, P.; Cassidy, T.; Cheung, T.; Evans, E. Fabric and garment drape measurement—Part 2. *J. Fiber Bioeng. Inform.* **2013**, *6*, 1–22. [CrossRef]
10. Capdevila, X.; Carrera-Gallissà, E. Application of discriminant analysis to parameters describing the drape for two types of woven fabrics. *J. Text. Inst.* **2015**, *107*, 784–790. [CrossRef]

11. Ragab, A.; Fouda, A.; El-Deeb, H.; Hemdan, A. A simple method for measuring fabric drape using digital image processing. *J. Text. Sci. Eng.* **2017**, *7*, 1000320.
12. Pan, N.; Lin, C.; Xu, J. A new method for measuring fabric drape with a novel parameter for classifying fabrics. *Fibers* **2019**, *7*, 70. [CrossRef]
13. Hussain, A.; Xin, W.; Naveed, T.; Waqar, I.; Iqbal, D.; Zhong, Y.; Yu, Z. Optimization of fabric drape measurement based on 3D model. *Industriatextila* **2019**, *70*, 318–323. [CrossRef]
14. Alikhanov, J.; Penchev, S.; Georgieva, T.; Moldazhanov, A.; Shynybay, Z.; Daskalov, P. An indirect approach for egg weight sorting using image processing. *Food Meas. Charact.* **2017**, *12*, 87–93. [CrossRef]
15. Petrak, S.; Naglic, M.; Rogale, D.; Geršak, J. Analysis of polygonal computer model parameters and influence on fabric drape simulation. *Materials* **2021**, *14*, 6259. [CrossRef] [PubMed]
16. Rudolf, A.; Zadravec, M.; Stjepanović, Z. Investigations regarding the effects of simulating parameters during 3D garments' drape simulations. *Fibres Text. East. Eur.* **2016**, *24*, 143–150. [CrossRef]
17. Han, M.-G.; Chang, S.-H. Draping simulation of carbon/epoxy plain weave fabrics with non-orthogonal constitutive model and material behavior analysis of the cured structure. Composites Part A. *Appl. Sci. Manuf.* **2018**, *110*, 172–182. [CrossRef]
18. Efendioglu, N.; Mutlu, M.; Pamuk, O. An investigation on usability of 3D visualization and simulation programs in leather apparel. *J. Text. Inst.* **2021**, *112*, 1–16. [CrossRef]
19. Ju, E.; Choi, M. Estimating cloth simulation parameters from a static drape using neural networks. *IEEE Access* **2020**, *8*, 195113–195121. [CrossRef]
20. Korger, M.; Bergschneider, J.; Lutsz, M.; Mahltig, B.; Finsterbusch, K.; Rabe, M. Possi ble Applications of 3D Printing Technology on Textile Substrates. *IOP Conf. Ser. Mater. Sci. Eng.* **2016**, *141*, 012011. [CrossRef]
21. Meyer, P.; Döpke, D.; Ehrmann, A. Improving adhesion of three-dimensional printed objects on textile fabrics by polymer coating. *J. Eng. Fibers Fabr.* **2019**, *14*, 1–7. [CrossRef]
22. Kramer, L. SFAE. Construction of an inexpensive dreapeometer. In Proceedings of the International Textile Conference 2016, Bangalore, India, 22 January 2016.
23. Mladenov, M.; Penchev, S.; Dejanov, M.; Mustafa, M. Automatic classification of grain sample elements based on color and shape properties. *Sci. Bull. Ser. C Univ. Politech. Bucharest* **2011**, *73*, 39–54.
24. Spahiu, T.; Fafenrot, S.; Grimmelsmann, N.; Piperi, E.; Shehi, E.A. Varying fabric drape by 3D-imprinted patterns for garment design. *IOP Conf. Ser.: Mater. Sci. Eng.* **2017**, *254*, 172023. [CrossRef]

machines

MDPI

Review

Assessment of Industry 4.0 for Modern Manufacturing Ecosystem: A Systematic Survey of Surveys

Fotios K. Konstantinidis [1,*], Nikolaos Myrillas [1], Spyridon G. Mouroutsos [2], Dimitrios Koulouriotis [1] and Antonios Gasteratos [1]

[1] Department of Production and Management Engineering, Democritus University of Thrace, 12 Vas. Sophias, GR-671 32 Xanthi, Greece
[2] Department of Electrical and Computer Engineering, Democritus University of Thrace, Kimmeria, GR-671 00 Xanthi, Greece
* Correspondence: fokonsta@pme.duth.gr; Tel.: +30-6945-24-3307

Abstract: The rise of the fourth industrial revolution aspires to digitize any traditional manufacturing process, paving the way for new organisation schemes and management principles that affect business models, the environment, and services across the entire value chain. During the last two decades, the generated advancements have been analysed and discussed from a bunch of technological and business perspectives gleaned from a variety of academic journals. With the aim to identify the digital footprint of Industry 4.0 in the current manufacturing ecosystem, a systematic literature survey of surveys is conducted here, based on survey academic articles that cover the current state-of-the-art. The 59 selected high-impact survey manuscripts are analysed using PRISMA principles and categorized according to their technologies under analysis and impact, providing valuable insights for the research and business community. Specifically, the influence Industry 4.0 exerts on traditional business models, small and medium-sized enterprises, decision-making processes, human–machine interaction, and circularity affairs are investigated and brought out, while research gaps, business opportunities, and their relevance to Industry 5.0 principles are pointed out.

Keywords: Industry 4.0; SME; digital footprint; business model; survey of surveys

1. Introduction

At the turn of the 21st Century, a revolutionary trend was inaugurated that digitized production processes and transformed the business world. This trend triggered the onset of the fourth industrial revolution, impacting the entire production life cycle, from first-tier production and business organization to recycling processes. The "Industrie 4.0" (I4.0) initiative, coined by the German Ministry of Education and Research and presented at the Hanover Fair in 2011, constitutes a flagship concept that promotes factory digitalization and supply chain interconnection [1]. These ideas, presented in what some consider the most established German initiative, cascaded other relevant approaches around the globe, namely Smart Manufacturing, Factories 4.0, Industrial Internet, Manufacturing 3.0, and Made-in-China 2025 [2]. These ground-breaking endeavours contributed to the development of cutting-edge technologies in both the industrial sector area and major industries such as construction [3] and medical [4].

The impact of the fourth industrial revolution is reflected in the fusion of cutting-edge technologies that directly affect the operations and functions of industries. Billions of interconnected intelligent devices, edge processing power, smart storage services, self-adaptable sliced networks, and knowledge transfer platforms offer an unprecedented spread of innovation and multi-disciplinary implementation [5]. Published research and best practice reports have shown that I4.0 offers the potential to provide zero-fault production and controlled environments, as facilities and infrastructures have become more and more digitized without technological silos [6]. These advancements and increased flexibility have turned existing challenges into valuable opportunities that support the digital transition. On the other hand, innovations such as re-configurable manufacturing systems (RMSs) aim to enhance the factory's response to fluctuating markets and enable speedy and cost-effective competition in dynamic market environments [7]. During this change, in addition to the technological challenges of interoperability and compatibility issues, knowledge challenges have also arisen from organizational assemblies, such as the reluctance of manufacturers to adopt I4.0 solutions or the suitability of those in traditional business models [8–11].

The advancements and innovation provided by I4.0 should attract the attention of industrialists and decision-makers, as rapid changes are also expected to affect obsolete business and management models, based on real-time data and analysis, derived from the manufacturing facilities and the product's life cycle [12,13]. In every management stage, the decision-making process becomes more challenging, due to an abundance of available information. Note that the information validity process stands as a critical hurdle in corporate C-suite decisions, when inaccurate data can contribute to unprofitable strategic plans, costing appreciable amounts of money. As the technological dependencies of I4.0 are adopted in decision-making processes by large enterprises, business leaders can strengthen their strategies and increase their profit [14]. On a smaller scale, dominated by small and medium-sized enterprises (SMEs), data-based decisions lead to rapid tech-savvy solutions, accelerating the digital transformation and ensuring power supply safety within the factory [15], such that small and big businesses have equal opportunities to remain competitive in the global market [16].

Since the first industrial revolution, manufacturers have been challenged to produce more and more goods to meet the ever-increasing demands, initially adopting mass production techniques, while most recently, lean production schemes. As we move forward into the fourth industrial revolution, the smart factory concept has become a reality, which utilizes technologies (e.g., Industrial Internet of Things (IIoT), machine vision, digital twins) to enable holistic visibility of operations and production flexibility [14]. Fully automated facilities, machines, and material flow, within the manufacturing area, are managed by intelligent networks and smart execution systems, as a result of data exchange among all elements and intelligent functions [17]. Although a significant reduction in human intervention is to be anticipated, human–machine collaboration technologies provide the common "integration point" between workers and machines, thereby harmonizing a smooth interaction while enabling effective collaboration [18].

This new era has metamorphosed the conventional supply chain ecosystem, where stakeholders have traditionally reported to sales departments to inquire about factory supplies or customers' deliveries, into a digitally independent supply chain network, providing horizontally integrated operations between customers and suppliers [19]. Advanced planning processes and (semi-)autonomous vehicles, empowered by technological pillars (5G, big data, cloud and edge computing, IIoT, blockchain), provide intelligence and real-time traceability within the supply chain network, while adaptability, speed, and service quality are maintained [20]. As the supply chain network grows, the personalised product availability is increased, creating the need for sustainable supply chains. The key link between them is provided by the circular economy policies that have been advanced through the Industry 4.0 ecosystem [21]. The closed circular loop model and industrial symbiosis platforms are central components of I4.0 and drive the *Fifth Industrial Revolution* (Industry 5.0) [22], in which end-of-life products, scraps, or by-products are recycled to be used as raw materials in the same (or different) manufacturing processes, thereby extending a product's life cycle and reducing waste mass that is harmful for the environment [23].

Generally, the major purpose of modern manufacturing is to speed production and related industries, such as logistics, while creating new business opportunities and models by going beyond automation and optimization, taking also into account sustainability and human centring aspects. Even established sectors are expected to shift their focus from products to services for their enterprises' success [24]. For example, the growth of advanced technologies immediately fostered the buildup of subsidiary R&D capabilities, since it raised the complexity of processing processes and led to the emergence of new technological issues [25] or the employment of re-configurable ways in first-tier materials, enhanced diagnostics, and cyber–physical manufacturing systems, improving the overall design and operation of modern manufacturing [7].

Although an increasing number of surveys have been conducted on the aforementioned topics resulting in finding silos, surveys to provide multidimensional evidence are missing. Specifically, a substantial number of existing surveys examine the gaps the characteristics of I4.0 exhibit, while other surveys deal with the implementation of the fourth industrial revolution in SMEs. However, these implementation surveys miss considering the above-mentioned characteristics in conjunction with the human element, as well as with a series of critical concepts, namely sustainability, the circular economy, and supply chains. Missing such a multifaceted approach, there is no other survey in the current literature that can provide answers to the reader about how to:

- Investigate the influence in existing business models of SMEs;
- Identify the decision-making procedures within the manufacturing ecosystem;
- Clarify the interaction between humans and machines;
- Offer circularity-based insights to the stakeholders;
- Provide multidisciplinary results from existing surveys.

Having said that, it is acknowledged that there are several studies on these topics, yet a pivotal objective of the paper in hand is to strengthen the existing literature by providing answers to research questions using a fundamental systematic literature review (SLR) technique.

In an attempt to facilitate and accelerate this process, this manuscript aims to highlight the contribution of I4.0 in traditional manufacturing and smart factory schemes, by investigating high-impact journal literature. Specifically, a systemic survey of surveys' methodology [26] has been conducted to provide decision-makers and researchers in manufacturing brief insights from current systematic studies, summarising existing research or highlighting the gaps, while the PRISMA protocol has been followed [27]. By presenting and analysing the existing literature, this paper strives to offer a comprehensive compilation of relevant data that will empower both the research and business communities. This manuscript will provide insightful I4.0 knowledge, in terms of circularity, business model adaptability, and factory intelligence, for entrepreneurs and policymakers. Considering this unique opportunity to clarify the relation between the aforementioned concepts, we are attempting to answer the following research questions (*RQs*):

RQ1: How will Industry 4.0 adoption challenge traditional business models?
RQ2: Can the existing Industry 4.0 business models be applied in SMEs?
RQ3: What is the role of decision-making in Industry 4.0?
RQ4: What is the role of human–machine interaction in Industry 4.0?
RQ5: How are the circular economy, the smart factory, and the supply chain concepts connected under the framework of Industry 4.0?

The remainder of this paper is structured as follows. In Section 2, we present the SLR methodology followed to shape the paper in hand. Consequently, Section 3 contains the identified fields under discussion and the journals where the selected survey articles were published. In the following Section 4, the findings of our SLR are discussed, while in Section 5, we summarize our findings and draw conclusions.

2. Systematic Literature Review

One of the main advantages of systematic reviews over other types of literature analysis approaches is that explicit, systematic methods are utilised in order to minimize bias and provide reliable findings, from which conclusions can be drawn. In particular, a systematic review attempts to collate all empirical evidence that fits pre-specified eligibility criteria to answer specific research questions [28].

The main purpose of this manuscript is to investigate Scopus survey articles that have been reported on I4.0 between 2014 and 2022. Following the survey of surveys methodology, we intend to summarize and describe survey publications that primarily focus on key factors that should be adhered to by the modern industrial environment. There are different terms for "survey of surveys" in the literature such as overview of reviews, review of reviews, umbrella reviews, synthesis of reviews, and others, which synthesise data that are generated by other systematic reviews [28]. Furthermore, in order to build a knowledge database capable of easily locating, synthesizing, and summarizing data points, we utilized the SLR methodology based on the PRISMA protocol. SLR is considered a trustworthy tool that ensures the quality and quantity of the data being under analysis. The reliability and transparency of the SLR process are guaranteed by applying the widely accepted, five-stage methodology [29], as presented in Figure 1.

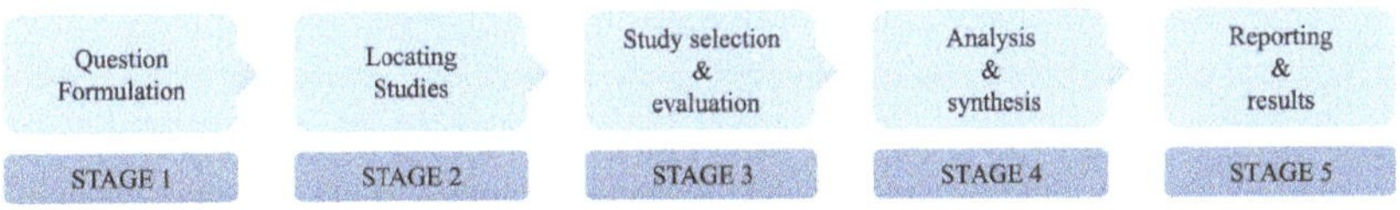

Figure 1. The figure illustrates the five sequence stages of the systematic literature review process, in which five connected blocks represent the research determination, the location of studies, the study selection and evaluation, the analysis, and the conclusion report [29].

2.1. *Question Formulation*

In the first stage, the scope of the literature review is defined and research questions are formulated. This manuscript explores how trending research fields are expected to affect the manufacturing I4.0 ecosystem. Consequently, the research questions were used to analyse the current technological stage, by comparing the challenges with the impact, in order to provide structured knowledge, in terms of business model adaptability, factory intelligence, human–machine interaction, and circularity.

2.2. *Locating Studies*

As a second step, with respect to the academic literature, a search was performed on review titles and abstracts, to gather the required publications. In order to answer the research questions with data from high-quality publications, Scopus® was used. This database includes over 34,000 peer-reviewed journals in top-level subject fields. During the collection of the publications, we selected the simple term *"Industr* 4.0 AND (Review OR Survey)"*, aiming to collect review papers that provide a mainstream overview of I4.0 and do not concentrate on highly technical topics. The research protocol and the selection criteria are presented in Table 1.

Table 1. Summary of the literature search protocol including inclusion and exclusion criteria.

Research Protocol	Details Description
Research databases	Scopus®
Publication type	Peer-reviewed papers (indexed by Scopus®)
Language	Only papers in English
Time period	2014–2022
Search field	Title and Abstract
Search term	*Industr* 4.0 AND (Review OR Survey)*
Criteria for inclusion (SC)	Articles exclusively referred to Industry 4.0
Criteria for exclusion	Articles not referring to Industry 4.0
"	Articles using Industry 4.0 as a supportive concept
"	Conference proceeding papers
"	Non-English text

2.3. *Study Selection and Evaluation*

The following stage involved the selection and evaluation of the collected papers. To ensure the dependable evaluation of the survey articles, inclusion and exclusion criteria were applied to exclude the ones not being relevant for our analysis or of low scientific quality. The specific criteria are described in Table 1. Surveys that did not focus on the fourth industrial revolution specifically or referred to it as future research were excluded as well. The criteria also led to the exclusion of articles that did not address technological pillars, enablers, or concepts and did not provide innovative knowledge to readers, as presented in Figure 2.

Especially, during the abstract analysis, attention is placed on identifying surveys that are focused on the basic concepts of Industry 4.0, but not on specific technology enablers. This approach aided our study by highlighting crucial information and technologies that are widely referredto in surveys. Our SLR criteria yielded a total of 511 survey articles published between January 2014 and July 2022. After the first screening of the titles, 127 papers were excluded as referring to other domains (such as Construction 4.0 [30] and Health 4.0 [31] initiatives, among others). As a next step, we utilised our university's network to discover the accessible manuscripts, except the public ones, excluding the 45 manuscripts that we were unable to locate. After applying the exclusion criteria, articles that did not pertain to Industry 4.0 exclusively, were not peer-reviewed, or were not written in English were also eliminated. Finally, 59 academic literature publications were selected for analysis. A graphical representation of the review process based on the PRISMA methodology is shown in Figure 2.

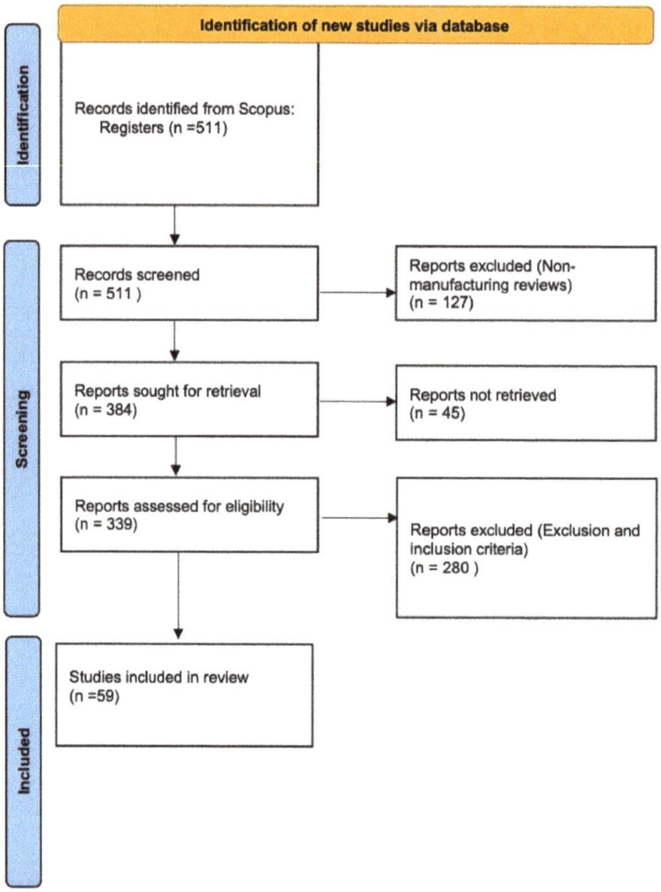

Figure 2. The figure shows different blocks being connected of the PRISMA methodology and represents how articles are decreased at each stage of the SLR process, selecting 59 out of 511 survey articles [27,32].

2.4. Analysis and Synthesis

In the data analysis and synthesis stage, the publications' content should be presented according to PRISMA in a way that answers the formulated research questions. In this paper, we firstly identified the key enablers and concepts of I4.0 that are totally related to the formulated question, as described in Section 3.1, and then identified the presence of these enablers in each analysed article, as shown in Table 4. The first high-level analysis proved the most famous technologies and concepts within the examined surveys. After that, in order to extract information regarding the research journal trends, we structured the selected survey articles into categories according to the thematic region and their occurrence (Table 3). In addition, the statistics of the publication dates were extracted, to identify the trend in which researchers tend to publish I4.0-related research according to our research questions.

2.5. Reporting and Using the Results

Thereafter, the results were categorized and discussed in the context of *traditional business models* (Section 4.1), *SMEs* (Section 4.2), *decision-making* (Section 4.3), *human–machine interaction* (Section 4.4), and *circularity* (Section 4.5). The aforementioned sections answer the research questions based on the identified key enablers and provide strong research guidance for researchers and entrepreneurs. These opportunities and suggestions are further described below in Section *Findings* (Section 4.6), specifically as it pertains to research and business opportunities, as well as the relevance to Industry 5.0 principles. In this manuscript, the occurrence of the fields under investigation is collocated in Table 2, while relationships between the survey articles and the fields are described in Table 4.

Table 2. Distribution of concepts and technologies obtained by survey articles.

Field	Number	Occurrence	Reference
Advanced Manufacturing Systems	26	44.83%	[5,8,10,14,33–54]
Artificial Intelligence	24	41.38%	[5,8,14,34,35,38–40,46,47,49,52,54–65]
Sustainability	20	34.48%	[10,37,43,45,48,59,61–74]
Challenges	16	27.59%	[8,11,14,39,47,48,50,51,60–62,65,66,75–77]
Impact	15	25.86%	[11,13,39,42,47,49,60,61,64–67,75,76,78]
Supply Chain	15	25.86%	[14,34,52,58,61–66,70,74–77]
SME	13	22.41%	[11,13,14,46,49,50,56,62,70,76,79–81]
Circularity	10	17.24%	[37,48,50,52,57,59,63,64,72,78]
Business model	10	17.24%	[11–13,34,37,53,76,80–82]
Decision Making	9	15.52%	[10,11,46,49,63,65,66,80,83]
Human-machine Collaboration	7	12.07%	[35,38,43,49,60,61,78]
Maturity	3	5.17%	[56,62,79]

3. Results

This section provides an overview of the most common areas identified in the selected publications, as well as an inspection of the journals in which researchers tend to publish I4.0-related research.

3.1. Research Field

During the full corpus investigation phase, several aspects affecting I4.0 adoption were encountered. We decided to use these as categories, around which we could structure our literature analysis, by investigating the impact of each category in I4.0. In order to cluster different aspects, we organized the key information in a contextual database, as presented in Table 4, while the survey articles were sorted by publishing date. The database contains technologies and concepts selected according to their direct or indirect reference to, and their importance in, the fourth industrial revolution, while a short description of their contribution is as follows:

- **Advanced manufacturing systems:** These survey articles promote cutting-edge technologies facilely adaptable to factory facilities, bringing about a broader transformation in operations and enterprises.

- **Artificial intelligence:** These survey articles address applications and scenarios influenced by digital transformation, as well as traditional concepts enhanced by artificial intelligence, which act as enablers for smart factories, products, networks, and production.
- **Industry 4.0 impact:** The survey articles falling into this field include analyses and conclusions concerning the positive or negative impact of the fourth industrial revolution in sectors such as business, supply chains, and management.
- **Sustainability:** These survey articles describe the concept and implementation procedure frameworks of sustainability, highlighting the benefits, value, and importance of the I4.0 ecosystem.
- **Human–machine collaboration:** The survey papers of this category discuss developments and applications that enhance human–machine collaborations in the I4.0 era, including Research and Development (R&D) and safety aspects.
- **Circularity:** These are survey articles that mention the concept and the transition of the circular economy in horizontal and vertical planes, as well as methodologies and techniques in the context of I4.0.
- **Business model:** The updated business models generated by the ongoing industrial transition, as well as their impact on the traditional models and the operations of the worldwide markets are mentioned in this cluster of survey articles.
- **Challenges:** The updated business models generated by the ongoing industrial transition, as well as their impact on the traditional models and the operations of the worldwide markets are mentioned in these survey articles.
- **Supply chain:** These survey articles analyse the role of the supply chain within the frame of I4.0, as well as the effect, challenges, and characteristics of the interconnected network.
- **Small and medium-sized enterprises:** The survey articles in this field include information about the current factory level and the requirements for their transition to the I4.0 ecosystem.
- **Decision-making:** These survey articles analyse how the data-driven, decision-making processes empower the strategic and management policies and drive the innovation research communities in the digital factories.
- **Maturity models:** These survey articles discuss innovative and upcoming maturity models, influenced by the context of smart manufacturing and factory of the future concepts.

The aforementioned topics constitute the smart factory and the level of their adoption should be considered in order to advise researchers and entrepreneurs. As the academic community can discover the research gaps through this manuscript, innovative business personnel are informed by the analysis of the value and opportunities provided by the aforesaid topics. Table 2 provides a summarised view of the examined survey articles regarding the aforementioned topics' occurrence in their full corpus, while the published journals and the detailed analysis is presented in Tables 3 and 4 accordingly.It is important to note that each topic was discussed in more than one survey article, with some topics being addressed more frequently than others. Therefore, Table 2 is sorted by topic, in descending order of occurrence and visualised in Figure 3.

From a high-level analysis viewpoint, we observed that advanced manufacturing systems and artificial intelligence technology were the fields appearing more often, as they are referred to in the biggest proportion of the selected papers in comparison to other concepts and technologies. Specifically, 15% of the papers included advanced manufacturing enablers, while artificial intelligence concepts appeared in 14% of the surveys. Discussions regarding the challenges and impact of I4.0, in terms of sustainability, circularity, human–machine collaboration, business models, etc., appeared to be less relevant. More specific, concepts such as maturity, human–machine collaboration, and decision-making were included in 2%, 4%, and 5% of the selected papers, respectively, while the frequency percentage of others such as business models, circularity, SMEs, supply chains, and impact was slightly increased at 6%, 6%, 8%, 9%, and 9%, respectively. Besides advanced manufacturing systems and artificial intelligence technology, only challenges and sustainability passed 10% in terms of frequency.

Table 3. This table includes the research areas of the journals and their occurrence.

Journal	Number	Research Area	Occurrence	Reference
Journal of Cleaner Production	6	Sustainable and environmental production	10.34%	[37,43,48,68,70,71]
Journal of Manufacturing Systems	5	Applied manufacturing system-based research	8.62%	[47,49,54,56,83]
International Journal of Production Research	4	Production ecosystem advancements	6.9%	[44,52,55,59]
Sustainability	4	Environmental, cultural, economic sustainability	6.9%	[65,67,72,73]
Applied Sciences	3	Applied natural sciences	5.17%	[14,69,79]
Journal of Manufacturing Technology Management	3	Digital operations management	5.17%	[12,57,62]
Benchmarking	2	Quality management	3.45%	[66,76]
Benchmarking: An International Journal	2	Operations and organization management	3.45%	[39,58]
Computers in Industry	2	Industrial ICT innovation and application	3.45%	[34,78]
Journal of Enterprise Information Management	2	Information systems	3.45%	[63,64]
Business Process Management Journal	1	Business process management	1.72%	[11]
Cogent Engineering	1	Engineering and technology	1.72%	[10]
Designs	1	Engineering design	1.72%	[41]
Economies	1	Economics, macroeconomics	1.72%	[13]
Electronics	1	Electronics advancements	1.72%	[46]
Electronics (Switzerland)	1	Science of electronics	1.72%	[80]
Engineering Science and Technology, an International Journal	1	R&D engineering advancements	1.72%	[40]
Enterprise Information Systems	1	Information systems and management	1.72%	[77]
Future Internet	1	Internet technologies	1.72%	[82]
IEEE Communications Surveys & Tutorials	1	Telecommunication research	1.72%	[38]
Industrial Robot	1	Industrial engineering, design, and manufacturing	1.72%	[60]
International Journal of Automation Technology	1	Engineering	1.72%	[35]
International Journal of Industrial and Systems Engineering	1	Industrial and systems engineering (ISE)	1.72%	[53]
International Journal of Precision Engineering	1	Green Technology aspects of precision engineering	1.72%	[61]
Journal of Engineering	1	R&D engineering advancements	1.72%	[42]
Journal of Industrial Information Integration	1	Industrial ICT innovation and application	1.72%	[8]
Journal of Industrial Integration and Management	1	Innovation and entrepreneurship	1.72%	[75]
Journal of Intelligent Manufacturing	1	Intelligent digital manufacturing	1.72%	[5]
Journal of Science and Technology Policy Management	1	Technology management/strategy	1.72%	[81]
Jurnal Teknologi	1	Industrial ICT innovation and application	1.72%	[33]
DAAAM International Vienna	1	Manufacturing	1.72%	[36]
Metals	1	Metallurgy engineering	1.72%	[45]
Supply Chain Forum: An International Journal	1	Supply chain management	1.72%	[74]
The International Journal of Advanced Manufacturing TecN.	1	Advanced manufacturing applications	1.72%	[50]
The Royal Society	1	Mathematical, physical, and engineering sciences	1.72%	[51]

Table 4. Overview of the review papers about Industry 4.0, while their impact on the identified concepts and technologies is marked. The acronyms AMS, AI, I, and S reflect Advance Manufacturing Systems, Artificial Intelligence, Impact, and Sustainability. The terms Sustainability Human Machine collaboration, Circularity, Business Model, and Challenges are described as S, HM, Ci, BM, and Ch. Lastly, the columns SC, SME, DM, and MM refer to the Supply Chain, Small–Medium Enterprises, Decision-Making, and Maturity Models.

Title	Year	AMS	AI	I	S	HM	Ci	BM	Ch	SC	SME	DM	MM
Industry 4.0: A review on industrial automation and robotic [33]	2016	yes	no	no	no	no	no	no	no	no	no	no	no
Understanding the implications of digitisation and automation in the context of Industry 4.0: A triangulation approach and elements of a research agenda for the construction industry [34]	2016	yes	yes	no	no	no	no	yes	no	yes	no	no	no
"Industrie 4.0" and Smart Manufacturing—A Review of Research Issues and Application Examples [35]	2017	yes	yes	no	no	yes	no	no	no	no	no	no	no
Industry 4.0: A survey on technologies, applications and open research issues. [8]	2017	yes	yes	no	no	no	no	no	yes	no	no	no	no
Past, present and future of Industry 4.0—a systematic literature review and research agenda proposal [55]	2017	no	yes	no	no	no	no	no	no	no	no	no	no
Industry 4.0 and supply chain sustainability: framework and future research directions [66]	2018	no	no	yes	yes	no	no	no	yes	yes	no	yes	no
A critical review of smart manufacturing & Industry 4.0 maturity models: Implications for small and medium-sized enterprises. [56]	2018	no	yes	no	no	no	no	no	no	no	yes	no	yes
Consequences of Industry 4.0 in Business and Economics [13]	2018	no	no	yes	no	no	no	yes	no	no	yes	no	no
Setting an Industry 4.0 research and development agenda for simulation-a literature review [36]	2018	yes	no	no	no	no	no	no	no	no	no	no	no
A review of emerging industry 4.0 technologies in remanufacturing [37]	2019	yes	no	no	yes	no	yes	yes	no	no	no	no	no
A Survey on Information and Communication Technologies for Industry 4.0: State-of-the-Art [38]	2019	yes	yes	no	no	yes	no	no	no	no	no	no	no
Exploring Industry 4.0 technologies to enable circular economy practices in a manufacturing context [57]	2019	no	yes	no	no	no	yes	no	no	no	no	no	no
Industry 4.0: Emerging themes and future research avenues using a text mining approach [78]	2019	no	no	yes	no	yes	yes	no	no	no	no	no	no
Industry 4.0: key findings and analysis from the literature arena. [39]	2019	yes	yes	yes	no	no	no	no	yes	no	no	no	no
Key ingredients for evaluating Industry 4.0 readiness for organizations: a literature review [58]	2019	no	yes	no	no	no	no	no	no	yes	no	no	no
Scanning the Industry 4.0: A Literature Review on Technologies for Manufacturing Systems [40]	2019	yes	yes	no	no	no	no	no	no	no	no	no	no
A review of research relevant to the emerging industry trends: Industry 4.0, IoT, blockchain, and business analytics [75]	2020	no	no	yes	no	no	no	no	yes	yes	no	no	no
A Strategic Roadmap for the Manufacturing Industry to Implement Industry 4.0 [41]	2020	yes	no	no	no	no	no	no	no	no	no	no	no
A survey on decision-making based on system reliability in the context of Industry 4.0 [83]	2020	no	no	no	no	no	no	no	no	no	no	yes	no
Analysis and synthesis of Industry 4.0 research landscape: Using latent semantic analysis approach [12]	2020	no	no	no	no	no	no	yes	no	no	no	no	no
Exponential Disruptive Technologies and the Required Skills of Industry 4.0 [42]	2020	yes	no	yes	no	no	no	no	no	no	no	no	no

Table 4. *Cont.*

Title	Year	AMS	AI	I	S	HM	Ci	BM	Ch	SC	SME	DM	MM
Impact of Industry 4.0 on Sustainability— Bibliometric Literature Review [67]	2020	no	no	yes	yes	no	no	no	no	no	no	no	no
Industry 4.0, digitization, and opportunities for sustainability [68]	2020	no	no	no	yes	no	no	no	no	no	no	no	no
Industry 4.0: How it is defined from a sociotechnical perspective and how much sustainability it includes-A literature review [43]	2020	yes	no	no	yes	yes	no	no	no	no	no	no	no
Literature review of Industry 4.0 and related technologies [5]	2020	yes	yes	no	no	no	no	no	no	no	no	no	no
Literature search of key factors for the development of generic and specific maturity models for Industry 4.0 [79]	2020	no	no	no	no	no	no	no	no	no	yes	no	yes
Production scheduling in the context of Industry 4.0: review and trends [44]	2020	yes	no	no	no	no	no	no	no	no	no	no	no
Supply chain integration and Industry 4.0: a systematic literature review [76]	2020	no	no	yes	no	no	no	yes	yes	yes	yes	no	no
Systematic Literature Review: Integration of Additive Manufacturing and Industry 4.0 [45]	2020	yes	no	no	yes	no	no	no	no	no	no	no	no
A review of data-driven decision-making methods for industry 4.0 maintenance applications [46]	2021	yes	yes	no	no	no	no	no	no	no	yes	yes	no
A Systematic Literature Review of Successful Implementation of Industry 4.0 Technologies in Companies: Synthesis of the IPSI Framework [69]	2021	no	no	no	yes	no	no	no	no	no	no	no	no
Big Data analytics in Smart Grids for renewable energy networks: Systematic review of information and communication technology tools [10]	2021	yes	no	no	yes	no	no	no	no	no	no	yes	no
Cyber-physical systems architectures for industrial internet of things applications in Industry 4.0: A literature review [47]	2021	yes	yes	yes	no	no	no	no	yes	no	no	no	no
Effective cloud resource utilisation in cloud ERP decision-making process for industry 4.0 in the united states [80]	2021	no	no	no	no	no	no	yes	no	no	yes	yes	no
Industry 4.0 and business models: a bibliometric literature review [11]	2021	no	no	yes	no	no	no	yes	yes	no	yes	yes	no
Industry 4.0 and sustainability: Towards conceptualization and theory Industry 4.0 and sustainable development: A systematic mapping of triple bottom line, Circular Economy and Sustainable Business Models perspectives [70]	2021	no	no	no	yes	no	no	no	no	yes	yes	no	no
Industry 4.0 smart reconfigurable manufacturing machines [49]	2021	yes	yes	yes	no	yes	no	no	no	no	yes	yes	no
Industry 4.0 technologies as enablers of collaboration in circular supply chains: a systematic literature review [59]	2021	no	yes	no	yes	no	yes	no	no	no	no	no	no
Industry 4.0 ten years on: A bibliometric and systematic review of concepts, sustainability value drivers, and success determinants [71]	2021	no	no	no	yes	no	no	no	no	no	no	no	no
Industry 4.0, transition or addition in SMEs? A systematic literature review on digitalization for deviation management [50]	2021	yes	no	no	no	no	yes	no	yes	no	yes	no	no

Table 4. *Cont.*

Title	Year	AMS	AI	I	S	HM	Ci	BM	Ch	SC	SME	DM	MM
Making smart manufacturing smarter—a survey on blockchain technology in Industry 4.0 [77]	2021	no	no	no	no	no	no	no	yes	yes	no	no	no
Six-gear roadmap towards the smart factory [14]	2021	yes	yes	no	no	no	no	no	yes	yes	yes	no	no
Stochastic model predictive control framework for resilient cyber–physical systems: Review and perspectives [51]	2021	yes	no	no	no	no	no	no	yes	no	no	no	no
The applications of Industry 4.0 technologies in manufacturing context: a systematic literature review [52]	2021	yes	yes	no	no	no	yes	no	no	yes	no	no	no
Towards a conceptual development of Industry 4.0, servitisation, and circular economy: A systematic literature review [72]	2021	no	no	no	yes	no	yes	no	no	no	no	no	no
Twin transition through the implementation of industry 4.0 technologies: Desk-research analysis and practical use cases in Europe [73]	2021	no	no	no	yes	no	no	no	no	no	no	no	no
Adopting open innovation for SMEs and industrial revolution 4.0 [81]	2022	no	no	no	no	no	no	yes	no	no	yes	no	no
An overview of Industry 4.0 in manufacturing industries [53]	2022	yes	no	no	no	no	no	yes	no	no	no	no	no
Augmented reality—an important aspect of Industry 4.0 [60]	2022	no	yes	yes	no	yes	no	no	yes	no	no	no	no
Business Models for the Internet of Services: State of the Art and Research Agenda [82]	2022	no	no	no	no	no	no	yes	no	no	no	no	no
Cyber-Physical Systems as an Enabler of Circular Economy to Achieve Sustainable Development Goals: A Comprehensive Review [61]	2022	no	yes	yes	yes	yes	no	no	yes	yes	no	no	no
Drivers and barriers of Industry 4.0 technology adoption among manufacturing SMEs: a systematic review and transformation roadmap [62]	2022	no	yes	no	yes	no	no	no	yes	yes	yes	no	yes
Integrating Industry 4.0 and circular economy: a review [63]	2022	no	yes	no	yes	no	yes	no	no	yes	no	yes	no
Mapping the links between Industry 4.0, circular economy and sustainability: a systematic literature review [64]	2022	no	yes	yes	yes	no	yes	no	no	yes	no	no	no
The Application of Industry 4.0 Technological Constituents for Sustainable Manufacturing: A Content-Centric Review [65]	2022	no	yes	yes	yes	no	no	no	yes	yes	no	yes	no
The individual and integrated impact of Blockchain and IoT on sustainable supply chains:a systematic review [74]	2022	no	no	no	yes	no	no	no	no	yes	no	no	no
Towards edge computing in intelligent manufacturing: Past, present and future [54]	2022	yes	yes	no	no	no	no	no	no	no	no	no	no

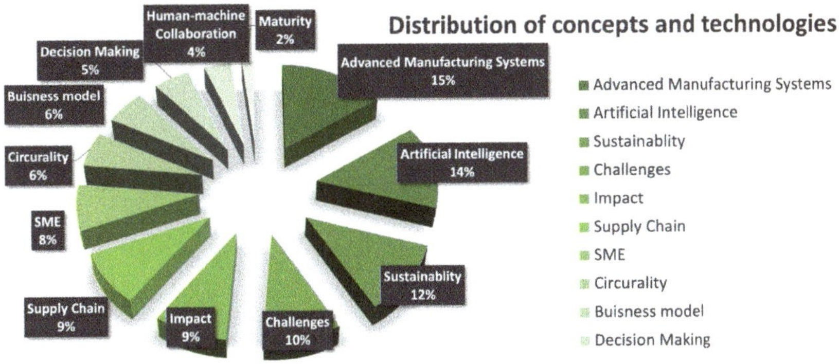

Figure 3. The figure presents the occurrence of the technologies and research fields within the 59 reviewed surveys.

This low percentage of survey articles is justified, as the aforementioned fields depend not only on the developments provided by advanced manufacturing technologies, but also on multi-disciplinary areas, such as financial and social sciences, which all need to be taken into consideration in order to extract accurate and valuable data. Lastly, our SLR confirms the trend towards the limited use of the I4.0 schema in the management of the factory and supply chain, as fields such as SMEs, decision-making, maturity models, and supply chains have not been well analysed, which resulted in the occurrence rates below 15%.

3.2. *Journal Publication*

Due to the plethoric availability of journals, researchers publish manuscripts with various publishers. Thus, in order to extract information regarding the research journal trends, we structured the selected survey articles into categories according to the journal during the analysis phase. The occurrence and their research area are shown in Table 3. It is easily distinguished that journals devoted to industry-related topics scored at the top of the preference, while sustainability, economic, engineering, and information journals have also attracted a great number of Industry 4.0 survey journal papers. A timeline analysis of the publications that were chosen for the study is visualised in Figure 4. During the period under investigation (2016–2022), there was an increasing research interest and activity compared to previous years. The proportion of publishing surveys climbed steadily over the course of the five years, beginning at four percent in 2016 and reaching a high of thirty-two percent in 2021. Besides, by the end of 2022, we expected to have an increase in published journals, as shown in Figure 4. This verifies the argument that, as time passes, the research community is able to provide much more information, while researchers massively review them, in order to summarise the knowledge within review journals, thus providing well-defined answers and results.

The distribution of the publishers taking into account the research results are illustrated in Figure 5. It is clear that the ranking is topped by Elsevier Ltd. , Emerald Group Holdings Ltd., and Multidisciplinary Digital Publishing Institute (MDPI). Despite the fact that there are more publishers, it is evident that the three at the top of this list are the most sought-after by authors for publishing Industry 4.0 review journals. It is remarkable that just two of the remaining twelve publishers indicated in Figure 5, notably Taylor & Francis Ltd. and Springer, are the most popular inside the subset. Finally, the rest of them show no difference as far as their distribution.

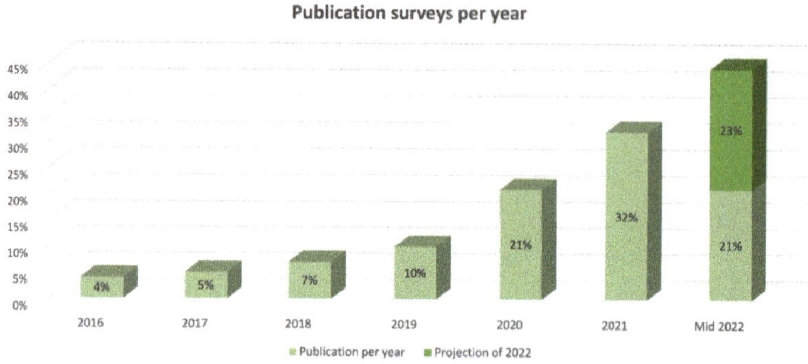

Figure 4. The figure presents the occurrence of the publication years within the examined period.

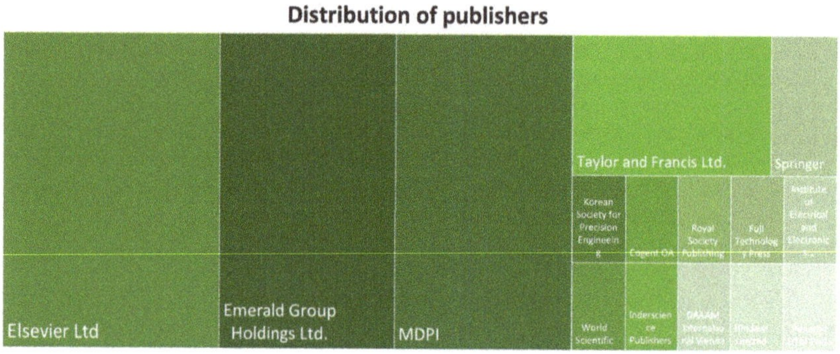

Figure 5. Distribution of publishers.

Our SLR identified that the majority of the published articles appeared in the *Journal of Cleaner Production* [37,43,48,68,70,71], the *International Journal of Production Research* [47,49,54,56,83], the *International Journal of Manufacturing Systems* [44,52,55,59] and *Sustainability* [65,67,72,73], with more than four articles in each of them. On the other hand, *Applied Sciences* [14,69,79] and the *Journal of Manufacturing Technology Management* [12,57,62] journals accounted for six publications. As concerns the research interests of these journals, the analysis revealed that the most preferable is focused not only on applied manufacturing [36,47,49,54,56,83] and the production ecosystem [44,51,52,55,59], but also on sustainability areas [37,43,48,65,67,68,70–73].

A significant number of I4.0 survey papers [35,40–42] were submitted in engineering-related journals such as *Engineering*, *Designs*, and the *Journal of Engineering* among others, which are mainly attracting R&D engineering advancements, architectures, and methodologies. In addition, industrial-based journals (*Computers in Industry*, *Industrial Robot*, *Jurnal Teknologi*) were identified, which are mainly focused on ICT design [34,53,78] and related manufacturing operations [33,60]. Apart from them, industry-related organisational [39,58], economic [13], and digital quality management [50,66,76] journals are also interesting to the researchers as proven by our analysis, while there is an increased interest in business and technology management journals [11,77,81].

However, there are also influential journals such as the *International Journal of Advanced Manufacturing Technology* [50], the *Journal of Intelligent Manufacturing* [5], and the *IEEE Communications Surveys* [38], the main theme of which lies in industrial electronics, intelligent manufacturing, and telecommunication topics, respectively. It is noteworthy to highlight that there is a number of Industry 4.0 survey papers in the computer science [10], integration [8], and Internet [82] areas in some dedicated journals such as *Computer Science Review*, *Journal of Industrial Information Integration*, and *Future Internet*. As a last observation, we identified some manuscripts being published in journals with a broad spectrum of topics, such as *Expert Systems* [75], *Electronics* [80], the *Journal of Precision Engineering* [61], or in journals with themes not directly related to I4.0, such as *Metals* [45], *Supply Chain Forum: An International Journal* [74], and *Electronics* [46].

Summarising, our study identified the trend that the academic community provides developments in the industrial manufacturing landscape with a close look into environmental and sustainable aspects. Furthermore, it is noticeable that our SLR confirms the global direction that Industry 4.0 papers are not only published in production-related journals, but in ones covering the whole spectrum of the value chain and business topics, including journals in financial, organizational, and management areas.

4. Discussion

The SLR has highlighted some aspects that help to clarify how I4.0 is transforming the existing manufacturing environment and how cutting-edge technologies are employed in the new era. These indications are grouped and generated based on identified research fields and formulated research questions. The relevance among them, as well as the findings of this survey of surveys procedure are visualised in Figure 6, while they are further described by focusing on the effects of I4.0 on traditional business models, the transformation of SMEs, the decision-making process, the human–machine interaction, and the circularity aspects.

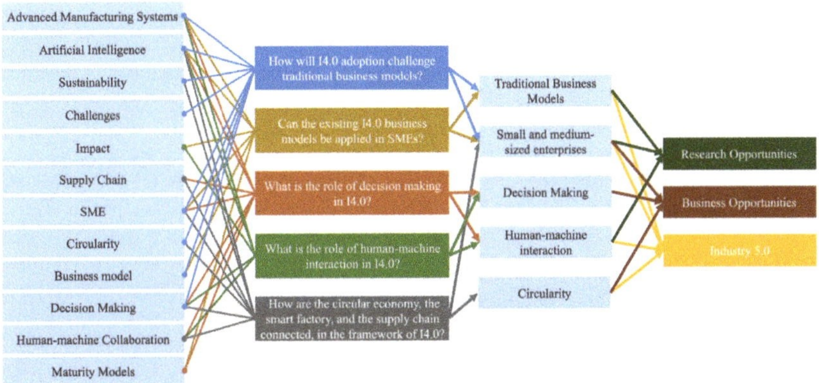

Figure 6. In this figure, the relation among identified research fields, research questions, and discussion points, in addition to the findings are visually represented.

4.1. Traditional Business Models

During the adoption of the Industry 4.0 schema in the production environment, several challenges have arisen either as obstacles to the implementation phase or as enablers that were eventually partnered with the innovations, thereby expanding the technological capabilities [47]. Ten percent of the survey articles refer to the challenges of I4.0 or significant advanced technologies and factors involved. The challenges of technological pillars, such as the Internet of Things, big data, etc., were mentioned [10], while the parameters of the achievements of Industry 4.0 have also been explored [39]. The interdependence between the level of business model adoption and actual deployed technologies was spotted by Nayernia et al. [73]. The horizontal and vertical integration capabilities through interoperability was highlighted by Lu et al. [8], who proposed a conceptual evaluation framework. In other disciplines, issues related to the circular economy and sustainable production were reported, while the challenges of Industry 4.0 in relation to business models were reported by [10,11].

Business models are heavily influenced by I4.0, and this trend will facilitate the creation of collaborative environments, interconnecting humans, processes, and production [53].

These new business models are mainly derived from organizations that adopt I4.0 to utilize more efficient and reliable production systems and innovative technologies [12,50]. Customers benefit from the fourth industrial revolution by improving their experience, as Industry 4.0 improves integration and collaboration across the value chain, which are key factors in all business models [13]. The Internet of Services (IoS) could be characterized as a new ecosystem where service providers and consumers explore their business networks for service provision and consumption. In the research community, though, there is a lack of a detailed view of an IoS-based business model to support the concept in creating, delivering to, and capturing value for customers [82]. Business models are key pillars of success in many areas such as the circular economy. More specifically, the success of a circular economy depends on new business models, which are built upon reusing the value of products at the end of their life cycle [37]. Different business model proposals consider additive manufacturing as a factor of sustainable technology that supports the circular economy [45]. Especially, additive manufacturing technology transforms the traditional sectors (footwear, textile, and wood industries), reducing the material that would be wasted otherwise [65]. More circular-based business models provide high-quality services or solutions to achieve technology cycles exceeding their life expectancy and, finally, allowing them to offer both short-lived and durable products through *"remedial methods"* [37]. In business models, sustainability plays an important role as well. In order to achieve it, organisational capabilities and corporate social responsibilities must be seriously taken into account [70]. Last, it is worth noting that despite the emergence of new business models, due to I4.0, challenges affecting traditional business models, such as interconnection, personalisation, pricing, smart services, value chain fragmentation, decentralized facilities, integrated production, and human ingenuity, have yet to be completely addressed and cannot be ignored [11].

4.2. Small and Medium-Sized Enterprises

According to the European Commission, enterprises with less than 250 employees and an annual turnover of less than EUR 50 million are classified as Small and Medium-sized Enterprises (SMEs) [56]. Due to the limited financial budget and the knowledge gap about the benefits provided by I4.0 technologies, SMEs' innovation investments are limited. As a result, multinational corporations are still pioneers in the adoption of emerging technologies, and SMEs should learn from their paradigms to quickly start their journey into smart manufacturing and Industry 4.0 [56]. Apart from the technological effects, SMEs are also facing organisational challenges, such as competitive strategies, business models, organizational architecture, etc. [79], which should be solved by embracing digitalisation and having an open mindset.

The literature study revealed that I4.0-related standards, architectures, and business models for the industrial management of SMEs have already been proposed. They are supported by technical transformation solutions from existing companies or new start-ups, which provide more flexible and interoperable systems when compared to those offered by traditional enterprise IT [13]. However, the existing industrial production assets, manufactured in the past 25 years, cannot completely support the new opportunities, in terms of scheduling and production control. The challenges and deployment constraints were explored by [50], concluding that SMEs invest in digital transformation without being fully aware of the value of the digital shift. As a consequence, the development of maturity models is a challenging task, creating barriers to the assessment of the factory's effectiveness and the development of strategic management to achieve its goals [79].

The need for a dedicated future model for SMEs was highlighted by [56], indicating that the existing maturity models have been developed for general purposes and are difficult to apply in SMEs. The dedicated maturity model would demonstrate the organization's readiness for smart production and would provide a tailored strategic plan to enable the realization of the factory of the future [56]. In addition, it would also provide specific tools and technologies to enrich the organizational dimensions, thereby guiding

SMEs into the fourth industrial revolution. Open innovation is also extremely important. Having collaboration as the main component is actually considered a new paradigm of a business innovation model that allows SMEs to reduce cost and time [81]. SMEs, though, surely struggle with a variety of problems such as upskilling challenges or poor cybersecurity regulations. That is why it is time for the research community to focus more on explaining how governments can speed up, simplify, and assist with SMEs' Industry 4.0 transformation [62].

4.3. Decision-Making

Decision-making constitutes a cornerstone process, upon which any production system is based. Automation systems gather information, assess situations, and proceed to specific actions, according to predefined procedures. As the volume of uncertain and incomplete information increases, decision support systems mainly utilize statistical, machine, or deep learning techniques to provide decision support, across a wide range of areas [83]. In an I4.0 ecosystem, decision support systems' technology is a vital enabler, as it empowers not only automation, but managers and strategy makers as well. Big data, in combination with artificial intelligence technology, analyses available data in a way that helps the system to proceed with more accurate, data-driven decisions, reducing the need for human experience in the decision-making process [83]. Besides, blockchain technology secures the data transactions used for decisions while the data costs are decreased [74]. Within the manufacturing ecosystem, intelligent data-based systems are used to investigate existing or upcoming problems in the whole value chain (viz. defective products, equipment failures, resource allocation, and energy needs) [75]. There are also enablers derived from IT (e.g., law and policy regarding employment, improved IT standards and security, corporate governance, etc.), which will actually help to streamline information flow across the supply chain network. Information flow and transparency will aid in better decision-making [66]. The decision-making process varies between an SME to a large company. The SMEs tend to adopt a quick and straightforward decision as the ownership lies in one person's hands, while the complexity of ERP in a large enterprise requires a more in-depth analysis supported by an analysis of different options and choosing the best alternative [80].

Based on the systematic, bibliographic review by Osterrieder et al. [14], a smart factory research model was proposed, which identifies the research pillars capable of enabling and improving the smart factory concept. One such pillar is the decision-making procedures, which include data-driven decisions that occur in manufacturing and rely on visualization, machine learning, and storage techniques. In particular, maintenance-based decision-making methods were investigated by [46] emphasizing the importance of cyber–physical systems in facilitating preventive and reliable decisions. Among other manufacturing-based decisions, the main ones are resource optimization, process planning, production equipment control, quality control operations, organizational-based predictions, and performance measurement [14].

4.4. Human–Machine Interaction

I4.0 aims to develop a work environment where teamwork is a key feature, not only for businesses, but also for individuals, allowing collaboration throughout the entire production ecosystem [38]. As the I4.0 and smart factory concepts unfold, human operators face increasing complexity in their day-to-day tasks, which illustrates the need to be extremely flexible and adaptable in this dynamic work environment [78]. The smart factory is a highly flexible and intelligent factory that actually empowers human–machine interaction since humans, machines, sources, and products communicate just like a social network [60]. In order to empower the operators, tools and approaches should be developed that should be easily: (i) integrated into existing everyday practices and (ii) combined within complex methodologies with high usability [78]. In the I4.0 ecosystem, humans and assets will collaborate using cognitive technologies in industrial environments, with smart machines being able to assist personnel with most of their daily tasks, by using speech recognition, computer vision, sensing systems, and machine learning techniques [54].

Augmented Reality (AR) technology is used in a wide variety of applications within the digital factory, allowing operators to stay informed about critical events and collaborate with digital assets [43]. For instance, in the maintenance procedures, operators use their mobiles to allocate and monitor their jobs throughout the day, interact with production equipment [84], and read step-by-step digitized procedures using AR architectures [18]. Besides, mechanical intelligence, enabled by artificial intelligence technologies, plays an important role in supporting human–machine collaboration. This way, machines' ineptitude in understanding and managing their environment can still be of use in a modern manufacturing environment. Moreover, advanced learning models for machines, such as robots, are needed so that humans and machines can develop complementary skills [54]. Even machine learning facilitated by CPSs provides a new means of people–machine interactions in the effort to create a smoother and more friendly user experience [61].

Human–machine collaboration belongs in a zero-fault environment, where human flexibility in combination with machine accuracy can achieve error-free and maximized production performance [35]. This should be considered a valuable enabler to accelerate the adoption of the factory of the future in the most complex production operations [54]. However, the overall performance of human–machine interaction applications can be improved by integrating environmental information and interaction models into a decision-making element that acts as a monitoring process for interaction control [51]. In addition, smart technologies should be employed to track humans in the manufacturing environment and notify stakeholders in the event of an emergency, creating a safer working environment [85]. Nevertheless, the safety principles and regulations should be taken into comprehensive consideration when adopting the human–machines collaboration techniques [73].

4.5. Circularity

The circular economy focuses on environmental performance improvements rather than taking a holistic view of sustainability dimensions. Among the various definitions of the circular economy, an insightful one is an *industrial economy that is restorative or regenerative by intention and design* [57], emphasizing the society-based economic methodology that restores waste and uses it as feedstock in circular chains. However, the large-scale implementation of the circular economy cannot be achieved since cultural, technical, marketing, and regulatory barriers need to be overcome [72]. The Circular Economy (CE) is considered a critical solution to global problems because of the considerable adverse effects of the current linear economic models [64]. Adopting combined factors of I4.0 and CE such as barriers, drivers, and enablers plays a vital role in decision-making and effective implementation [63].

One of the most critical enablers is information and communications technology, which provides interaction and integration capabilities in both existing and new applications, such that customers, organizations, and enterprises can economically benefit from the waste and unused resources [37]. In order to move forward for efficient and widespread implementations, forums, forces, and industrial symbiosis initiatives should be expanded, such that participants embrace innovative goals, key activities, and methodologies, to gain an advantage in the competitive, global market [33,78].

In the I4.0 ecosystem, where digitization and cyber–physical systems are ubiquitous throughput the value chain, the circular economy has also benefited and circular-based management models have been developed [58,71]. As stated in [78], there is a clear trend towards promoting sustainable production and consumption reduction, given the role of supply chain networks in sustainability. In spite of the low integration of I4.0 technologies in the circular economy, the limited adoption provides positive results when used to address sustainability issues, while more impressive results, such as economies of scale and knowledge transfer, can be obtained using technological advancements [59,67]. For instance, blockchain technology has offered valuable opportunities to promote the sustainability of the energy sector [68,86] and the traceability of the supply chain. It is worth mentioning that [37,43] also highlighted important sustainability issues, in relation to I4.0, such as its *consciousness* and the level at which it is implemented. The concept of the smart factory is evolving alongside the smart supply chain, which is also activated by the implementation of I4.0 [52,59], in favour of a circular economy. This transformation has led to a growing emphasis on customer importance and environmental sustainability, promoting a focus on *customer centricity* [52].

4.6. Findings

This SLR analysis spotlighted evidence that helps to provide an overview of the opportunities generated by the I4.0 ecosystem. We identified how the fourth industrial revolution affects the current state-of-the-art in production and circularity issues, while the main technological enablers were identified as contributing to the transformation of manufacturing operations and management. It is worth noting that the examined topics were directly or indirectly related to at least a third one. This relation confirms the importance of horizontal and vertical integration when using technological pillars in the context of smart manufacturing. In addition, the study proved that trending technology topics, such as artificial intelligence or advanced manufacturing systems (Table 2), are comprehensively addressed by scholar survey articles, providing a stable sense of I4.0 outcomes. Nonetheless, despite the fact that experts are continuously contributing to the I4.0 research area, some topics have not progressed as much as the aforementioned ones (viz. artificial intelligence) and should be further explored in the near future. These opportunities and suggestions are further described below, specifically as it pertains to research and business opportunities. The limitations of our SLR are also discussed.

4.6.1. Research Community

One of the most critical issues that requires further exploration is the interpretation of the term "Industry 4.0", based on its contribution at the manufacturing and business level, the value of the futuristic use cases, and the available technological enablers. It is widely known that different technological initiatives, developed around the globe, have been described as I4.0 advancements. These various forms of ingenuity have obscured the original I4.0 inspiration, and as a result, the methodology needed to achieve key developments and business visions has not yet been clearly defined. Despite the fact that the first definition of I4.0 promotes horizontal and vertical integration, by utilizing through-engineering solutions [1], the proposed concept is not fully applied in other disciplines that are directly related to the factory, such as sustainability, management policies, and product distribution, as well as those within the manufacturing area. In addition, we believe that future research will showcase the application of new technology enablers, such as blockchain or edge computing, and their impact on smart factories, circularity, and SMEs [54].

An opportunity for further research into the adoption of I4.0 in SMEs was observed. As the latest factories facilitate the I4.0 solutions [13], a vast number of traditional plants cannot support them due to their ageing infrastructure. Nevertheless, a limited number of add-on systems have been developed (e.g., product tracking [87], image-based vibration monitoring [88], etc.) in order to retrofit the aged machines and assets, thereby providing integration and cognitive capabilities. Furthermore, the existing shop floors are in need of systems and architectures that enable data sharing, not only for the in-house assets and departments, but also for collaborative factories, such as first-tier suppliers and wholesale customers. As we identified a lack of survey articles that referred to retrofitting policies and systems, we believe that future research should be conducted to provide opportunities and insights for the implementation of I4.0 add-on systems.

The interplay between technological pillars used in smart manufacturing, across the value chain, including the supply chain and circularity, is not fully understood. Our SLR identified that artificial intelligence technology and advanced manufacturing systems were discussed in the majority of the survey articles, highlighting the capabilities of the digital brain in the manufacturing ecosystem [10] (Table 2). On the contrary, technologies such as machine vision, digital twins, and additive manufacturing have not been analysed, in spite of their value. For example, machine vision is used not only to identify the quality of the products, but also to guide unmanned vehicles inside and outside the factory [89]. Additive manufacturing promotes sustainability by reducing the volume of raw materials, while digital twins predict upcoming production faults, allowing the machines to undergo self-configuration. Furthermore, the adoption of 5G in manufacturing networks via the tactile Internet has been analysed in existing case studies [90], but the regulatory framework should be further explored. The analysis and discussion of such technologies, in the entire value chain, through survey papers, will be valuable for researchers and technology adopters.

The regulatory and governmental perspectives of I4.0, especially for circularity and sustainability topics, have not been clearly addressed by recent survey articles. While regulatory directions have already been stated for trending technologies (i.e., artificial intelligence) [91]), a systematic review analysis to correlate the regulatory directions and provide meaningful insights has not been performed yet. Furthermore, an interesting, yet unexplored, topic is the barriers and challenges of current regulations in the adoption of I4.0, when humans are called to collaborate and interact with machines [73]. Generally, the regulatory and government states stand as a barrier to the success of the upcoming revolutions, and their comprehensive analysis, through future studies, would greatly contribute to this complex field.

A lack of review papers regarding educational strategies to prepare industrial workers for revolutionary changes, especially in human–machine collaboration, was observed [68,71]. Until now, the published literature has covered technologies that empower the worker and enable the Operator 4.0 concept [43,54,84], such as augmented reality, the Internet of Things, and artificial intelligence. It is widely accepted that the ongoing digital transformation has created new jobs, which require data-centric soft skills, open mindsets, and digital awareness [24]. Nevertheless, education and human-centred upskilling procedures have not been identified, which will allow us to reduce the skills gap and provide practical insights.

4.6.2. Business Community

From a business community point of view, the present manuscript summarizes the results of high-impact I4.0 survey articles that assist entrepreneurs in their understanding of the influence of digital transformation in factory and business operations. It is essential for entrepreneurs to grasp the potentialities of the technologies provided within the I4.0 bunch before adopting them within their facilities. For example, as discussed in Sections 4.1 and 4.2, recently developed business models also take into account sustainable production policies, combined with circularity-based techniques and technology enablers across the value chain [37]. The adoption of these models in SMEs requires careful consideration of the maturity level of the existing infrastructure [56], maximizing the strategic impact, and reducing non-profitable investments. Apart from that, their retrofication strategy should be also explored in other disciplines of the value chain such as warehouses, where intelligent techniques started to be used to automate packaging- or human-related procedures [92,93].

Decision-makers in business and technology adoption should prioritize horizontal integration, with the aim to remove silos and enable interoperability [8]. The interconnected factory provides zero-fault operational management, as the decision-making processes rely on real-time data streams [83]. Apart from that, new technologies, namely artificial intelligence, digital twins, edge computing, and cloud computing, can be broadly implemented in the factory, as the interconnections provide accessible and structured data [40,54,94]. For instance, as stated in Section 4.4, the interaction between a smart machine and a human requires data from the management system and the surrounding environment. Finally, advanced techniques, such as speech recognition and computer vision, analyse humans' movements and the machine acts in tandem, based on the system's state, enabling human–machine collaboration.

4.6.3. Industry 5.0

This survey of surveys study demonstrates that, over the past several years, a substantial number of surveys have been conducted on the technological elements, productivity, and flexibility, as well as the difficulties and consequences associated with their implementation (Figure 3). This result validates the design concepts and aims of the German Industry 4.0 strategy [1]. In contrast, during the past two years (Table 4), the sustainability element has attracted a great deal of attention from the academic community, which is attempting to discover methods for implementing circular policies and mapping the relationships among circularity, Industry 4.0, and sustainability. This trend aligns with the *Fifth Industrial Revolution* (Industry 5.0) strategy of the European Commission, which promotes sustainable, human-centred, and resilient aspects [22]. Nevertheless, as stated in Section 4, some aspects have already been mentioned in surveys without being noted as Industry 5.0 principles, highlighted as follows:

The *human-centric* aspect places the human at the centre of manufacturing activities, utilising technology solutions and innovative approaches to upskill and reskill them. According to Agostini et al. [11], open innovation initiatives democratise human knowledge, hence decreasing the cost and implementation time for SMEs. In another study, cognitive technologies such as voice recognition and computer vision can be utilised to increase cooperation between humans and machines by boosting their complementary skills [54], while augmented reality assists humans in their everyday operations in a user-friendly manner [60,95].

The *resilience* aspect focuses on the production's adaptation and resilience in the face of crucial unforeseen changes. Intelligence can be added to traditional systems through the use of flexible, interoperable technologies and new solutions developed by startups or established enterprises [13]. Big data and artificial intelligence enable machines to manage accurate decisions by merging data from multidisciplinary sources, such as urban demands, supply chains, and environmental changes [62]. On the manufacturing site, the application of predictive maintenance solutions extends the life of machines, decreasing direct expenses [46], which can further be invested in the digital transformation.

The *sustainability* aspect refers to the effectiveness of the circular economy through the manufacturing and supply chain processes. According to Kerin et al. [37], business models based on reused products enhance the reusability of natural resources. Societies play crucial roles in this principle by promoting waste restoration policies to collect waste for reuse as feedstock in circular chains [57]. Reliable communication technologies provide interconnections among stakeholders (customers, organisations, and businesses) in order to collect data, hence facilitating the effective management of waste and underused resources [63]. This change emphasises the value of humans, promoting *customer centricity* in favour of environmental sustainability [52].

4.6.4. Limitations

The time limitations in this research could be characterized as non-critical, as I4.0 encompasses relatively new, constantly evolving concepts, which is evident from the general research activity—most of which was published in recent years. The main limitations were in the coverage of topics and selected sources, as the selection process and the criteria that were defined at the beginning of this meta-analysis of the review articles limited its scope to journals written in English that were archived in the Scopus® database.

Although the time limitation may not have been a critical issue, other limitations, such as the breadth of coverage, may have had a substantial impact on our analysis. Survey articles published between January 2016 and July 2022 that addressed important I4.0-related problems were the focal point of our investigation. To access the information that we needed regarding the digital footprint of Industry 4.0, we followed a survey of surveys approach based on PRISMA guidelines, as explained in Section 2. The different stages are illustrated in Figure 1. During those stages, the inclusion and exclusion criteria further limited the number of articles that could enter the analysis phase. These decisions were based mostly on practical considerations, which, in many cases, were unavoidable. As a result, some very recent and interesting publications were excluded.

5. Conclusions

Industry 4.0 is expected to give rise to various changes in different areas. Based on our analysis, we should note that the I4.0 initiative started as an enabler to digitise industrial-related operations, but advanced technologies have already been adapted in several sectors, developing a I4.0 philosophy that composes a world view for tomorrow's society. Within the SMEs, it is evident that a powerful interaction between I4.0 and concepts such as decision-making, supply chain, circular economy, and business models exists. Actually, the 59 academic publications (Table 4) that were selected to account for the body of our analysis prove that such concepts (e.g., cyber–physical systems, artificial intelligence, human–machine collaboration, etc.) have a strong impact on Industry 4.0. All these topics were examined by scholar articles surveying different facets of the Industry 4.0 ecosystem, thus providing a distilled concept about the fourth industrial revolution results and the need for Industry 5.0.

Specifically, circular-based business models should lead the modern manufacturing ecosystem by providing high-quality services and durable products that are driven by the Internet of Services (IoS) principles. In order to accelerate the transformation, maturity assessment models could assist SMEs in evaluating their technology level and investing in digital transformation. As a significant enabler, data-driven decision systems aid humans or robots in optimising production planning, operating equipment, predicting breakdowns, and measuring factory performance. Furthermore, human–machine collaboration enables a zero-fault environment in which human flexibility and machine precision can accomplish error-free and optimal production performance. Finally, in the modern manufacturing ecosystem, SMEs restore waste from customers and use it as feedstock in circular manufacturing chains, promoting the sustainability and circularity principles.

Moreover, opportunities and suggestions have been provided focusing on the research and business communities. Research gaps such as the adoption of Industry 4.0 in SMEs, the regulatory and governmental policies, as well as the educational strategies and managerial insights such as sustainable-based and human centring strategies were highlighted in Section 4.6. Yet, this is not enough, since continuous uninterrupted effort is surely needed. Sustainability, business models, and the adaptation of advancements to humans should be further analysed so as for Industry 4.0 to grow further and be transmitted to the Industry 5.0 era. This in turn should result in new challenges and prospects that shall enrich future studies, which will be based on developing transition frameworks.

Author Contributions: Most of the presented work was performed by F.K.K. and N.M. Conceptualization, D.K.; methodology, F.K.K., N.M., A.G. and D.K.; formal analysis, F.K.K. and N.M.; investigation, F.K.K. and N.M.; resources, N.M.; data curation, N.M., F.K.K., D.K. and S.G.M.; writing—original draft preparation, F.K.K. and N.M.; writing—review and editing, F.K.K., A.G. and D.K.; visualization, F.K.K. All authors have read and agreed to the published version of the manuscript.

Funding: Part of this research was co-financed by the European Union and Greek national funds through the Operational Program Competitiveness, Entrepreneurship and Innovation, under the call RESEARCH—CREATE—INNOVATE (project code: T2EDK-01658).

Institutional Review Board Statement: Not applicable.

Informed Consent Statement: Not applicable.

Data Availability Statement: Not applicable.

Conflicts of Interest: The authors declare no conflicts of interest.

Abbreviations

The following abbreviations are used in this manuscript:

Abbreviation	Full Form	Definition
I4.0	Industry 4.0	Industry 4.0, which refers to the fourth industrial revolution, is the cyber–physical transformation of manufacturing. The name is inspired by Germany's Industrie 4.0, a government initiative to promote connected manufacturing and a digital convergence between industry, businesses, and other processes
I5.0	Industry 5.0	The Industry 5.0 term, known as the fifth industrial revolution, referring to the development of a more human-centric, sustainable, and resilient industry. This is linked with the upcoming development of novel technology, industrial or supply chain processes, or new business models
SMEs	Small and Medium-sized Enterprises	According to the European Commission, enterprises with less than 250 employees and an annual turnover of less than EUR 50 million are classified as small and medium-sized enterprises
PRISMA	Preferred Reporting Items for Systematic Reviews and Meta-Analyses	PRISMA is an evidence-based minimal set of components for the systematic review and meta-analysis reporting. It is primarily intended for the reporting of reviews assessing the effects of interventions, but it may also be used to record systematic reviews with goals other than evaluating intervention effects
SLR	Systematic Literature Review	The SLR is a sort of literature review that collects and critically examines various research projects or publications in a systematic manner. The SLR's objective is to give a comprehensive overview of the existing literature related to a research issue
RMS	Re-configurable Manufacturing System	The RMS is a system that is intended from the start for rapid modification of its structure, as well as its hardware and software components, in order to rapidly adapt its production capacity and functionality, in reaction to market changes
IoS	Internet of Services	The IoS could be characterized as a new ecosystem where service providers and consumers explore their business networks for service provision and consumption
IoT	Internet of Things	The IoT refers to physical items equipped with sensors, processing power, software, and other technologies that connect to and exchange data with other devices and systems over the Internet or other communications networks
CE	Circular Economy	The CE is a production and consumption paradigm that emphasises sharing, leasing, reusing, repairing, refurbishing, and recycling of existing resources and goods for as long as feasible

References

1. Kagermann, H. Change through digitization—Value creation in the age of Industry 4.0. In *Management of Permanent Change*; Springer: Berlin/Heidelberg, Germany, 2015; pp. 23–45.
2. Xu, L.D.; Xu, E.L.; Li, L. Industry 4.0: State of the art and future trends. *Int. J. Prod. Res.* **2018**, *56*, 2941–2962. [CrossRef]

3. Michalis, P.; Konstantinidis, F.; Valyrakis, M. The road towards Civil Infrastructure 4.0 for proactive asset management of critical infrastructure systems. In Proceedings of the 2nd International Conference on Natural Hazards & Infrastructure (ICONHIC), Chania, Greece, 23–26 June 2019; pp. 23–26.
4. Stewart, C.E.; Kan, C.F.K.; Stewart, B.R.; Sanicola, H.W.; Jung, J.P.; Sulaiman, O.A.; Wang, D. Machine intelligence for nerve conduit design and production. *J. Biol. Eng.* **2020**, *14*, 1–19. [CrossRef] [PubMed]
5. Oztemel, E.; Gursev, S. Literature review of Industry 4.0 and related technologies. *J. Intell. Manuf.* **2020**, *31*, 127–182.
6. Wichmann, R.L.; Eisenbart, B.; Gericke, K. The direction of industry: A literature review on Industry 4.0. In *Proceedings of the Design Society: International Conference on Engineering Design*; Cambridge University Press: Cambridge, UK, 2019; Volume 1, pp. 2129–2138.
7. Koren, Y.; Gu, X.; Guo, W. Reconfigurable manufacturing systems: Principles, design, and future trends. *Front. Mech. Eng.* **2018**, *13*, 121–136. [CrossRef]
8. Lu, Y. Industry 4.0: A survey on technologies, applications and open research issues. *J. Ind. Inf. Integr.* **2017**, *6*, 1–10. [CrossRef]
9. Konstantinidis, F.K.; Mouroutsos, S.G.; Gasteratos, A. The Role of Machine Vision in Industry 4.0: An automotive manufacturing perspective. In Proceedings of the 2021 IEEE International Conference on Imaging Systems and Techniques (IST), Kaohsiung Taiwan, 21–23 June 2021; pp. 1–6.
10. Colmenares-Quintero, R.F.; Quiroga-Parra, D.J.; Rojas, N.; Stansfield, K.E.; Colmenares-Quintero, J.C. Big Data analytics in Smart Grids for renewable energy networks: Systematic review of information and communication technology tools. *Cogent Eng.* **2021**, *8*, 1935410. [CrossRef]
11. Agostini, L.; Nosella, A. Industry 4.0 and business models: A bibliometric literature review. *Bus. Process. Manag. J.* **2021**, *27*, 1633–1655. [CrossRef]
12. Wagire, A.A.; Rathore, A.; Jain, R. Analysis and synthesis of Industry 4.0 research landscape: Using latent semantic analysis approach. *J. Manuf. Technol. Manag.* **2019**,*31*,31–51. [CrossRef]
13. Maresova, P.; Soukal, I.; Svobodova, L.; Hedvicakova, M.; Javanmardi, E.; Selamat, A.; Krejcar, O. Consequences of industry 4.0 in business and economics. *Economies* **2018**, *6*, 46. [CrossRef]
14. Sufian, A.T.; Abdullah, B.M.; Ateeq, M.; Wah, R.; Clements, D. Six-gear roadmap towards the smart factory. *Appl. Sci.* **2021**, *11*, 3568. [CrossRef]
15. Lee, H.G.; Huh, J.H. A cost-effective redundant digital excitation control system and Test Bed Experiment for safe power supply for process industry 4.0. *Processes* **2018**, *6*, 85. [CrossRef]
16. Kostavelis, I.; Gasteratos, A. Conceptualizing Industry 4.0 for Greek Manufacturing Sector. *J. Eng. Sci. Technol. Rev.* **2022**,*15*,1–7. [CrossRef]
17. Koulinas, G.; Paraschos, P.; Koulouriotis, D. A machine learning-based framework for data mining and optimization of a production system. *Procedia Manuf.* **2021**, *55*, 431–438. [CrossRef]
18. Konstantinidis, F.K.; Kansizoglou, I.; Santavas, N.; Mouroutsos, S.G.; Gasteratos, A. MARMA: A Mobile Augmented Reality Maintenance Assistant for Fast-Track Repair Procedures in the Context of Industry 4.0. *Machines* **2020**, *8*, 88. [CrossRef]
19. Büyüközkan, G.; Tüfekçi, G.; Uztürk, D. Evaluating Blockchain requirements for effective digital supply chain management. *Int. J. Prod. Econ.* **2021**, *242*, 108309. [CrossRef]
20. Caiado, R.G.G.; Scavarda, L.F.; Gavião, L.O.; Ivson, P.; de Mattos Nascimento, D.L.; Garza-Reyes, J.A. A fuzzy rule-based industry 4.0 maturity model for operations and supply chain management. *Int. J. Prod. Econ.* **2021**, *231*, 107883. [CrossRef]
21. Rajput, S.; Singh, S.P. Connecting circular economy and industry 4.0. *Int. J. Inf. Manag.* **2019**, *49*, 98–113. [CrossRef]
22. Breque, M.; De Nul, L.; Petridis, A. *Industry 5.0: Towards a Sustainable, Human-Centric and Resilient European Industry*; European Commission, Directorate-General for Research and Innovation: Luxembourg, 2021.
23. Akrivou, C.; Łekawska-Andrinopoulou, L.; Tsimiklis, G.; Amditis, A. Industrial symbiosis platforms for synergy identification and their most important data points: A systematic review. *Open Res. Eur.* **2021**, *1*, 101. [CrossRef]
24. Lasi, H.; Fettke, P.; Kemper, H.G.; Feld, T.; Hoffmann, M. Industry 4.0. *Bus. Inf. Syst. Eng.* **2014**, *6*, 239–242. [CrossRef]
25. Szalavetz, A. Industry 4.0 and capability development in manufacturing subsidiaries. *Technol. Forecast. Soc. Chang.* **2019**, *145*, 384–395. [CrossRef]
26. Aromataris, E.; Fernandez, R.; Godfrey, C.M.; Holly, C.; Khalil, H.; Tungpunkom, P. Summarizing systematic reviews: Methodological development, conduct and reporting of an umbrella review approach. *JBI Evid. Implement.* **2015**, *13*, 132–140. [CrossRef] [PubMed]
27. Liberati, A.; Altman, D.G.; Tetzlaff, J.; Mulrow, C.; Gøtzsche, P.C.; Ioannidis, J.P.; Clarke, M.; Devereaux, P.J.; Kleijnen, J.; Moher, D. The PRISMA statement for reporting systematic reviews and meta-analyses of studies that evaluate health care interventions: Explanation and elaboration. *J. Clin. Epidemiol.* **2009**, *62*, 1–34. [CrossRef] [PubMed]
28. Hunt, H.; Pollock, A.; Campbell, P.; Estcourt, L.; Brunton, G. An introduction to overviews of reviews: Planning a relevant research question and objective for an overview. *Syst. Rev.* **2018**, *7*, 1–9. [CrossRef] [PubMed]
29. Denyer, D.; Tranfield, D. *Producing a Systematic Review*; Sage Publications Ltd.: Newbury Park, CA, USA, 2009; pp. 671–689.
30. Wang, M.; Wang, C.C.; Sepasgozar, S.; Zlatanova, S. A systematic review of digital technology adoption in off-site construction: Current status and future direction towards industry 4.0. *Buildings* **2020**, *10*, 204. [CrossRef]
31. Cavallone, M.; Palumbo, R. Debunking the myth of industry 4.0 in health care: Insights from a systematic literature review. *TQM J.* **2020**, *32*, 849–868. [CrossRef]

32. Page, M.J.; McKenzie, J.E.; Bossuyt, P.M.; Boutron, I.; Hoffmann, T.C.; Mulrow, C.D.; Shamseer, L.; Tetzlaff, J.M.; Akl, E.A.; Brennan, S.E.; et al. The PRISMA 2020 statement: An updated guideline for reporting systematic reviews. *Syst. Rev.* **2021**, *10*, 1–11. [CrossRef]
33. Bahrin, M.A.K.; Othman, M.F.; Azli, N.H.N.; Talib, M.F. Industry 4.0: A review on industrial automation and robotic. *J. Teknol.* **2016**, *78*, 6–13.
34. Oesterreich, T.D.; Teuteberg, F. Understanding the implications of digitisation and automation in the context of Industry 4.0: A triangulation approach and elements of a research agenda for the construction industry. *Comput. Ind.* **2016**, *83*, 121–139. [CrossRef]
35. Thoben, K.D.; Wiesner, S.; Wuest, T. "Industrie 4.0" and smart manufacturing-a review of research issues and application examples. *Int. J. Autom. Technol.* **2017**, *11*, 4–16. [CrossRef]
36. Vieira, A.A.C.; Dias, L.S.; Santos, M.Y.; Pereira, G.; Oliveira, J.A. Setting an industry 4.0 research and development agenda for simulation-a literature review. *Int. J. Simul. Model.* **2018**, *17*, 377–390. [CrossRef]
37. Kerin, M.; Pham, D.T. A review of emerging industry 4.0 technologies in remanufacturing. *J. Clean. Prod.* **2019**, *237*, 117805. [CrossRef]
38. Aceto, G.; Persico, V.; Pescapé, A. A survey on information and communication technologies for industry 4.0: State-of-the-art, taxonomies, perspectives, and challenges. *IEEE Commun. Surv. Tutor.* **2019**, *21*, 3467–3501. [CrossRef]
39. Rejikumar, G.; Arunprasad, P.; Persis, J.; Sreeraj, K.M. Industry 4.0: Key findings and analysis from the literature arena. *Benchmarking Int. J.* **2019**,*26*,2514–2542.
40. Alcácer, V.; Cruz-Machado, V. Scanning the industry 4.0: A literature review on technologies for manufacturing systems. *Eng. Sci. Technol. Int. J.* **2019**, *22*, 899–919. [CrossRef]
41. Butt, J. A strategic roadmap for the manufacturing industry to implement industry 4.0. *Designs* **2020**, *4*, 11. [CrossRef]
42. Bongomin, O.; Gilibrays Ocen, G.; Oyondi Nganyi, E.; Musinguzi, A.; Omara, T. Exponential disruptive technologies and the required skills of industry 4.0. *J. Eng.* **2020**, *2020*, 17. [CrossRef]
43. Beier, G.; Ullrich, A.; Niehoff, S.; Reißig, M.; Habich, M. Industry 4.0: How it is defined from a sociotechnical perspective and how much sustainability it includes–A literature review. *J. Clean. Prod.* **2020**, *259*, 120856. [CrossRef]
44. Parente, M.; Figueira, G.; Amorim, P.; Marques, A. Production scheduling in the context of Industry 4.0: Review and trends. *Int. J. Prod. Res.* **2020**, *58*, 5401–5431. [CrossRef]
45. Hernandez Korner, M.E.; Lambán, M.P.; Albajez, J.A.; Santolaria, J.; Ng Corrales, L.d.C.; Royo, J. Systematic literature review: Integration of additive manufacturing and industry 4.0. *Metals* **2020**, *10*, 1061. [CrossRef]
46. Bousdekis, A.; Lepenioti, K.; Apostolou, D.; Mentzas, G. A Review of Data-Driven Decision-Making Methods for Industry 4.0 Maintenance Applications. *Electronics* **2021**, *10*, 828. [CrossRef]
47. Pivoto, D.G.; de Almeida, L.F.; da Rosa Righi, R.; Rodrigues, J.J.; Lugli, A.B.; Alberti, A.M. Cyber-physical systems architectures for industrial internet of things applications in Industry 4.0: A literature review. *J. Manuf. Syst.* **2021**, *58*, 176–192. [CrossRef]
48. Beltrami, M.; Orzes, G.; Sarkis, J.; Sartor, M. Industry 4.0 and sustainability: Towards conceptualization and theory. *J. Clean. Prod.* **2021**, *312*, 127733. [CrossRef]
49. Morgan, J.; Halton, M.; Qiao, Y.; Breslin, J.G. Industry 4.0 smart reconfigurable manufacturing machines. *J. Manuf. Syst.* **2021**, *59*, 481–506. [CrossRef]
50. Chavez, Z.; Hauge, J.B.; Bellgran, M. Industry 4.0, transition or addition in SMEs? A systematic literature review on digitalization for deviation management. *Int. J. Adv. Manuf. Technol.* **2021**, *2021*, 1–20. [CrossRef]
51. Chen, J.; Shi, Y. Stochastic model predictive control framework for resilient cyber–physical systems: Review and perspectives. *Philos. Trans. R. Soc. A* **2021**, *379*, 20200371. [CrossRef] [PubMed]
52. Zheng, T.; Ardolino, M.; Bacchetti, A.; Perona, M. The applications of Industry 4.0 technologies in manufacturing context: A systematic literature review. *Int. J. Prod. Res.* **2021**, *59*, 1922–1954. [CrossRef]
53. Nimawat, D.; Gidwani, B. An overview of industry 4.0 in manufacturing industries. *Int. J. Ind. Syst. Eng.* **2022**, *40*, 415–454. [CrossRef]
54. Nain, G.; Pattanaik, K.; Sharma, G. Towards edge computing in intelligent manufacturing: Past, present and future. *J. Manuf. Syst.* **2022**, *62*, 588–611. [CrossRef]
55. Liao, Y.; Deschamps, F.; Loures, E.d.F.R.; Ramos, L.F.P. Past, present and future of Industry 4.0-a systematic literature review and research agenda proposal. *Int. J. Prod. Res.* **2017**, *55*, 3609–3629. [CrossRef]
56. Mittal, S.; Khan, M.A.; Romero, D.; Wuest, T. A critical review of smart manufacturing & Industry 4.0 maturity models: Implications for small and medium-sized enterprises (SMEs). *J. Manuf. Syst.* **2018**, *49*, 194–214.
57. Nascimento, D.L.M.; Alencastro, V.; Quelhas, O.L.G.; Caiado, R.G.G.; Garza-Reyes, J.A.; Rocha-Lona, L.; Tortorella, G. Exploring Industry 4.0 technologies to enable circular economy practices in a manufacturing context: A business model proposal. *J. Manuf. Technol. Manag.* **2019**, *30*, 607–627. [CrossRef]
58. Sony, M.; Naik, S. Key ingredients for evaluating Industry 4.0 readiness for organizations: A literature review. *Benchmarking Int. J.* **2019**,*27*,2213–2232. [CrossRef]
59. Gebhardt, M.; Kopyto, M.; Birkel, H.; Hartmann, E. Industry 4.0 technologies as enablers of collaboration in circular supply chains: A systematic literature review. *Int. J. Prod. Res.* **2021**, *2021*, 1–29. [CrossRef]

60. Sharma, A.; Mehtab, R.; Mohan, S.; Shah, M.K.M. Augmented reality—An important aspect of Industry 4.0. *Ind. Robot. Int. J. Robot. Res. Appl.* **2021**,*49*, 428–441. [CrossRef]
61. Ahmed, A.A.; Nazzal, M.A.; Darras, B.M. Cyber-physical systems as an enabler of circular economy to achieve sustainable development goals: A comprehensive review. *Int. J. Precis. Eng. Manuf.-Green Technol.* **2021**,*9*, 955–975. [CrossRef]
62. Ghobakhloo, M.; Iranmanesh, M.; Vilkas, M.; Grybauskas, A.; Amran, A. Drivers and barriers of Industry 4.0 technology adoption among manufacturing SMEs: A systematic review and transformation roadmap. *J. Manuf. Technol. Manag.* 2022, *ahead-of-print*. [CrossRef]
63. Sahu, A.; Agrawal, S.; Kumar, G. Integrating Industry 4.0 and circular economy: A review. *J. Enterp. Inf. Manag.* **2021**, *35*, 885–917. [CrossRef]
64. Patyal, V.S.; Sarma, P.; Modgil, S.; Nag, T.; Dennehy, D. Mapping the links between Industry 4.0, circular economy and sustainability: A systematic literature review. *J. Enterp. Inf. Manag.* **2022**, *35*, 1–35. [CrossRef]
65. Ng, T.C.; Lau, S.Y.; Ghobakhloo, M.; Fathi, M.; Liang, M.S. The Application of Industry 4.0 Technological Constituents for Sustainable Manufacturing: A Content-Centric Review. *Sustainability* **2022**, *14*, 4327. [CrossRef]
66. Bag, S.; Telukdarie, A.; Pretorius, J.C.; Gupta, S. Industry 4.0 and supply chain sustainability: Framework and future research directions. *Benchmarking Int. J.* **2018**, *28*, 140–1450. [CrossRef]
67. Ejsmont, K.; Gladysz, B.; Kluczek, A. Impact of industry 4.0 on sustainability—bibliometric literature review. *Sustainability* **2020**, *12*, 5650. [CrossRef]
68. Ghobakhloo, M. Industry 4.0, digitization, and opportunities for sustainability. *J. Clean. Prod.* **2020**, *252*, 119869. [CrossRef]
69. Cardin, O. A Systematic Literature Review of Successful Implementation of Industry 4.0 Technologies in Companies: Synthesis of the IPSI Framework. *Appl. Sci.* **2021**, *11*, 8917. [CrossRef]
70. Khan, I.S.; Ahmad, M.O.; Majava, J. Industry 4.0 and sustainable development: A systematic mapping of triple bottom line, Circular Economy and Sustainable Business Models perspectives. *J. Clean. Prod.* **2021**, *297*, 126655. [CrossRef]
71. Ghobakhloo, M.; Fathi, M.; Iranmanesh, M.; Maroufkhani, P.; Morales, M.E. Industry 4.0 ten years on: A bibliometric and systematic review of concepts, sustainability value drivers, and success determinants. *J. Clean. Prod.* **2021**, *302*, 127052. [CrossRef]
72. Atif, S.; Ahmed, S.; Wasim, M.; Zeb, B.; Pervez, Z.; Quinn, L. Towards a conceptual development of Industry 4.0, servitisation, and circular economy: A systematic literature review. *Sustainability* **2021**, *13*, 6501. [CrossRef]
73. Ortega-Gras, J.J.; Bueno-Delgado, M.V.; Cañavate-Cruzado, G.; Garrido-Lova, J. Twin Transition through the Implementation of Industry 4.0 Technologies: Desk-Research Analysis and Practical Use Cases in Europe. *Sustainability* **2021**, *13*, 13601. [CrossRef]
74. Dutta, P.; Chavhan, R.; Gowtham, P.; Singh, A. The individual and integrated impact of Blockchain and IoT on sustainable supply chains: A systematic review. In *Proceedings of the Supply Chain Forum: An International Journal*; Taylor & Francis: Oxfordshire, UK, 2022; pp. 1–24.
75. Zhang, C.; Chen, Y. A review of research relevant to the emerging industry trends: Industry 4.0, IoT, blockchain, and business analytics. *J. Ind. Integr. Manag.* **2020**, *5*, 165–180. [CrossRef]
76. Tiwari, S. Supply chain integration and Industry 4.0: A systematic literature review. *Benchmarking Int. J.* **2020**, *28*, 990–1030. [CrossRef]
77. Zuo, Y. Making smart manufacturing smarter—A survey on blockchain technology in Industry 4.0. *Enterp. Inf. Syst.* **2021**, *15*, 1323–1353. [CrossRef]
78. Galati, F.; Bigliardi, B. Industry 4.0: Emerging themes and future research avenues using a text mining approach. *Comput. Ind.* **2019**, *109*, 100–113. [CrossRef]
79. Jesus, C.d.; Lima, R.M. Literature search of key factors for the development of generic and specific maturity models for Industry 4.0. *Appl. Sci.* **2020**, *10*, 5825. [CrossRef]
80. Marinho, M.; Prakash, V.; Garg, L.; Savaglio, C.; Bawa, S. Effective cloud resource utilisation in cloud erp decision-making process for industry 4.0 in the united states. *Electronics* **2021**, *10*, 959. [CrossRef]
81. Anshari, M.; Almunawar, M.N. Adopting open innovation for SMEs and industrial revolution 4.0. *J. Sci. Technol. Policy Manag.* **2021**, *13*, 405–427. [CrossRef]
82. Reis, J.Z.; Gonçalves, R.F.; Silva, M.T.d.; Kazantsev, N. Business Models for the Internet of Services: State of the Art and Research Agenda. *Future Internet* **2022**, *14*, 74. [CrossRef]
83. Souza, M.L.H.; da Costa, C.A.; de Oliveira Ramos, G.; da Rosa Righi, R. A survey on decision-making based on system reliability in the context of Industry 4.0. *J. Manuf. Syst.* **2020**, *56*, 133–156. [CrossRef]
84. Liu, C.; Zheng, P.; Xu, X.(Accepted/In press) Digitalisation and servitisation of machine tools in the era of Industry 4.0: A review. *Int. J. Prod. Res.* **2021**, 1–13.
85. Park, S.; Huh, J.H. Effect of cooperation on manufacturing it project development and test bed for successful Industry 4.0 project: Safety management for security. *Processes* **2018**, *6*, 88. [CrossRef]
86. Wlazlak, P.; Säfsten, K.; Hilletofth, P. Original equipment manufacturer (OEM)-supplier integration to prepare for production ramp-up. *J. Manuf. Technol. Manag.* **2019**, *30*, 506–530. [CrossRef]
87. Konstantindis, F.K.; Gasteratos, A.; Mouroutsos, S.G. Vision-Based Product Tracking Method for Cyber-Physical Production Systems in Industry 4.0. In Proceedings of the 2018 IEEE International Conference on Imaging Systems and Techniques (IST), Beijing, China, 18–20 October 2017; pp. 1–6.

88. Bampis, L.; Mouroutsos, S.G.; Gasteratos, A. A Product Pose Tracking Paradigm Based on Deep Points Detection. *Machines* **2021**, *9*, 112. [CrossRef]
89. Konstantinidis, F.K.; Kansizoglou, I.; Tsintotas, K.A.; Mouroutsos, S.G.; Gasteratos, A. The Role of Machine Vision in Industry 4.0: A textile manufacturing perspective. In Proceedings of the 2021 IEEE International Conference on Imaging Systems and Techniques (IST), Kaohsiung Taiwan, 21–23 June 2021.
90. Mourtzis, D.; Angelopoulos, J.; Panopoulos, N. Smart Manufacturing and Tactile Internet Based on 5G in Industry 4.0: Challenges, Applications and New Trends. *Electronics* **2021**, *10*, 3175. [CrossRef]
91. Bogoviz, A.V. Perspective directions of state regulation of competition between human and artificial intellectual capital in Industry 4.0. *J. Intell. Cap.* **2020**, *21*, 583–600. [CrossRef]
92. Balaska, V.; Folinas, D.; Konstantinidis, F.K.; Gasteratos, A. Smart counting of unboxed stocks in the Warehouse 4.0 ecosystem. In Proceedings of the 2022 IEEE International Conference on Imaging Systems and Techniques (IST), Kaohsiung, Taiwan, 21–23 June 2022; pp. 1–6.
93. Konstantinidis, F.K.; Balaska, V.; Symeonidis, S.; Mouroutsos, S.G.; Gasteratos, A. AROWA: An autonomous robot framework for Warehouse 4.0 health and safety inspection operations. In Proceedings of the 2022 30th Mediterranean Conference on Control and Automation (MED), Vouliagmeni-Athens, Greece, 28 June–1 July 2022; pp. 494–499.
94. Huh, J.H.; Seo, Y.S. Understanding edge computing: Engineering evolution with artificial intelligence. *IEEE Access* **2019**, *7*, 164229–164245. [CrossRef]
95. Katika, T.; Konstantinidis, F.K.; Papaioannou, T.; Dadoukis, A.; Bolierakis, S.N.; Tsimiklis, G.; Amditis, A. Exploiting Mixed Reality in a Next-Generation IoT ecosystem of a construction site. In Proceedings of the 2022 IEEE International Conference on Imaging Systems and Techniques (IST), Kaohsiung, Taiwan, 21–23 June 2022; pp. 1–6.

machines

Article

A Novel Computational-Based Visual Brand Identity (CbVBI) Product Design Methodology

Athanasios Manavis [1,*], Anastasios Tzotzis [1], Apostolos Tsagaris [2] and Panagiotis Kyratsis [1,*]

1 Department of Product and Systems Design Engineering, University of Western Macedonia, GR50100 Kila Kozani, Greece
2 Department of Industrial Engineering and Management, International Hellenic University, GR57400 Thessaloniki, Greece
* Correspondence: manavis.athanasios@gmail.com (A.M.); pkyratsis@uowm.gr (P.K.)

Abstract: Product design is a promising field for the application of new technologies and methodologies emerging from the digital evolution of Industry 4.0. A great number of tools have been developed in order to accentuate the use of modern Computer-Aided Design (CAD) systems and computational design techniques for design customization in product applications. The present paper deals with the development of two different applications for designing furniture based on the Computational-based Visual Brand Identity (CbVBI) design methodology. For the first case study, the Application Programming Interface (API) Solidworks™ (VBA event-driven programming language) is used. The second case study focuses on the visual programming language of Grasshopper™, which is incorporated within Rhinoceros3D™. The proposed case studies offer a great deal of flexibility in both design and manufacturing, while many design alternatives could become available in a very short period.

Keywords: CAD; computational design; Application Programming Interface; brand identity; mass customization; furniture design

1. Introduction

Automated product design is a topic receiving increasing interest from both industry and academia. Modern industries tend to automate every aspect related to the development of a consumer product. Hence, the conceptual design, the testing and the finalization of a product are only some of the processes considered for automatization. An Application Programming Interface (API) in CAD systems is a tool that can facilitate the realization of the aforementioned tasks. Parallelly, computational design affords a great number of benefits to CAD systems that can level up common CAD-based design procedures. Specifically, computational design is a novel strategy of using visual programming aspects to create and transform shapes, forms and structures. This allows the end-users to act as co-designers and take part in creating alternative options for the original products. Finally, the increase in the productivity and the automation of daily routines has enabled designers to effectively allocate their time to more creative tasks.

2. State of the Art

2.1. Automated Product Design and Mass Customization

CAD-based programming is commonly used for the simulation of several industrial processes, such as the kinematics of machining and robotics, the prediction of machining components, the extraction of manufacturing data [1–4], measurements, quality assessment [5,6] and many more. The computational design methodology derives from similar procedures in mass customization. The usability of these techniques has been expanded to multiple fields of application, such as the design and assembly of products, which is the topic of the two case studies presented in this article, the generation of engineering drawings and assembly instructions and high-quality renderings and illustrations.

2.2. CAD-Programming Design and APIs

Tzotzis et al. [7] and Kyratsis et al. [8] used similar techniques to develop applications for the automated accomplishment of a set of engineering processes related to the development of standardized mechanical parts and systems. Hong et al. [9] managed to simplify the assembly process of complex systems, optimizing in this way the time-consuming and resource-intensive tasks that are required for the assembly of such parts. The majority of the case studies from the proposed references list that are related to the design and assembly of consumer products [10] include the use of material textures, coloring and high-quality renderings that provide the consumer with a complete experience regarding the presentation of the product, as well as its framework. CAD-based programming can be applied to finite element modeling as well.

A number of studies [11,12] exist in the literature in which the automated design of machine tools was realized with the aid of CAD-based coding. The implementation of such techniques can lead to the generation of tool models that can be imported to finite element analysis (FEA) software for simulation purposes. Finally, the work of Nam et al. takes a different approach to programming in relation to CAD systems [13]. The authors developed a web-based material database aiming at the facilitation of the material selection process with respect to the functionality, the manufacturability and cost of the designed product.

2.3. Graphical Algorithmic Design and Web-Based APIs

Castro e Costa et al. [14] proposed an online application that promotes the development of a parametric design system for the mass customization of ceramic tableware. The implementation of the final digital 3D forms accelerated with the use of shape grammar rules according to the specific parametric procedure of computational design. Manavis and Kyratsis [15] managed to blend the principles of brand product identity with the automated procedures of computational design. Specifically, the authors presented a web-based application for the re-creation of novel souvenirs under the umbrella of the "Cycladic marble figurines from the Early Ages" concept. The whole process of the previous application was based on the Computational-based Visual Brand Identity (CbVBI) strategy. Furthermore, Kyratsis et al. [16] defined a unique non-conventional procedure for the form generation of furniture and decorative artifacts. The proposed procedure was based on graphical algorithmic design rules. More specifically, the crucial rules of the aforementioned case study are linked to the morphology and shape parametrization of the produced furniture and decorative artifacts. The proposed approach concerns an ergonomic point of view of product development. Finally, the ChairDNA Design Tool [17] is a great example of a mass customization procedure for product design under the influence of computational design parameterization.

2.4. Product Design as Brand Element

The "product shape generation to support brand identity elements" [18] relates to CbBVI design strategy. The current CbVBI model tries to effectively offer an automatic way of designing alternative products under a specific brand (e.g., morphological specifications, aesthetic criteria, etc.). This unique CbVBI methodology approach connects two different design areas according to state-of-the-art research. Specifically, CbVBI methodology is a bridge between the different design areas of branding and product design under the main design procedure.

The present paper introduces two case studies based on the programming of CAD systems, namely SolidWorks™ (Dassault Systèmes SE, Vélizy-Villacoublay, France) and Rhinoceros3D™ (Robert McNeel & Associates, Seattle, WA, USA). In the first case study, an API is utilized to develop an application for the automated design and assembly of a piece of furniture with a wide range of options that can be commercially distributed. The application comprises three tabs, one for the design options, one for the material selection and one for the product summary and preview, so that the user can select the desired dimensions and construction options, the colors and materials, as well as obtain an instant cost estimation. The second application deals with the design of a desk workstation. Specifically, it introduces a web-based application that offers specific brand parameters to end-users, allowing them to design their own version of the prototype desk workstation. The parameters are related to the main ontology of the desk, e.g., dimensions, materials and colors. The development of the application is based on both the Rhinoceros3D™ and the Grasshopper™ software.

3. The Proposed CbVBI Methodology

Figure 1 illustrates the proposed CbVBI design methodology used to create specific branded pieces of furniture. The workflow is divided into three different sections. The first section of CbVBI deals with the branding elements and references specification. The Do-It-Yourself (DIY) branding theme relates to the furniture ontology (e.g., dimensions variables, design, assembly parts, etc.) [19]. The implementation of the design procedure forms part of the second section of the workflow. More specifically, the second step is composed of two parallel directions: (a) designing with CAD programming tools (textual programming) and (b) designing with computational tools (graphical algorithmic design). These directions are considered in the following procedure according to the digitalized furniture design parametrization: (a) the designer creates the design options according to the furniture ontology, which is related to the object's morphology, and (b) a material database is created by the designer in order to offer the aesthetic parameters in the design procedure of customizing DIY furniture with a 3D-rendered representation. The final section involves the digital export of the new furniture ideas created from the end-users' point of view (e.g., rendered images, assembly instructions, etc.). The basic objectives of the CbVBI methodology (as the final result of the research) are summarized in three specific directions: (a) the parametrization of the branding criteria (visual elements), (b) the digitalization of the product development according to computational strategies and finally, (c) the automatization of the product development.

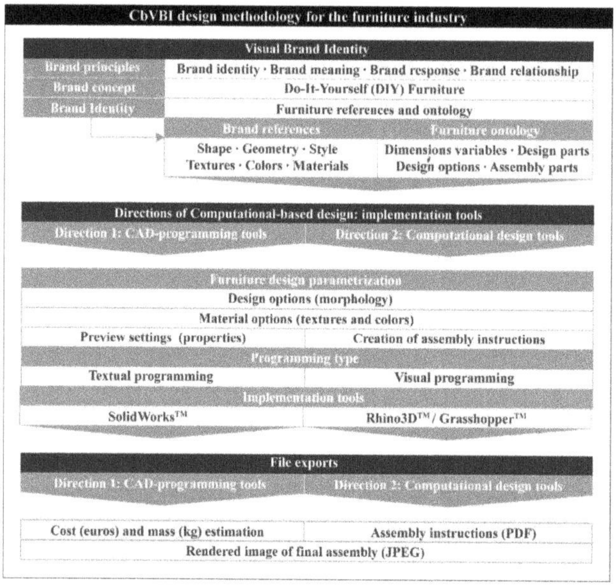

Figure 1. Computational-based Visual Brand Identity design methodology for the furniture industry.

4. Case Studies and Applications

4.1. CAD-Based Design and Assembly of a DIY Coffee Table

Engineering task automation is a key factor in the effective allocation of working time and the creativity of engineers. In particular, design and assembly are two important stages during the development of a product that benefit greatly from such techniques. It is seen more often than not that companies implement CAD-based automation in their projects with the aim of increasing their productivity.

In this case study, a typical example of CAD-based programming is presented, in relation to the design and assembly of a coffee table. Such furniture is common in coffee shops and cafeterias, as well as in similar shops. Despite the fact that this paper does not focus on the design aspect of the product, but rather on the implementation of CAD-based programming for the design procedures, it must be noted that these tables are usually compact, lightweight, somewhat tall and sturdy to allow them to be carried easily, take up only a small area and provide adequate ergonomics for the user. The user interface is considered to be the bridge between the user and the CAD system [20]; therefore, it is important that it is clear and understandable, as well as easy to navigate.

The application is divided into three tabs, as seen in Figure 2. The first tab "Morphology" includes all the necessary fields and lists, where the user can input the dimensions of the table and select the model as well, which corresponds to three different board shapes (rectangle/square, circle and rhombus) and three different chassis types. These parameters yield nine unique table combinations. In addition, the variety of the board materials and colors hugely increases the number of possible combinations. The second tab, namely "Material/Color", includes a number of buttons, which are used to select one of the available materials and colors.

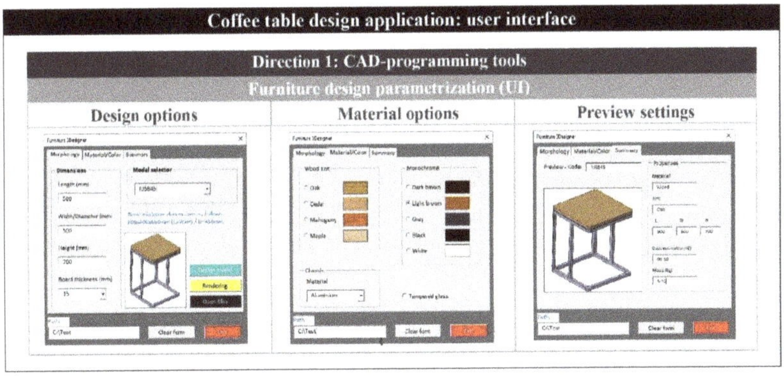

Figure 2. The three tabs of the coffee table design application.

Additionally, an option is included for a glass board, instead of a wooden one. Furthermore, the user can select the chassis material in this tab, ranging from plain steel for a more economic final product to aluminum, which offers a more stylish look and a lightweight final product. The third tab (Summary) provides the user with auxiliary functions, such a preview of the final product, its weight and cost, as well as a summary of its basic attributes (material, color and dimensions).

The application can be operated with the three command buttons "Design model", "Rendering" and "Open files", found in the first tab, which are used to perform the complete design and assembly process, the generation of a rendering of the final product and the compiling of a folder containing the files, respectively.

According to Figure 3, the code follows a straightforward execution. First of all, the necessary variables, relating to the dimensions, the material and the color, are declared. Next, these variables are linked to the corresponding design options that are available in the interface. For instance, when the user types a number in the "Length (mm)" field, the variable that corresponds to the length of the board changes its value to the typed one. As soon as the user pushes the "Design model" button, the design process begins. Therefore, the design loop is executed for every part comprising the table. This number varies depending on the selected design options. Regarding the design procedure, the creation of a new part document, the selection of a plane, the insertion of a new sketch, the design of a contour, the extrusion of a solid model, the application of a material texture/color and the storage of the model are tasks that are performed on every iteration of the loop. The only difference is that the contour sketched on each iteration varies according to the part being designed. Hence, a rectangle/square, a rhombus or a circle will be sketched for the design of the board. Similar shapes will be sketched for the design of the chassis parts. To facilitate the design code for the sketched contours of the parts, the "If-Then-Else" structure was used.

Coffee table design application: workflow		
Direction 1: CAD-programming tools		
Furniture design parametrization		
Step	**Name**	**Procedures**
01	Decleration of variables	Lenght, width, height, texture, color, etc.
02	Linking of design options with the variables	**01.** Board shape **02.** Lenght, width/diameter **03.** Board thickness **04.** Board material **05.** Chassis height **06.** Chassis type **07.** Chassis material
03	Design of parts ('for' loop)	**01.** Creation of new part document **02.** Selection of pre-defined plane and insertion of new sketch **03.** Design according to options **04.** Extrusion of solid model **05.** Application of material properties **06.** Storage of part model
04	Assembly of parts	**01.** Creation of new assembly document **02.** Insertion of components **03.** Application of mate features **04.** Storage of assembly
05	Export of properties	**01.** Mass and volume **02.** Cost **03.** Material, texture and color
06	Rendering of assembly and creation of preview	

Figure 3. Workflow of the coffee table design application.

The assembly procedure is carried out in a similar manner. Thus, a loop is utilized to repeat the execution of the assembly code. First of all, a new assembly document is created; next, the components comprising the assembly are inserted; and finally, the appropriate mates are applied to the components. In contrast to the design process, the loop only contains the mates, rather than the whole process. With the completion of the assembly, several useful properties, such as the cost, the mass and the volume, are automatically exported to the fields that are available under the "Summary" tab in order to inform the user.

Finally, it is possible to generate a rendering of the final product by pressing the "Rendering button" found under the first tab. The image is saved in a standard image format, JPG. An image showing the user a preview of the final product is automatically generated as soon as the rendering is complete. This way, the user acquires a complete view of the full product.

The two first steps of the workflow do not require any special API methods, but rather standard VBA™ statements. The methods and the properties used in the step where the design of the parts is performed are summarized in Figure 4. Most of the methods in the VBA™ language are self-explanatory, and the syntax requires a number of elements that fully define a method, function or statement. Therefore, for the design process, the "SelectByID2" method is used to select named entities, such as edges and faces and especially planes, for sketching. The "InsertRefPlane" method is used to create a number of extra planes on the designed parts that can be used later during the assembly process.

API methods and properties

Design procedure		
API method/property	**Explanation**	**Return/property value**
NewDocument	Creates a new document based on a specified template	Newly created document
ActiveDoc	Connects to the currently active document	Model document
SelectByID2	Selects the specified entity	True/false
Select4	Selects an entity and marks it	True/false
InsertRefPlane	Inserts a constraint-based reference plane using the selected reference entities	Reference plane
InsertSketch	Inserts a new sketch in the current part or assembly document	New sketch
CreateCenterRectangle	Creates a center rectangle	Array of sketch segments that represent the edges and diagonals created for this center rectangle
CreateCircleByRadius	Creates a circle based on a center point and a specified radius	Sketch segment for the circle
FeatureExtrusion3	Creates an extruded feature	Access to the feature
FeatureCut4	Creates a cut extrude feature	Cut extrude feature
FeatureFillet3	Creates the specified fillet feature for selected edges or faces	New fillet feature
InsertFeatureChamfer	Inserts a chamfer	New chamfer feature
SetMaterialPropertyName2	Sets the name of the material property for the specified configuration	Applied material
SaveAs3	Saves the active document to the specified name in the specified format	True/false

Figure 4. API methods and properties used for the design procedure.

The "InsertSketch" method enables the sketching of a plane, whereas the "CreateCenterRectangle" and the "CreateCircleByRadius" are used to sketch a rectangle/square and a circle, respectively. Additional methods of the same type are available for the sketching of lines, polygons and other shapes. Once the contour is sketched, the "FeatureExtrusion3" and the "FeatureCut4" methods can be utilized to generate solids. With the appropriate syntax, it is possible to create a variety of solids representing the individual parts of the assembly, which can then receive attributes such as mass, volume, texture, color and other visual and physical properties. "SetMaterialPropertyName2" is a method utilized to apply a material name to a part, conferring on it various properties that correspond to the material.

During the stage of the assembly process, a new assembly document is opened and all the parts that comprise the assembly are opened in the background. The "ActivateDoc3" and the "AddComponent5" methods are responsible for the activation of the newly created document and the insertion of each one of the opened parts into this document. "SelectByID2" is used in this stage as well for the same purpose: to select the already named entities for the application of the mate features with the "CreateMate" method. As has been already discussed, it is important that the parts are designed with predefined planes and other geometric entities in order to facilitate the mating process during the assembly of its components. The most commonly used API methods are shown in Figure 5.

API methods and properties		
Assembly procedure		
API method/property	**Explanation**	**Return/property value**
NewDocument	Creates a new document based on a specified template	Newly created document
OpenDoc6	Opens an existing document and returns a pointer to the document object	Newly loaded model document
ActivateDoc3	Activates a loaded document and rebuilds it as specified	Model document
AddComponent5	Adds the specified component for the specified configuration options to this assembly	Addition of component to assembly
SelectByID2	Selects the specified entity	True/false
CreateMate	Creates a mate with the specified feature data object	Application of mate
SaveAs3	Saves the active document to the specified name in the specified format	True/false

Figure 5. API methods and properties used for the assembly procedure.

Figure 6 includes the necessary methods and properties for the export of the product's properties, as well as the realization of the rendering. The method "CreateMassProperty" calculates the mass and volume of an individual part or an assembly by utilizing the physical properties of the applied material. Therefore, this method functions only if a material is set for each of the components.

API methods and properties		
Rendering / exprorts procedures		
API method/property	**Explanation**	**Return/property value**
CreateMassProperty	Obtains mass property	Mass properties object
GetRayTraceRenderer	Get a ray-trace rendering engine	Ray trace renderer
RayTraceRendererOptions	Gets a ray-trace rendering engine's options	Ray-trace rendering engine's options
InvokeFinalRender	Invokes the final render window	True/false
RenderToFile	Saves the current view of the rendered model as an image to the specified file	True/false

Figure 6. API methods and properties used for the rendering and the property export procedures.

Moreover, to carry out the rendering, "GetRayTraceRenderer" enables the use of the renderer that is available in the CAD system, whereas the "RayTraceRendererOptions" property sets the rendering options, such as image resolution, dimensions and format. The final rendered image of the product is generated and stored with the methods "InvokeFinalRender" and "RenderToFile", respectively.

The previous code snippet (Figure 7) presents the lines required for the design of a rectangular board, including the creation of a new part document, the selection of a plane, the insertion of a new sketch, the sketch of rectangle, the extrusion of the sketch and the application of the texture and color of a material. The choice of material is not only important for cosmetic purposes; it is also necessary for the extraction of the physical properties.

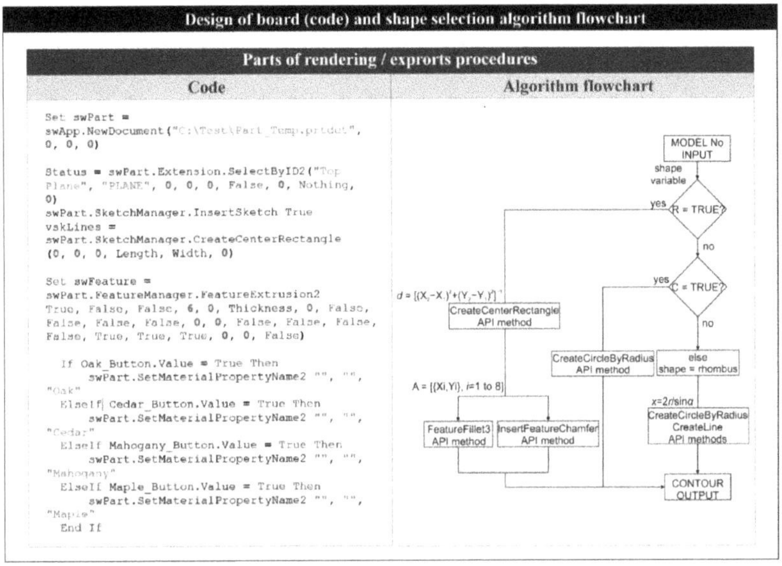

Figure 7. Code sample and algorithm flowchart used.

As soon as the user selects a model number, an algorithm (Figure 7) performs a check on the shape of the table so that the geometric measurements can be applied accordingly. If the user inputs dimensions that correspond to a rectangle or a square, the algorithm uses the corresponding API method to generate a rectangle or a square using the diagonal $d = \left[(x_2 - x_1)^2 + (y_2 - y_1)^2\right]^{-1}$. Next, the fillet or the chamfer feature is applied based on the model selection. A matrix containing the coordinates (x, y) of the eight points comprising the four chamfered or filleted corners of the rectangle or square is used for the application of the equivalent feature. In the case of a circular table, the algorithm skips the rectangular or square shape and sketches a circle instead via the respective API method, whereas in the case of a rhombic table board, the methods used to sketch a circle and a set of lines are used. Therefore, if the selected model corresponds to a circular board, a circle is sketched with a radius equal to the width given by the user. Finally, the lines are sketched tangential to the circle, forming a tangential quadrilateral with an edge length equal to $x = 2r/sina$.

4.2. Graphical Algorithmic Design and Assembly of DIY Desk Workstation

Graphical algorithmic editors are important for developing parametric designs through visual programming [21]. The numerous designs that are generated by the proposed Cb-VBI application are digital forms of a desk workstation. This procedure was based on Grasshopper^{TM} and Rhinoceros3D^{TM} software in order to offer a fully useful design application for branded furniture under the concept of DIY [22].

The present application is divided into three different design areas, as seen in Figure 8. The first area includes all the important parameters according to the design options (e.g., basic dimensions of the furniture, number of parts, etc.). Some of the design options are the total height of the product (TH), the length and the width of Desk1&2 and the number of the shelves (LS1 to LS5). Similar to the previous case study, the second area includes a number of options for the material and texture of the furniture. All the materials and skin types are related to wood textures. Finally, the third area provides the export files for the assembly instructions and the rendered representations of the final DIY furniture.

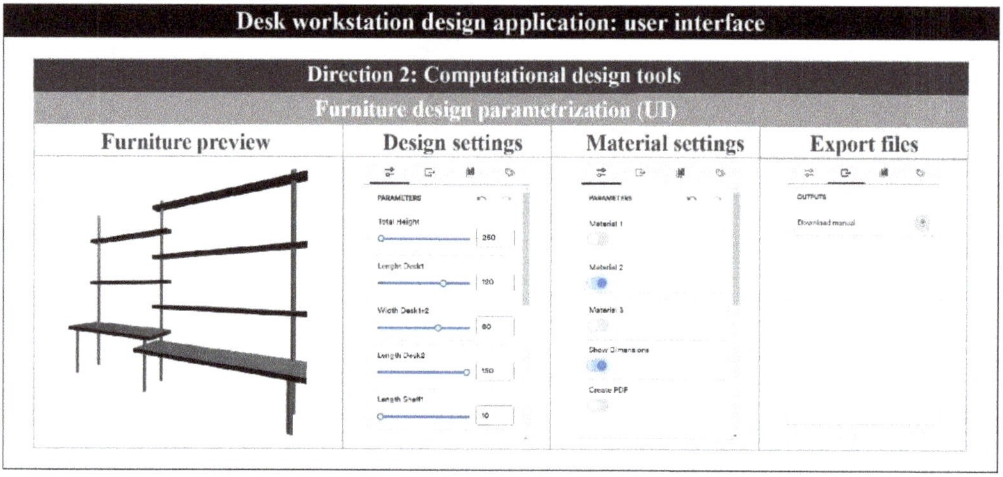

Figure 8. Desk workstation design application as created with Shape Diver™.

The proposed application was developed with the web-based platform Shape Diver™. Shape Diver™ allows end-users to design their own objects under specific parameters, which are developed from the programmer/designer's point of view.

Figure 9 depicts the design procedure for the computational-based visual code, which follows a specific hierarchy of steps. Every step of the proposed workflow is related to a specific node in the graphical algorithm of the whole code. The algorithm has five significant steps/nodes:

- First of all, the necessary variables relating to the dimensions and colors are declared.
- Next, these parameters are linked to the corresponding design options. Every aesthetic design criterion is related to a specific variable of the visual programming code.
- Following this, the third part of total hierarchy is the visual representation of the produced parts of the furniture. With both Grasshopper™ and Rhinoceros3D™, a final assembled desk is produced with all the design and manufacturing attributes. As a result of the third step, the application is capable of exporting a *.gh file with all the variables and parameters of the design object.
- The fourth stage of the workflow involves the linking procedure of the *.gh file within the web-based platform Shape Diver™. At this stage, the programmer/designer organizes the application interface for the furniture parametrization.
- Finally, the end-user is able to create his own desk workstation according to the "Do-It-Yourself" brand identity. In this case, the user has the responsibility of filling out all the variables in the online CbVBI application.

Figure 10 summarizes the group of the applications, pieces of software and add-ons that were used for the development of the CbVBI application. Apart from the aforementioned software Grasshopper™, Rhinoceros3D™ and Shape Diver™, the authors used two kinds of special add-ons for the application implementation.

Desk workstation design application: workflow			
Direction 2: Computational design tools			
Furniture design parametrization			
Step	**Name**	**Procedures**	**Application**
01	Decleration of variables	**01.** Total Height (TM) **02.** Lenght Desk1 (LD1) **03.** Lenght Desk2 (LD2) **04.** Width (W) **05.** Lenght Shelf1 (LS1) **06.** Lenght Shelf2 (LS2) **07.** Lenght Shelf3 (LS3) **08.** Lenght Shelf4 (LS4) **09.** Lenght Shelf5 (LS5) **10.** Material1 (M1) **11.** Material2 (M2) **12.** Material3 (M3)	Grasshopper™
02	Linking of design attributes with the variables	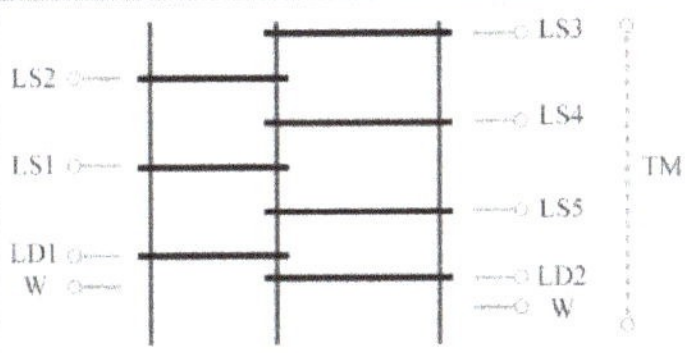	Grasshopper™ Rhinoceros™
03	Design and assemply of parts	**01.** Selection of pre-defined design attributes. **02.** Design and assemby according to Grasshopper™ options. **03.** Export of **.gh* file format. 	Rhinoceros™ Grasshopper™
04	Parametrization of furniture	**01.** Insertion of **.gh* file into Shape Diver™ platform. **02.** Application of customized furnitures according to end-user's choices. 	Shape Diver™
05	Exporting of assembly instructions and rendered images	**01.** Assembly instructions export (**.pdf* file format). **02.** Rendered image of final desk workstation (**.jpeg* file format) 	Shape Diver™

Figure 9. Workflow of the desk workstation design application.

Computational design applications		
Design and assembly procedures for parameterization capability		
Application	**Explanation**	**Result**
Rhinoceros™	A freeform surface modeler that utilizes the NURBS mathematical model.	CAD models visualization during the design procedure.
Grasshopper™	A visual programming language that runs within the Rhinoceros application	CAD-based 3D forms creation based on a number of parameters incorporated.
Shape Diver™	An online application that it provides customized parameters from end user's point of view.	Final product alternatives and creation of manufacturing file exports for each product.
Pufferfish™	A plugin set of a great number of components which focuses on special shape changing.	Methods for tween / blend / morph / lattice operations.
Squid™ (ShapeDiver edition)	A Grasshopper™ plugin for drawing bitmap images.	Outputs Grasshopper™ bitmap objects as mesh textures (for the PDF manual.)

Figure 10. Implementation tools based on computational design techniques.

Pufferfish™ is a plugin set with a great number of components relating to shape transformations (e.g., tween, blend and morph operations). Additionally, Squid™ is an important plugin for drawing bitmap images. This is a crucial code operation for the final export of the assembly instructions in a PDF file format. Specifically, Squid™ outputs Grasshopper™ bitmap objects as mesh textures in order to represent the final object's morphology as an image.

According to the previous procedure, the end-user is able to create his own desk workstation following the DIY brand identity principles. Specifically, the designer has the responsibility of filling all the numerical values for the design and material settings in the online application of Shape Diver™. The proposed Computational-based Visual Brand Identity (CbVBI) methodology exports two types of files in order to support the development of specific applications (e.g., technical drawings and assembly instructions). Finally, the production of the rendered images is based on the main characteristics of rendering synthesis: materials, textures, shadows, lights and the environment elements.

Figure 11 presents parts of the visual codes for each stage of the application workflow. More specifically, the whole code is divided into three programming areas: (a) design and assembly of the furniture, (b) application for the end-user's design contribution and (c) export files. The first area of the graphical algorithm includes the necessary methods and properties for the design and assembly of the product's morphology. At this stage, the designer is able to create the parts of furniture according to Grasshopper™'s commands and procedures. Three kinds of parameters define the design procedure: the number of parts, type of parts and part specification.

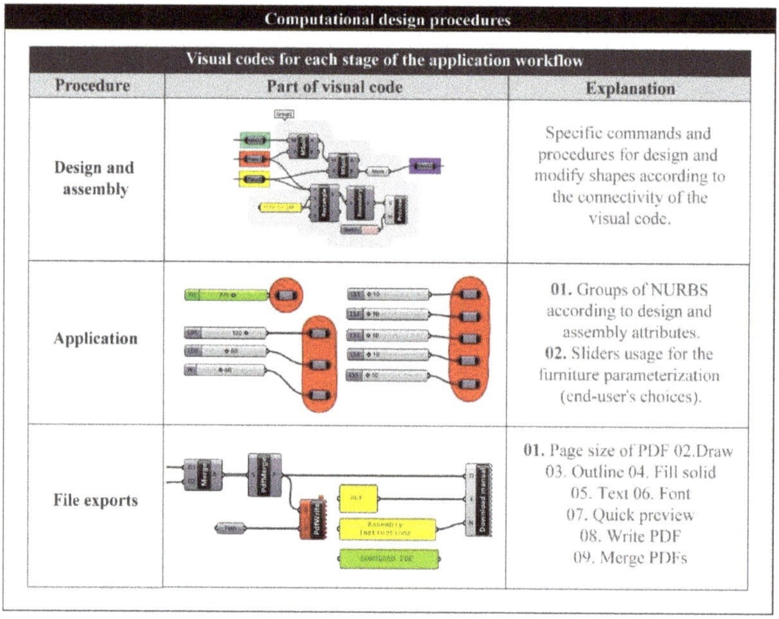

Figure 11. Computational design procedures.

Moreover, there are two more kinds of variables in the assembly definition: the types of connectivity and the technical requirements, such as total height. Following that, the designer groups the NURBS according to design and assembly attributes. More specifically, the designer uses special sliders with numerical values for the final furniture parameterization. The aforementioned sliders are connected to the parameters of the previous stage (number of parts, type of parts, part specification, etc.). It is important to note that the numerical values of the proposed parameters are made to be specific in order to protect the branded image of the prototype furniture.

The final code area is related to the procedure of exporting files. At this stage, the authors created nine operations for the final production of a PDF file that includes the assembly instructions. These significant operations deal with the (a) page size of the PDF (e.g., A4, A3, etc.), (b) visual representation of each furniture part, (c) outline creation based on wireframe models, (d) solid view of the whole object, (e) text and font specifications and finally (f) the merging of the produced pages. This means that the application automatically produces a full set of instructions and a product description for all the alternative designs requested by the user. The full automation of this procedure offers a significant cost reduction, while at the same time providing the final user with the ability to contribute towards the furniture design via choosing attributes of his preference.

The present case study shows an application for the automatic creation of a unique branded furniture with the aid of Grasshopper™ and Rhinoceros3D™. Figure 12 illustrates an image of the desk workstation created based on the strong concept theme of DIY followed by the representation of two indicative desk design alternatives, which are based on the CbVBI methodology. The final desks are the result of a combination of a number of parameters.

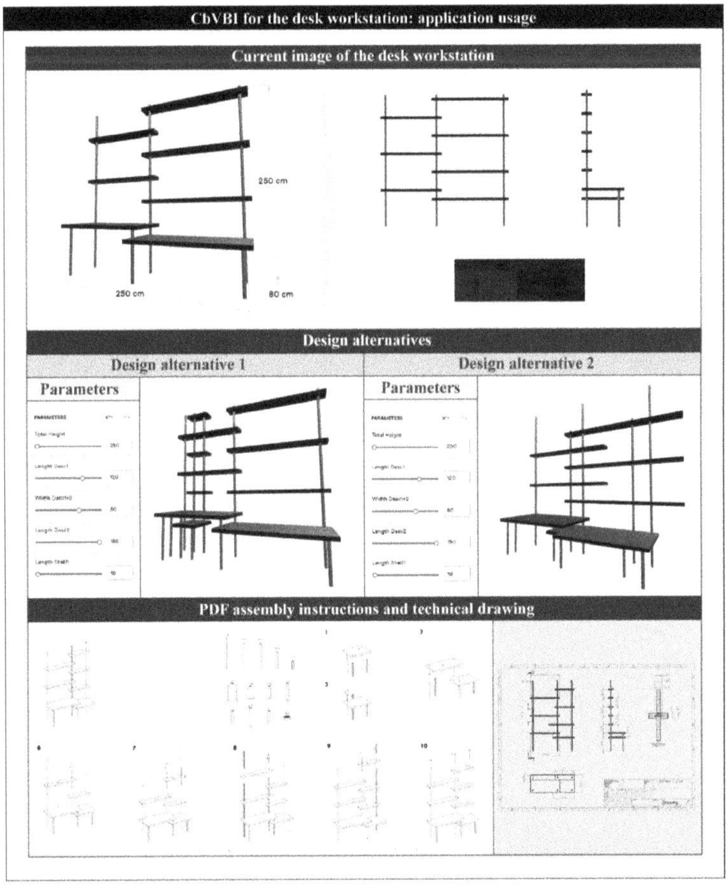

Figure 12. Desk workstation design alternatives and export files.

Design alternative 1 includes the following values: TH = 250, LD1 = 120, WD1&2 = 80, LD2 = 150 and LS = 10. On the other hand, design alternative 2 is shaped by the values: 300, 200, 100, 150 and 50. Finally, according to the CbVBI methodology, a PDF manual with all the appropriate instructions for the furniture assembly is produced. The current manual is similar and looks like the instruction books found in IKEA^{TM} DIY products.

Finally, it is important to note the time saved with the use of the CbVBI methodology in comparison with traditional design procedures. More specifically, Figure 13 presents the final times of the two aforementioned case studies: (a) CAD-based design and assembly of a DIY coffee table and (b) graphical algorithmic design and assembly of a DIY desk workstation. More specifically, the time needed for the development of a coffee table design (first case study) using a traditional design method, Solidworks^{TM}, is estimated to be about 85 min (including design and rendering procedures). On the other hand, the time needed for the development of the same coffee table via the use of CbVBI methodology was demonstrated to be 3.4 s. It is important to note that this time is relative to a computer's sensitivity. The second case study involved the development of a workstation design. The time that a designer needs to develop a CAD model of a workstation with all design and assembly instructions via RhinoTM is estimated at about 270 min. The CbVBI methodology is capable of generating the same files (i.e., the CAD model and assembly instructions) in 8.2 s.

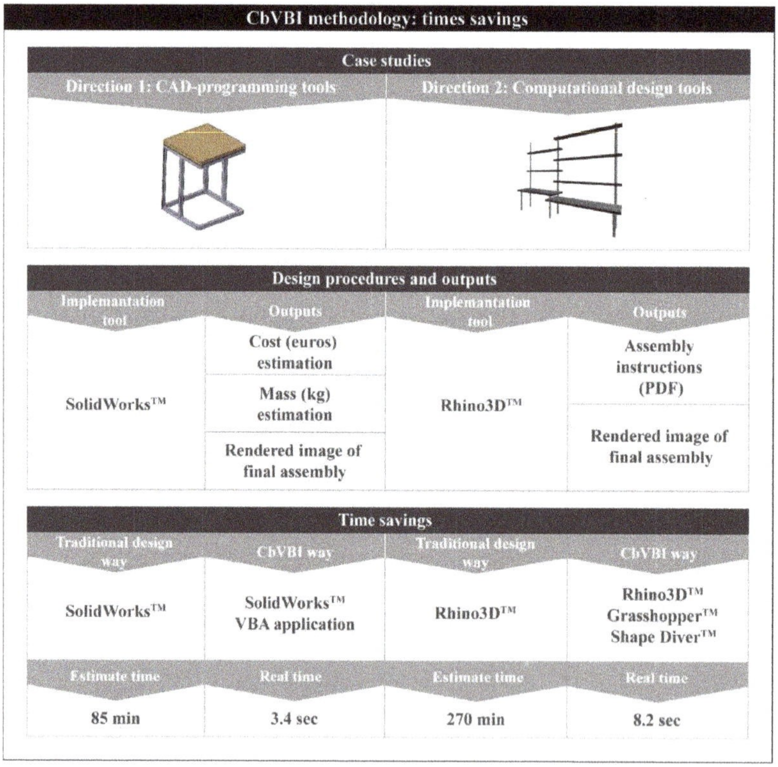

Figure 13. CbVBI time savings.

5. Conclusions

The present work deals with novel ideas for product development. The perception of the aforementioned CbVBI methodology deals with the parametrization of the visible elements of DIY design strategy (e.g., number of product parts, types of assembly, kinds of materials, etc.) according to a computational design approach. The core concept of the proposed design strategy is the automatic generation of furniture under the theme of DIY. The main contribution from the authors' point of view is linked to the practical connection between the visual elements of branding identity and the design parameters according to automated and/or computational design approaches. The proposed case studies of the two different furniture pieces underline the digitalization of the aesthetic and ergonomic characteristics of the DIY design strategy. The authors suggest two different strategies for the implementation of the CbVBI methodology: (a) an application based on the textual script using Solidworks™ and VisualBasic™ and (b) an application based on graphical algorithmic design using Rhinoceros3D™ and Grasshopper™.

The visual programming languages allow designers to build graphic-based programs rapidly, simply and flexibly. On the other hand, textual programming language (VBA) uses an event-driven approach to achieve the final outcome. Furthermore, textual languages follow three specific principles: primitive elements, combination mechanisms and abstraction mechanisms.

The result of these two different directions is an automated way to design a great number of furniture variations (coffee table and desk workspace) according to the end-users' needs. The user becomes a design contributor via their participation in selecting the design parameters and thus strengthens the brand identity principles through product shape and morphology.

Author Contributions: Conceptualization, A.M. and P.K.; methodology, A.M. and P.K.; software, A.M., A.T. (Anastasios Tzotzis) and A.T. (Apostolos Tsagaris); validation, A.M., A.T. (Anastasios Tzotzis), A.T. (Apostolos Tsagaris) and P.K.; formal analysis, A.M., A.T. (Anastasios Tzotzis) and A.T. (Apostolos Tsagaris); investigation, A.M. and A.T. (Anastasios Tzotzis); resources, P.K.; data curation, A.M. and A.T. (Anastasios Tzotzis); writing—original draft preparation, A.M. and P.K.; writing—review and editing, A.M., A.T. (Anastasios Tzotzis), A.T. (Apostolos Tsagaris) and P.K.; visualization, A.M. and A.T. (Anastasios Tzotzis); supervision, P.K.; project administration, A.M. and P.K.; funding acquisition, P.K. All authors have read and agreed to the published version of the manuscript.

Funding: This research received no external funding.

Institutional Review Board Statement: Not applicable.

Informed Consent Statement: Not applicable.

Data Availability Statement: Not applicable.

Conflicts of Interest: The authors declare no conflict of interest.

References

1. Neto, P.; Mendes, N.; Arajo, R.; Pires, J.N.; Moreira, A.P. High-level robot programming based on CAD: Dealing with unpredictable environments. *Ind. Rob.* **2012**, *39*, 294–303. [CrossRef]
2. Tapoglou, N.; Antoniadis, A. 3-Dimensional kinematics simulation of face milling. *Meas. J. Int. Meas. Confed.* **2012**, *45*, 1396–1405. [CrossRef]
3. Tzotzis, A.; Manavis, A.; Efkolidis, N.; Kyratsis, P. Cad-based automated g-code generation for drilling operations. *Int. J. Mod. Manuf. Technol.* **2021**, *13*, 177–184. [CrossRef]
4. Kyratsis, P.; Tapoglou, N.; Bilalis, N.; Antoniadis, A. Thrust force prediction of twist drill tools using a 3D CAD system application programming interface. *Int. J. Mach. Mach. Mater.* **2011**, *10*, 18–33. [CrossRef]
5. Gella-Marín, R.; Tzotzis, A.; García-Hernández, C.; Huertas-Talon, J.L.; Kyratsis, P. CAD software integration with programming tools for modelling, measurement and verification of surfaces. In *Experiments and Simulations in Advanced Manufacturing*; Kyratsis, P., Davim, J.P., Eds.; Springer: Berlin/Heidelberg, Germany, 2021; pp. 91–116.
6. García-Hernández, C.; Gella-Marín, R.; Huertas-Talón, J.L.; Berges-Muro, L. Algorithm for measuring gears implemented with general-purpose spreadsheet software. *Meas. J. Int. Meas. Confed.* **2016**, *85*, 1–12. [CrossRef]
7. Tzotzis, A.; Garcia-Hernandez, C.; Huertas-Talon, J.L.; Tzetzis, D.; Kyratsis, P. Engineering applications using CAD based application programming interface. *MATEC Web Conf.* **2017**, *94*, 01011. [CrossRef]
8. Kyratsis, P.; Tzotzis, A.; Tzetzis, D.; Sapidis, N. Pneumatic cylinder design using cad-based programming. *Acad. J. Manuf. Eng.* **2018**, *16*, 107–113.

9. Xiao, H.; Li, Y.; Yu, J.F.; Cheng, H. Dynamic assembly simplification for virtual assembly process of complex product. *Assem. Autom.* **2014**, *34*, 1–15. [CrossRef]
10. Kyratsis, P.; Gabis, E.; Tzotzis, A.; Tzetzis, D.; Kakoulis, K. CAD based product design: A case study. *Int. J. Mod. Manuf. Technol.* **2019**, *11*, 110–115.
11. Tzotzis, A.; García-Hernández, C.; Huertas-Talón, J.L.; Kyratsis, P. Cad-based automated design of fea-ready cutting tools. *J. Manuf. Mater. Process.* **2020**, *4*, 104. [CrossRef]
12. Vijayaraghavan, A.; Dornfeld, D.A. Automated drill modeling for drilling process simulation. *J. Comput. Inf. Sci. Eng.* **2007**, *7*, 276–282. [CrossRef]
13. Chun, D.M.; Kim, H.J.; Lee, J.C.; Ahn, S.H. Web-Based Material Database for Material Selection and Its Application Programming Interface (API) for CAD. *Key Eng. Mater.* **2007**, *345–346*, 1593–1596. [CrossRef]
14. E Costa, E.C.; Jorge, J.; Duarte, J. Comparing digital tools for implementing a generative system for the design of customized tableware. *Comput. Aided Des. Appl.* **2019**, *16*, 803–821. [CrossRef]
15. Manavis, A.; Kyratsis, P. A computational study on product shape generation to support brand identity. *Int. J. Mod. Manuf. Technol.* **2021**, *13*, 115–122.
16. Kyratsis, P.; Manavis, A.; Gianniotis, P.; Ghiculescu, D. A non-conventional methodology for interior product design using conceptual design principles and parametric tools. *Nonconv. Technol. Rev. Tehnol. Neconv.* **2019**, *23*, 16–22.
17. Garcia, S.; Menezes Leitão, A. Shape grammars as design tools: An implementation of a multipurpose chair grammar. *Artif. Intell. Eng. Des. Anal. Manuf. AIEDAM* **2018**, *32*, 240–255. [CrossRef]
18. Khan, S.; Awan, M.J. A generative design technique for exploring shape variations. *Adv. Eng. Inform.* **2018**, *38*, 712–724. [CrossRef]
19. Hatton-Jones, S.; Teah, M. Case analysis of the do-it-yourself industry. *Asia Pac. J. Mark. Logist.* **2015**, *27*, 826–838. [CrossRef]
20. Kyratsis, P.; Tzotzis, A.; Manavis, A. Computational design and digital fabrication. In *Advances in Manufacturing Systems*; Kumar, S., Rajurkar, K.P., Eds.; Springer: Singapore, 2021; pp. 1–17.
21. Mengoni, M.; Peruzzini, M.; Raffaeli, R.; Raponi, D. A Web-based Platform to Support Contract Furniture Design. *Comput. Aided Des. Appl.* **2014**, *11*, 533–543. [CrossRef]
22. Bernal, M.; Haymaker, J.R.; Eastman, C. On the role of computational support for designers in action. *Des. Stud.* **2015**, *41*, 163–182. [CrossRef]

machines

Article

Optimization of Selective Laser Sintering/Melting Operations by Using a Virus-Evolutionary Genetic Algorithm

Nikolaos A. Fountas [1], John D. Kechagias [2] and Nikolaos M. Vaxevanidis [1,*]

[1] Department of Mechanical Engineering, School of Pedagogical and Technological Education (ASPETE), GR 151 22 Amarousion, Greece
[2] Design & Manufacturing Lab (DML), Department of FWSD, University of Thessaly, 43100 Karditsa, Greece
* Correspondence: vaxev@aspete.gr; Tel.: +30-210-2896841

Abstract: This work presents the multi-objective optimization results of three experimental cases involving the laser sintering/melting operation and obtained by a virus evolutionary genetic algorithm. From these three experimental cases, the first one is formulated as a single-objective optimization problem aimed at maximizing the density of Ti6Al4V specimens, with layer thickness, linear energy density, hatching space and scanning strategy as the independent process parameters. The second one refers to the formulation of a two-objective optimization problem aimed at maximizing both the hardness and tensile strength of Ti6Al4V samples, with laser power, scanning speed, hatch spacing, scan pattern angle and heat treatment temperature as the independent process parameters. Finally, the third case deals with the formulation of a three-objective optimization problem aimed at minimizing mean surface roughness, while maximizing the density and hardness of laser-melted L316 stainless steel powder. The results obtained by the proposed algorithm are statistically compared to those obtained by the Greywolf (GWO), Multi-verse (MVO), Antlion (ALO), and dragonfly (DA) algorithms. Algorithm-specific parameters for all the algorithms including those of the virus-evolutionary genetic algorithm were examined by performing systematic response surface experiments to find the beneficial settings and perform comparisons under equal terms. The results have shown that the virus-evolutionary genetic algorithm is superior to the heuristics that were tested, at least on the basis of evaluating regression models as fitness functions.

Keywords: virus-evolutionary genetic algorithm; optimization; selective laser sintering; heuristics; additive manufacturing; hardness; density; tensile strength

1. Introduction

Setting advantageous values for the process parameters in manufacturing processes constitutes a crucial task in process planning, requiring extensive experience in manufacturing engineering operations, deep knowledge of establishing and conducting systematic experiments, a mathematical background concerning the statistical analysis of experimental results and comprehensive know-how for formulating machining modeling problems for optimization [1–4]. Unfortunately, such elements can hardly be implemented in real-world applications owing to the tedious efforts of process planners under pressing timespans and high productivity/quality requirements. Consequently, industrial practices still suggest the usage of handbooks, manuscripts and tool catalogues in order to examine the operational ranges for different process parameters which try to determine advantageous—if not optimal—settings for them as a combination of parameter values that will affect the whole process. This approach continues to be applied since systematic experimentation and parameter examination are time-consuming processes that also contribute to higher costs.

To set an "optimum" combination for the parameter values, several researchers have developed and implemented various algorithmic modules following the principles of dynamic programming [5,6], goal programming [7], integer linear programming [8] and fuzzy environments [9]. Even though such systems have succeeded in solving several engineering optimization problems, they come with certain drawbacks in regards to their objective-dependent searching mechanisms, as well as their constraint-type functions (i.e., non-linear and/or linear ones). In addition, they are unable to maintain the same efficiency when they are altering the size of the solution domain or its structure (i.e., convex, concave and saddle ones, etc.).

To solve the numerous engineering optimization problems with the highest rate of success, researchers have proposed the implementation of algorithms that adhere to either population or swarm intelligence. In the work presented in [10], the authors took advantage of an optimization problem related to the abrasive water jet machining, which was originally studied in [11]. In their work, they implemented the Greywolf algorithm [12] to maximize the material removal rate (*MRR*), with the nozzle diameter, the feed of nozzle, the mass flow rate of the abrasives, the water mass flow rate and the water pressure at the nozzle as the independent process parameters. In the same work presented in [10], a second case was solved by the Greywolf algorithm related to a bi-objective problem that was originally examined in [13]. In this case, the objectives of *MRR* and *Ra* were simultaneously optimized, exhibiting their trade-off, owing to *MRR* maximization and *Ra* minimization requirement. The authors in [14] implemented their Teaching/Learning optimization algorithm (TLBO) to solve three non-conventional machining processes based on previous experimental works conducted by others, namely, ultrasonic machining (USM), abrasive jet machining (AJM) and wire electro-discharge machining (wEDM). The TLBO algorithm was also applied in [15] to solve three single-objective and two multi-objective problems concerning fused deposition modeling (FDM). All five case studies reported in this work were based on previous experimental works [16–20]. Under the same research concept, the authors of [21] implemented the MO-Jaya algorithm to optimize the previously investigated machining operations; wire electro-discharge machining [22], the laser cutting process [23], electrochemical machining [24] and focused ion beam (FIB) micro-milling [25]. Other machining and micro-machining processes based on the experiments of others were also optimized using the TLBO algorithm [26], while some other abrasive waterjet machining operations were optimized using the Jaya algorithm along with the PROMETHEE method [27].

The initial variant of the virus-evolutionary genetic algorithm (VEGA) was developed and implemented in several research works with emphasis to those that are presented in [28] and [29]. An enhanced version of this algorithm was later proposed in [30]. A noticeable characteristic of the original variant of the VEGA was the transformation of the multi-objective optimization problems to single-objective ones through the adaptation of the "weighted summation" strategy [31].

This paper implements a virus-evolutionary genetic algorithm that is capable of controlling both the techniques for solving optimization problems (single and multi-objective ones). To enable this control, a sub-routine that activates the functions related to single or multi criteria processing has been programmed. Emphasis is given to the multi-objective optimization selection, where a single simulation run can prompt the algorithm to evaluate an entire set comprising many non-dominated solutions existing in a Pareto front. In the case of a multi-objective problem represented by using the "weighted summation" approach, the proposed virus-evolutionary genetic algorithm investigates the problem with the determination of all the possible combinations among the weights applied to the objectives. The multi-objective virus-evolutionary genetic algorithm is called MOVEGA in the paper. The MOVEGA is applied to solve three problems related to a cutting-edge additive manufacturing process that is widely known as selective laser sintering—selective laser melting (SLS-SLM). All the problems have been formulated with reference to the regression equations that are derived from original experiments, and they are quadratic and parameter bounded. In the first case, the processing layer thickness, linear energy density, hatching space and scanning strategy are treated as the independent process parameters, with the maximum density of the manufactured specimens being the objective of interest.

The second case uses the laser power, scanning speed, hatch spacing, scan pattern angle and heat treatment temperature as the independent variables, while maximum hardness and maximum tensile strength are considered as the optimization objectives. Finally, the third case presents a three-objective optimization problem where the laser power, scanning speed and hatch spacing are the independent variables, whilst the minimum mean surface roughness, maximum density and maximum hardness are treated as the optimization objectives. The regression models used for optimizing the aforementioned problems have been previously validated with regard to the corresponding experiments conducted by previous researchers.

2. Virus-Evolutionary Genetic Algorithm for Multi-Objective Optimization (MOVEGA)

Two methods are distinguished for creating new candidate solutions to solve a combinatorial problem. The first method is based on a stochastic philosophy, whereas the second method is associated with local information that characterizes the problem. Even though genetic algorithms (GAs) are primarily considered as stochastic search modules, local information is very useful in order to escape from local optima and increase the possibility of finding the global optimum solution to a problem. Crossover and selection operators are used for creating new candidate solutions, and they provide local information for the search space. The mechanism of generating new effective individuals under the concept of producing robust schemata is prone to premature local convergence when one applies only a proportional selection. On the contrary, MOVEGA targets a promising substring from the binary encoded string of an individual and creates the virus individual. By combining the strings of the individuals and viruses, the algorithm manages to produce new candidate solutions in horizontal propagation. This strategy is quite promising since the coevolution among hosts and viruses (global and local information) facilitates fast optimization problem solving.

The virus theory of evolution suggests that virus transduction is of paramount importance for transporting genetic segments across species [32]. Transduction implies the genetic modification of a bacterium by the genes from another bacterium transported though a bacteriophage. The majority of the viruses found in nature may easily cross species barriers, and they are usually transferred directly from individuals of one phylum to another. This means that natural viruses may transfer their gene to the host populations in horizontal propagation. In addition, the viruses may be included into germ cells, and thus, transmitted from generation to generation, following vertical inheritance. As a result, this paper applies a virus-evolutionary genetic algorithm that implements two operators; one of them is used for simulating vertical inheritance, and the other one is used simulating horizontal propagation. Two populations are created and co-evolve in this algorithm: the host population (candidate solutions) and the virus population (substrings of the selected host individuals). Typical genetic operators (i.e., selection and crossover ones) are implemented for performing standard activities in genetic algorithms, whilst the viral infection operators, which are the "reverse transcription" and "transduction" ones, are applied to simulate the natural operations of viruses. Figure 1 illustrates the functions of the two operators for performing viral infection.

In the program developed to build the functions of the proposed algorithm, C_{Pop}^{init} is the initial chromosomes population; Var is an independent parameter; $N_b^{Var}{}_{i,j}$ is the number of bits necessary for representing the accuracy of each parameter Var for the i_{th} chromosome of the j_{th} population; $Lgth_i$ is the i_{th} length of the chromosome. By considering the independent parameter Var_i, along with its corresponding bound $D_{Var_i} = [Ub_i, Lb_i]$ and the i_{th} chromosome length $Lgth_i$, the formula given in Equation (1) can used for switching from binary to real encoding.

$$Var_i = Lb_i + fnc(BinStr) \times \frac{Ub_i - Lb_i}{2^{Lgth_i} - 1} \tag{1}$$

where Ub_i and Lb_i are the parameters' upper and lower bounds, respectively, whilst $fnc(BinStr)$ denotes the function for returning the decimal values referring to binary encoding based on their accuracy. When it comes to algorithmic simulations, there is a low possibility that an already evaluated individual from a previous run might be selected again. To avoid possibly selecting an already evaluated individual, a routine has been developed based on flag statements either 0 (a new individual) or 1 (an already examined individual from previous runs) in the *Microsoft® Visual Basic® for Applications* environment. Objective function computation is conducted for all the candidates in a population. Thereby, the results are examined by the ranking function, which ranks them based on their objective scores. The ranking is sorted into ascending order. The fitness evaluation follows next, and it is applied to every individual in a population. Elite individuals are those having attained the lowest scores in the ranking operation. The individuals are finally stored along with their corresponding fitness scores in a file ("fitnessScores.dat"), and they are sorted into descending order.

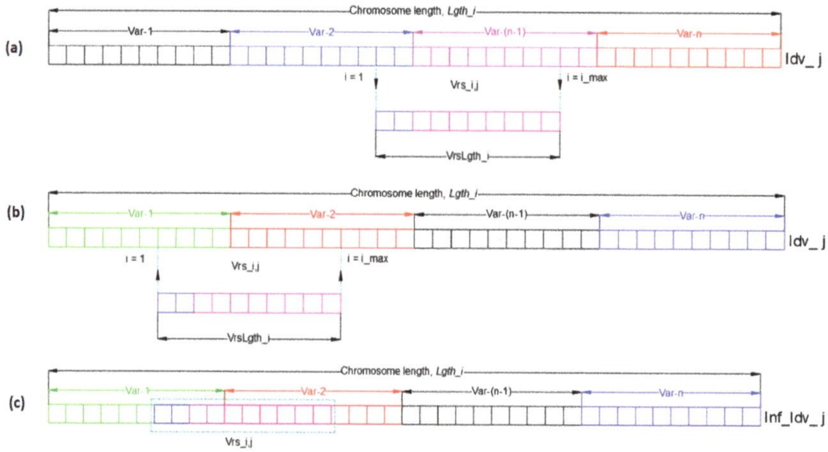

Figure 1. (**a**) Transduction operation for the generation of a virus individual. (**b**) Reverse transcription operation for infecting selected individuals with a virus. (**c**) Infected individual after the reverse transcription operation performed by a virus.

The process continues further by applying selection operator, where some individuals are selected for reproduction according to their fitness. Equation (2) represents the cumulative sum *Fitsum* of the entire population after importing the fitness scores of the individuals. The election range is formulated according to the lower and upper bounds, i.e., zero and *Fitsum*, respectively (0, *Fitsum*). Further, the individuals are subsequently ranked with regard to their fitness score; i.e., (0, Fit1), (Fit1, Fit2), ... , (FitN-1, *Fitsum*), whereas a random generator is used to create the values within this region. The random generator produces as many random values as the number of selected individuals for the crossover. Random values exist between one of the above sub-ranges that denote the individuals that are to be selected. Thereby, elite individuals are favored according to their fitness so that the elitist behavior is maintained during the entire operation. After the selection of a particular individual is achieved, both the selection range [0, *Fitsum*] and the individual's sub-range are redefined by applying Equation (2) to prevent the iterative selection of this particular outstanding individual.

$$Fit_i = Fit_i - \frac{Fitsum}{2N} \quad (2)$$

where Fit_i is the fitness function score of i_{th} individual; N is the population's size; *Fitsum* is the aggregate sum of the entire population. The selected individuals are stored in a dedicated file ("Selected.dat" file) to be further handled by the crossover operator. Pairs of individuals are randomly selected to produce offspring. When the crossover function is completed, all the individuals—the parents and offspring—are imported into a new candidate population (individual pool), whose size is twice as big as the initial population's size. The objective function is computed again, and the individuals are ranked accordingly by applying a ranking function. Based on their ranking N, the best individuals are deterministically selected, where N is the original number of initial individuals that constitute the first population. Finally, the individuals are mutated to give a new population of candidate solutions.

The next operation is mutation. Mutation involves the number of individuals, the number of mutated variables in terms of binary digits (i.e., 0 and 1) and the mutation rate. A population—as a result of selection and crossover—is imported and processed in order to determine the pointers for the selected offspring to indicate the locations in chromosomes where the change between "0" and "1" digits are to be made. Mutation begins with a rather high initial probability to preserve the diversity of exploration, while it gradually lowers via a linear expression that involves the initial mutation rate value and the number of generations to prevent a completely scattered random search, thus, it has a high computational cost.

Viruses are created after the fitness evaluation of all the individuals comprising the main population is conducted. The number of viruses is actually a fraction of the main population's magnitude. The MOVEGA algorithm manages to conduct both the targeted and random selection of individuals for infection. The former selection is applied to the elite of some outstanding individuals, whilst the latter one is normally applied to the rest individuals according to the probability. This ensures that there is an unbiased selection scheme. The successfully infected individuals appear as offspring. If their fitness score is improved after the infection, they replace their ancestors. Therefore, these individuals survive in the next generation. The initial chromosomes population C_{Pop}^{init} is randomly created, and then, a following transduction operation is applied to the fitted and randomly selected individuals for the creation of the virus population V_{Pop}^{init}. The viruses are stored in a binary representation of a related archive ("virus_population.dat"). A single virus individual Vrs_{ij} is created by transducing from the i_{th} chromosome of the j_{th} population. The cut substrings represent the viruses' chromosomes, whose length is $VrsLgth_i$. $i = 1$ is the starting point from which the $VrsLgth_i$ length is specified, while Locus $i_{\max}$ denotes the end point. These two points are randomly selected and constrained to the original host's chromosome length $Lgth_i$. The chromosome length ($Lgth_i$) of the individuals in the main population is constant, whereas the length of the virus individuals ($VrsLgth_i$) increases as the evolution process continues $\left(VrsLgth_i = V_{strlength}\max\right)$. The population index from which selected individuals have been attacked is stored in the "infected_host_population.dat" archive. The phenotypes of the individuals that are candidates for infection are stored in the "virus_phenotype.dat" archive. The objective results of the virus individuals are finally stored in the "virusobjvalues.dat" archive. Transduction and reverse transcription constitute the two fundamental operations of viral infection. The transduction operator is applied to the host individuals to generate the virus population. Viruses Vrs_{ij} attack to infect the individuals by applying reverse transcription to overwrite their own substrings to randomly selected segments of the individuals' Idv_j strings. The indices of Vrs_{ij} and Idv_j are "a priori" declared so as to perform the subsequent replacement of the binary digits according to the predetermined references.

An assessment of the virus individuals is achieved with the use of their fitness scores, which are denoted as $FitVrs_{i,j}$, reflecting their infection strength. This fitness is computed after the successful infection of Idv_j by Vrs_{ij}, as it is presented by Equation (3):

$$FitVrs_{i,j} = FitInfIdv_j - FitIdv_j \tag{3}$$

The value obtained by Equation (3) is the difference between the two fitness values of individual Idv_j before and after its infection by Vrs_{ij}. Given that Vrs_{ij} might infect more than a single individual (let S be the set of infected individuals), $FitVrs_{i,j}$ reveals the improvement of fitness values of all the infected individuals, and as Equation (4) determines:

$$FitVrs_i = \sum_{j \in S} FitVrs_{i,j} \tag{4}$$

A virus Vrs_{ij} has a maximum viral infection rate $V_{infRate}\max$ for controlling the number of viral infections satisfying the relation $1 \leq V_{infRate}\max \leq 10$. As a consequence, the number of reverse transcriptions that a single virus may perform will depend on its viral infection rate. The maximum viral infection rate $V_{infRate}\max$ is related also to its fitness value $FitVrs_{i,j}$. The higher the fitness $FitVrs_{i,j}$ is, then the higher the $V_{infRate}\max$ will be. Equation (5) gives the relation adopted to relate the viral infection parameters mentioned above and control $V_{infRate}\max$ with regard to virus fitness $FitVrs_{i,j}$. In Equation (5), $a(>0)$ is a fixed parameter that is used to improve or degrade $V_{infRate}\max$ with regard to either the positive or negative results referring to virus fitness Vrs_{ij}.

$$V_{infRate}\max_{i,G+1} = \left\{ \begin{array}{l} (1+a) \times V_{infRate}\max_{i,G} \\ (1-a) \times V_{infRate}\max_{i,G} \end{array} \right\}, \begin{array}{l} FitVrs_i \geq 0 \\ FitVrs_i < 0 \end{array} \tag{5}$$

A virus Vrs_{ij} has a life force indicator that represents its contribution to the main population in terms of its successful infections. The life force of a virus Vrs_{ij} is denoted as $VrsLiforce_{i,G}$, where i is the index of the virus Vrs_{ij} and G the current generation. $VrsLiforce_{i,G}$ depends also upon the virus fitness Vrs_{ij}, and it is compared to the one obtained by the virus Vrs_{ij} in a previous generation. It its value is negative, then a new *transduction* operation is applied by the virus Vrs_{ij} to alter its scheme by randomly selecting an individual. Otherwise, Vrs_{ij} cuts a partially new substring by a successfully infecting an individual. The magnitude of $VrsLiforce_{i,G}$ is computed in each generation according to the virus life reduction rate $rate \downarrow V_{life}$, satisfying $0.001 \leq rate \downarrow V_{life} \leq 1.0$. Therefore, the maximum viral infection rate $V_{infRate}\max$ and the virus life reduction rate $rate \downarrow V_{life}$ are related through the expression given in Equation (6).

$$VrsLiforce_{i,G+1} = rate \downarrow V_{life} \times VrsLiforce_{i,G} + FitVrs_i \tag{6}$$

The $V_{infRate}\max$ and $VrsLiforce_{i,G}$ parameters are initialized in MOVEGA as $V_{infRate}\max_{init} = V_{infRate}\max_{i,0}$ and $VrsLiforce_{i,0} = 0$, respectively. The operation of partial transduction in the case where $VrsLiforce_{i,G} < 0$ is presented in Figure 2 with reference to transduction and reverse transcription operations presented in Figure 1 above.

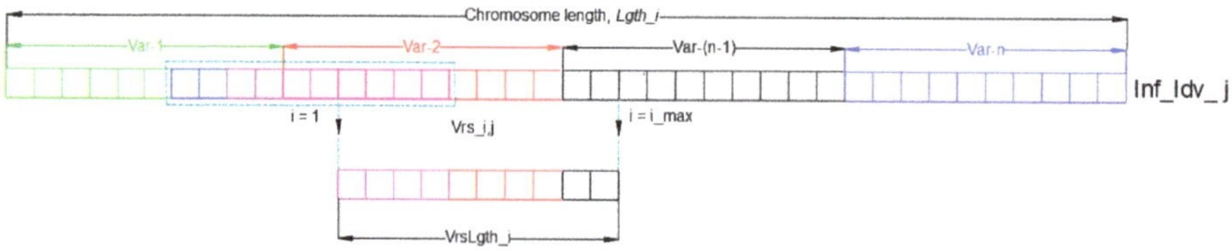

Figure 2. Partial transduction operation for changing the virus scheme.

The entire workflow (flow chart) for the proposed algorithm is presented in Figure 3.

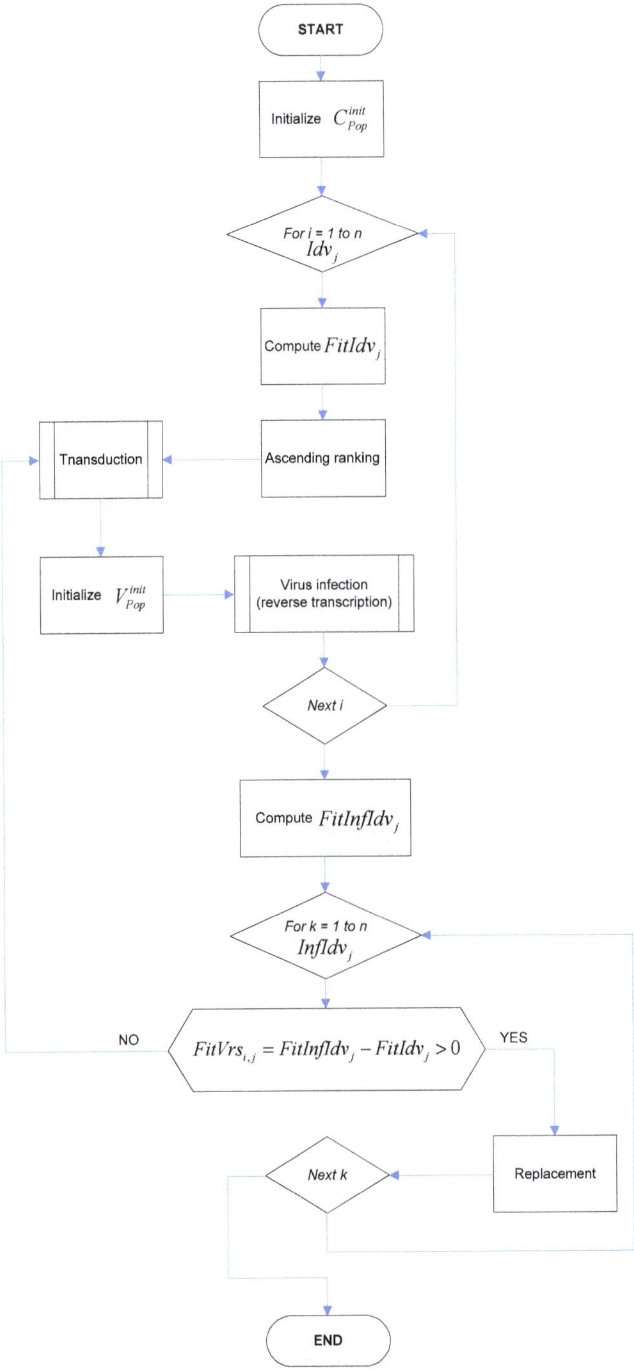

Figure 3. Flow chart of viral infection after the evaluation of main population's individuals.

3. Performance Metrics

When it comes to multi-objective optimization, a diverse group of multiple non-dominated solutions are obtained. To examine and characterize the performance of an algorithm, some discrete performance metrics can be used. Two of the most important and commonly implemented metrics are the coverage of two sets $Cvg(NDS_{i-1}, NDS_i)$ and the spacing S [33]. The coverage of two groups of non-dominated solutions (NDS_{i-1}, NDS_i) refers to the percentage of dominance exhibited by the individuals of one group on the solutions of the other group. This metric can be computed by using the relation given in Equation (7).

$$Cvg(NDS_{i-1}, NDS_i) = \frac{|\{b \in NDS_i | \exists\_a \in NDS_{i-1} : a \leq b\}|}{|NDS_i|} \quad (7)$$

where NDS_{i-1}, NDS_i represent two non-dominated solution sets to compare; $a \leq b$ implies that a either dominates it, or it is equal to b. $Cvg(NDS_{i-1}, NDS_i) = 1$ implies that all the non-dominated solutions in NDS_i are dominated, or they are equal to all the non-dominated solutions in NDS_{i-1}, whilst $Cvg(NDS_{i-1}, NDS_i) = 0$ implies that none of the non-dominated solutions of NDS_i are covered by those found in NDS_{i-1}. Note that both $Cvg(NDS_{i-1}, NDS_i)$ and $Cvg(NDS_i, NDS_{i-1})$ should be considered since there is a possibility that $Cvg(NDS_{i-1}, NDS_i) \neq 1 - Cvg(NDS_i, NDS_{i-1})$. In the case that $Cvg(NDS_{i-1}, NDS_i) = 1$ and $Cvg(NDS_i, NDS_{i-1}) = 0$, the solutions of NDS_{i-1} entirely dominate those of NDS_i, and thus, this would be the ideal result for NDS_{i-1}. Thereby, $Cvg(NDS_{i-1}, NDS_i)$ gives the percentage of the non-dominated points in NDS_i which are inferior or equal to the points in NDS_{i-1}, whereas $Cvg(NDS_i, NDS_{i-1})$ gives the percentage of non-dominated points in NDS_{i-1} which are inferior or equal to the points in NDS_i.

Spacing is a performance metric used for quantifying the spread of the non-dominated solutions, or equivalently, how uniform the distribution among the different solutions is. The spacing metric is determined according to Equation (8) as follows:

$$S = \sqrt{\frac{1}{|n-1|}\sum_{i=1}^{n}\left(\bar{d} - d_i\right)^2} \quad (8)$$

In Equation (8), n is the number of the different non-dominated solutions in a Pareto front, whilst $d_i = \min_{i, i \neq j} \sum_{Obj=1}^{ObjMax} \left| f_{Obj}^{i} - f_{Obj}^{j} \right|, i, j = 1, 2, \ldots, n$ is the distance variance. $ObjMax$ is the maximum number of objectives, Obj is an objective number, and f_{Obj} is the objective function that corresponds to Obj objective. Since the objectives of interest might have different magnitudes, it is important to remove their inherent bias by normalizing them. Normalization can be achieved by using Equation (9).

$$\bar{d} = \sum_{i=1}^{n}\left(\frac{d_i}{|n|}\right) \quad (9)$$

In the case where $S = 0$ (ideal and optimal case), all the non-dominated solutions are uniformly spread and equidistantly spaced to the entire Pareto front. The spacing indicator is suitable for evaluating the non-dominated solutions of the regular Pareto fronts, which are obtained by different algorithms. However, it is important that the solutions are unique, i.e., they cannot be duplicated.

4. Optimization Problems Related to Selective Laser Sintering/Melting—SLS/SLM

Selective laser sintering/melting (SLS/SLM) constitutes a very important, rapid prototyping process, where the materials in the form of powder are used for fabricating parts from computer-aided design (CAD) models. In the wide range of materials that are used for

SLS/SLM, one may distinguish engineering thermoplastics such as polymers, polyamides, ABS, nylons and polycarbonates, as well as metallic materials such as titanium and its alloys, stainless steel alloys and tool steels, etc. SLS/SLM is a very challenging operation owing to weak strength, dimensional inaccuracy and poor surface finish that characterize most of the products fabricated using SLS/SLM technology. As it occurs in any other manufacturing process, choosing the different settings for the values of process-related parameters is an important aspect. Since the problems presented in this research are examined in terms of optimization for the first time in the literature, no other reports were found to retrieve information about the algorithm-specific parameters and related settings. MOVEGA has been applied to solve all three problems with reference to preliminary research work concerning the tuning of the algorithm-specific parameters related to the proposed algorithm. Consequently, the settings applied for MOVEGA's parameters are $(C_{pop} = 20)$ for the population size, $(V_{pop} = 12)$ for the virus population size, the maximum variable number of bits in the virus chromosome substring, which is equal to 25 (given as a fraction of the 100-digit chromosome string of the individuals in the main population) $\left(V_{strlength}\max = 25 = \frac{1}{4}C_{strlength}\max\right)$, the maximum virus life reduction rate, which is equal to -0.5 $\left(rate \downarrow V_{life} = -0.5\right)$, and the maximum infection rate, which is equal to 70% of the maximum viral infectivity $(V_{\text{inf}Rate}\max = 7)$. The same problems have been solved by applying other intelligent algorithms, with their algorithm-related parameters having been tuned to the best possible extend for rigorous comparisons to be made. These algorithms are mentioned in each problem, and they represent three population-based ones and one swarm-based one.

4.1. Optimization of SLM Parameters for Forming Ti6Al4V Alloy Specimens with Maximum Density

The optimization procedure for the case study is based on the experimental results found in the work presented in [29]. In the aforementioned work, a custom SLM machine (DiMetal-280) was used for conducting the experiments. The major specifications of their machine are a continuous fiber laser with a wavelength of 1075 nm, the X-Y galvanometer scanning mode, which was focused by the *f*-θ lens with a scanning velocity ranging from 5 to 5000 mm/s, the beam quality factor $M^2 \leq 1.1$, the laser spot diameter of 70 µm and the thickness, which was layered by a roller and ranged from 20 to 80 µm. The material was a gas-atomized spherical Ti6Al4V powder, while the average diameter of 95% powder was under 20 µm. During the SLM operation, the oxygen content was below 0.02%. Based on the results in [29], a regression equation for estimating the mean density of produced parts was generated and adopted in this case for maximizing the mean density (%). However, the process parameters along with their corresponding ranges are the same as those considered by the authors of [29]. The parameters are: the processing layer thickness PT (mm), the linear energy density LED (J/mm) and the hatching space HS (mm). As a categorical factor, the scanning strategy was kept constant in the "*X-Y inter-layer stagger scanning*" mode. The ranges for the parameters were: $(0.02 \leq PT \leq 0.035)$, $(0.2 \leq LED \leq 0.5)$ and $(0.04 \leq HS \leq 0.07)$, according to [29]. Equation (10) gives the regression equation that was used as the objective function for maximizing the mean density $\max\overline{\rho}$ in this single-optimization case.

$$\max\overline{\rho} = -\Big(96.3 + 136 * PT + 13.1 * LED - 216 * HS - 630 * PT^2 - 46.4 * LED^2 - 1861 * HS^2 - 901 * PT * LED - 60 * PT * HS + 974 * LED * HS\Big) \quad (10)$$

The authors of [29] optimized (maximized) the objective of the mean density $\max\overline{\rho}$ by applying numerical optimization, and they achieved the value of 94.4424 (%) as the optimal solution. In [34] the optimal solution for the same problem was found equal to 94.4424. An attempt was conducted to further improve this result by applying the MOVEGA using its single-objective module. The same procedure was used for the rest of the optimizers used for comparison: the Greywolf algorithm (GWO), the Multiverse algorithm (MVO), the Antlion algorithm (ALO) and the Dragonfly algorithm (DA) [12,35–37]. The MOVEGA as well as the competitive algorithms ran with a population size that was equal to 20 for 200 iterations, thus, resulting in 4000 function evaluations. All the algorithms managed to obtain the same maximum result for the mean density max$\bar{\rho}$, which was equal to 94.4751 against the maximum value of 94.4424 reported in [34]. From the results obtained by the algorithms tested, in terms of the convergence speed and iteration number at which the best result was achieved, it is revealed that there is no need for setting such a large number of iterations, however, this has been intentionally considered in order to examine whether the algorithms become trapped in a local optimum solution or not. the MOVEGA (in this single-objective optimization case, it is referred as VEGA) exhibited the best convergence speed against the rest algorithms, whilst it obtained the maximum result for the mean density max$\bar{\rho}$ in the 6th iteration (120 function evaluations). The GWO, MVO, ALO and DA algorithms converged to the 17th, 53rd, 25th and 23rd iterations, respectively. The total execution timespan for the algorithms was approximately the same, ranging from 10 s to 13 s. Optimal values obtained were found that were equal to 0.02, 0.5 and 0.07 for the processing layer thickness *PT* (mm), linear energy density *LED* (J/mm) and hatching space *HS* (mm), respectively. These outputs are in full agreement with the experimental ones in [34]. Note that the algorithms were ran 20 times to examine their stochastic nature and repeatability for obtaining the best result. Figure 4 gives the best convergence trends exhibited by all the algorithms with regard to the total of 20 algorithmic simulation tests.

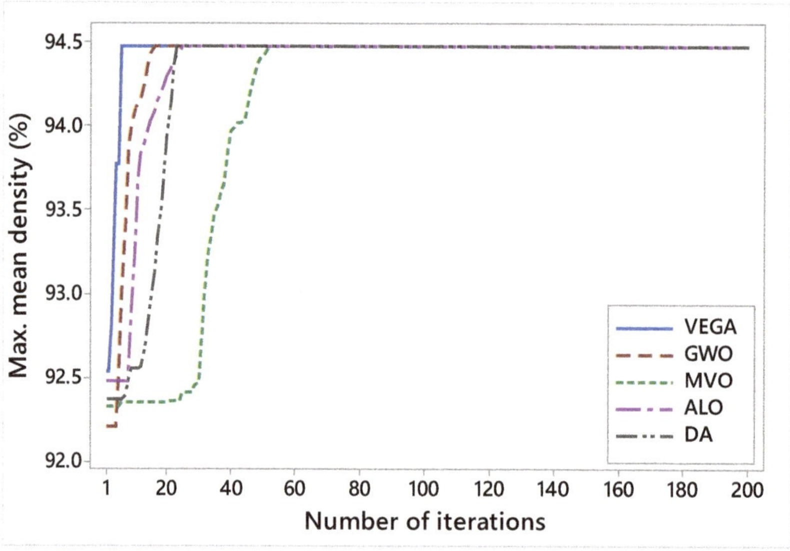

Figure 4. Convergence speed exhibited by VEGA, GWO, MVO, ALO and DA algorithms for maximizing mean density, with 200 iterations.

4.2. Case 2: Maximization of Density and Tensile Strength of Laser-Melted Ti6Al-4V Alloy Specimens

The optimization problem examined here is formulated by considering the experimental results presented in [38], where a Taguchi experimental design was established, with the laser power *LP* (W), scanning speed *SS* (mm/min), hatch spacing *HS* (μm), scan pattern angle *SPA* (°) and heat treatment temperature *HTT* (°C) being the independent process parameters, to characterize the Brinell hardness *HB* (HB) and tensile strength *TS* (MPa) of the laser-melted Ti6Al4V alloy specimens. The experimental design involved a total of 25 runs based on the number of parameters and their corresponding levels. The SLM equipment that was used was an SLM-125HL equipped with YLR-fiber-laser with a minimum spot size of 5μm, while the material used was a powdered Ti6Al4V Titanium alloy. The authors of [38] measured the hardness of their experimental specimens using a DuraJet® G5 apparatus for the Brinell hardness. The load and indenter values that were applied were 30 N and 1/30, respectively. The regression equations found in [38] were adopted to play the role of objective functions for the MOVEGA and the rest of the algorithms. The scanning strategy was kept constant. The ranges for the parameters were: $(90 \leq LP \leq 110)$, $(600 \leq SS \leq 800)$, $(65 \leq HS \leq 85)$, $(36 \leq SPA \leq 75)$ and $(20 \leq HTT \leq 1050)$. Equations (11) and (12) give the regression equations that were used as objective functions for simultaneously maximizing the hardness maxHB and the tensile strength maxTS in this two-objective optimization case.

$$\max HB = -(10376 - 152.9 * LP - 4.20 * SS - 52.2 * HS + 43.3 * SPA + 0.336 * HTT + 0.638 * LP^2 + 0.00264 * SS^2 + 0.204 * HS^2 + 0.0471 * SPA^2 - 0.000091 * HTT^2 + 0.0059 * LP * SS + 0.376 * LP * HS - 0.114 * LP * SPA - 0.00499 * LP * HTT - 0.00844 * SS * SPA + 0.000486 * SS * HTT - 0.376 * HS * SPA + 0.00175 * HS * HTT - 0.00465 * SPA * HTT) \quad (11)$$

$$\max TS = -(-11916 + 389 * LP + 1.3 * SS - 240 * HS + 92 * SPA + 4.23 * HTT - 1.93 * LP^2 + 0.00276 * SS^2 + 1.443 * HS^2 + 0.549 * SPA^2 - 0.000429 * HTT^2 - 0.0161 * LP * SS + 0.51 * LP * HS - 1.077 * LP * SPA - 0.0003 * LP * HTT - 0.0272 * SS * SPA - 0.00316 * SS * x(5) - 0.362 * HS * SPA - 0.0225 * HS * HTT + 0.0034 * SPA * HTT) \quad (12)$$

The two-objective optimization problem was solved by applying the MOVEGA, MOGWO, MOMVO, MOALO and MODA algorithms using a population size that was = equal to 20 for 200 iterations (4000 function evaluations). To examine the efficiency of the MOVEGA and the rest of antagonizing algorithms, 30 independent algorithmic simulations were conducted, resulting in 30 non-dominated Pareto fronts. From these 30 sets, the coverage has been computed among the pairs of two independent sets per two algorithms, while the spacing refers to the best non-dominated set obtained out of 30 trials. The standard deviation results refer to the 30 coverage results computed for the five multi-objective optimization algorithms. Figure 5 presents the best Pareto fronts obtained by the algorithms. Table 1 summarizes the results for the best 145 non-dominated solutions set obtained by the MOVEGA, whilst Table 2 gives the results for best results, the mean and the standard deviation for coverage values, as well as the spacing among the solutions for the best non-dominated set obtained by each algorithm. It is evident that the MOVEGA exhibited the best performance in terms of the metrics selected. As an example, the result of *Cvg(MOVEGA, MOGWO)* = 0.8328 implies that, with reference to the best values, 83.28% of MOGWO's non-dominated solutions are dominated by those obtained by MOVEGA. On the contrary, the result of *Cvg(MOGWO, MOVEGA)* = 0.7213 implies that, with reference to the best values, 72.13% of MOVEGA's non-dominated solutions are dominated by those obtained by MOGWO. With the same philosophy, the rest of the outputs for the coverage indicator can be similarly interpreted. The best results for the spacing indicator are 0.0257, 0.0912, 0.0922, 0.0697 and 0.0984 for the MOVEGA, MOGWO, MOMVO, MOALO and MODA optimizers, respectively. By considering all three statistical parameters (the best results, the mean and the St.Dev.), MOVEGA is superior to the rest algorithms. By reviewing the Pareto fronts given in Figure 5, it is evident that MOVEGA's Pareto front is more well-spread and continuous, with its 145 non-dominated points being uniformly distributed throughout the entire trend curve.

Table 1. Optimal non-dominated solutions set obtained by MOVEGA using the regression equations of case 2.

Sol. No.	LP (W)	SS (mm/min)	HS (μm)	SPA (°)	HTT (°C)	max*HB*	max*TS*
1	90.0000	600.0070	65.0000	74.9992	96.0606	724.1530	1676.6300
2	90.0000	600.0830	65.0000	75.0000	189.6659	716.3600	1768.1500
3	90.0002	600.8099	65.0002	74.9999	369.0359	696.4140	1922.2800
4	90.0004	600.2490	65.0015	74.9992	512.6129	677.2990	2025.6600

Table 1. *Cont.*

Sol. No.	LP (W)	SS (mm/min)	HS (μm)	SPA (°)	HTT (°C)	max*HB*	max*TS*
5	90.0004	600.0346	65.0005	74.9990	572.3014	668.2410	2063.6400
6	90.0004	600.0497	65.0005	74.9991	572.4180	668.2090	2063.7000
7	90.0002	600.0021	65.0014	74.9992	638.1319	657.2870	2101.8300
8	90.0002	600.0349	65.0013	74.9995	687.7818	648.4690	2128.1600
9	90.0013	600.4803	65.0001	74.9994	74.0921	725.1940	1654.4900
10	90.0020	600.1857	65.0008	74.9994	115.8210	722.4130	1696.6500
11	90.0001	600.7364	65.0004	74.9993	213.9841	713.4070	1790.9200
12	90.0002	600.0025	65.0002	74.9973	48.2313	727.4620	1626.9200
13	90.0001	600.0005	65.0009	74.9910	100.7152	723.7120	1681.0100
14	90.0002	600.0005	65.0009	74.9910	111.3896	722.9010	1691.7700
15	90.0007	600.0103	65.0001	74.9973	322.8056	702.6660	1885.1000
16	90.0003	600.0176	65.0007	74.9885	436.4406	688.3110	1972.6600
17	90.0006	600.0850	65.0001	74.9961	537.9085	673.5970	2042.0900
18	90.0010	600.0852	65.0000	74.9962	575.7634	667.6140	2065.6100
19	90.0037	600.0031	65.0004	74.9998	662.0168	653.0600	2114.7800
20	90.0001	600.0041	65.0001	74.9997	704.9853	645.3690	2136.8900
21	90.0002	600.0046	65.0001	74.9994	740.2719	638.7320	2153.8000
22	90.0055	600.0059	65.0003	74.9959	825.6876	621.5840	2190.0900
23	90.0030	600.0004	65.0006	74.9958	883.3804	609.4070	2211.2500
24	90.0001	600.0022	65.0018	74.9999	959.8236	592.3350	2235.0000
25	90.0000	600.0024	65.0009	74.9995	1049.9481	570.7760	2256.4100
26	90.0009	600.0032	65.0008	74.9997	72.2267	725.8430	1652.1300
27	90.0000	600.1076	65.0023	75.0000	211.0341	714.3050	1787.8900
28	90.0000	600.0121	65.0004	74.9986	300.1269	705.2500	1866.2800
29	90.0000	600.0203	65.0004	74.9984	304.2338	704.7820	1869.7200
30	90.0000	600.0722	65.0004	74.9967	362.5610	697.8850	1917.1000
31	90.0000	600.0714	65.0004	74.9970	362.6760	697.8740	1917.2000
32	90.0000	600.0657	65.0001	74.9993	423.4337	690.1280	1963.6000
33	90.0000	600.0660	65.0002	74.9993	431.3230	689.0670	1969.3800
34	90.0005	600.1135	65.0005	74.9996	482.8846	681.8100	2005.8100
35	90.0005	600.1110	65.0007	74.9997	531.7765	674.5350	2038.2600
36	90.0009	600.0268	65.0031	75.0000	608.8342	662.2050	2085.2100
37	90.0004	600.1641	65.0076	74.9999	689.8788	647.8650	2128.7900
38	90.0007	600.1651	65.0079	74.9999	710.0763	644.1420	2138.8300
39	90.0003	600.0472	65.0192	75.0000	811.1128	624.3400	2183.5100
40	90.0006	600.0391	65.0091	74.9999	875.5373	611.0010	2208.2300
41	90.0001	600.0028	65.0100	74.9999	936.3248	597.5660	2227.8300
42	90.0000	600.0027	65.0050	74.9998	1001.2611	582.5420	2245.5100
43	90.0043	600.0114	65.0043	74.9999	81.6854	725.0330	1661.7300
44	90.0002	600.0909	65.0436	74.9995	353.1494	698.1930	1907.9500
45	90.0002	600.0898	65.0461	74.9995	354.1915	698.0200	1908.6800
46	90.0000	600.0104	65.0133	75.0000	487.6573	680.9800	2008.5600
47	90.0001	600.3778	65.0084	75.0000	552.4658	670.9590	2050.8700
48	90.0003	600.4747	65.0014	74.9992	756.6315	635.1890	2160.6100
49	90.0002	600.4659	65.0014	74.9992	798.6468	626.8870	2178.7200
50	90.0005	600.7370	65.0019	74.9616	860.6133	613.5870	2200.6600
51	90.0000	600.0007	65.0044	74.9999	928.7629	599.3840	2225.8800
52	90.0000	600.0000	65.0019	74.9999	972.2506	589.4500	2238.3600
53	90.0000	600.0231	65.0003	75.0000	1021.0095	577.8650	2250.3200
54	90.0000	600.0224	65.0002	75.0000	1049.6006	570.8650	2256.3600
55	90.0004	600.6827	65.0039	74.9999	280.1177	706.7060	1849.3300
56	90.0000	600.1879	65.0004	74.9928	180.6262	716.9880	1759.4200
57	90.0008	600.1150	65.0002	74.9970	246.9769	710.7680	1820.2900
58	90.0003	600.0007	65.0000	74.9992	20.1417	729.2480	1596.9400

Table 1. *Cont.*

Sol. No.	LP (W)	SS (mm/min)	HS (μm)	SPA (°)	HTT (°C)	max*HB*	max*TS*
59	90.0003	600.0007	65.0001	74.9999	88.6743	724.7030	1669.1200
60	90.0002	600.0009	65.0001	74.9999	115.1837	722.7150	1695.9400
61	90.0001	600.0007	65.0001	74.9999	150.1976	719.8930	1730.4600
62	90.0001	600.0009	65.0001	75.0000	185.7119	716.8010	1764.3900
63	90.0042	600.0007	65.0015	74.9996	229.7076	712.5410	1804.7900
64	90.0005	600.0004	65.0026	74.9996	291.2892	706.1940	1858.7500
65	90.0018	600.0028	65.0023	74.9988	424.0614	690.0180	1963.9300
66	90.0018	600.0004	65.0023	74.9989	436.8278	688.3010	1973.2600
67	90.0075	600.0098	65.0006	74.9999	504.0405	678.6600	2020.0500
68	90.0073	600.0096	65.0004	74.9999	527.3985	675.1570	2035.4100
69	90.0028	600.0066	65.0001	74.9998	580.7397	666.8640	2068.7700
70	90.0031	600.0065	65.0001	74.9998	589.3602	665.4540	2073.9200
71	90.0101	600.0036	65.0005	74.9898	656.1425	653.8770	2111.1400
72	90.0000	600.0052	65.0013	74.9980	718.4244	642.8350	2143.3300
73	90.0000	600.0051	65.0013	74.9980	725.1892	641.5640	2146.5800
74	90.0004	600.0005	65.0003	74.9994	791.4698	628.6960	2176.4300
75	90.0003	600.0001	65.0005	74.9998	806.9547	625.5700	2182.8500
76	90.0009	600.0003	65.0003	74.9981	916.3233	602.2180	2222.1800
77	90.0286	600.0030	65.0050	74.9999	120.2652	721.6730	1700.3900
78	90.0041	600.0147	65.0003	74.9999	192.6370	716.0760	1770.8200
79	90.0041	600.0148	65.0002	74.9999	224.1196	713.1030	1799.8000
80	90.0019	600.0978	65.0022	74.9992	350.5860	699.2630	1907.5900
81	90.0018	600.0980	65.0013	74.9993	377.1215	696.0260	1928.5200
82	90.0010	600.0072	65.0009	75.0000	460.2781	685.1080	1990.1300
83	90.0020	600.0121	65.0008	74.9999	496.3530	679.9050	2014.9700
84	90.0023	600.0167	65.0009	75.0000	562.1374	669.8410	2057.4200
85	90.0002	600.0046	65.0001	74.9985	619.2623	660.5290	2091.2900
86	90.0002	600.0067	65.0000	75.0000	654.6460	654.4410	2110.9200
87	90.0004	600.0434	65.0002	74.9980	760.4103	634.7950	2162.8600
88	90.0006	600.0443	65.0017	74.9988	823.3227	622.1490	2189.2300
89	90.0003	600.0844	65.0009	74.9996	882.0087	609.7240	2210.8200
90	90.0003	600.0177	65.0015	74.9979	943.9839	595.9530	2230.4200
91	90.0002	600.0176	65.0015	74.9979	951.1902	594.3050	2232.4900
92	90.0004	600.0176	65.0004	74.9979	1018.4691	578.4620	2249.6600
93	90.0002	600.0319	65.0009	74.9998	48.8745	727.4000	1627.7000
94	90.0006	600.0274	65.0009	74.9999	101.5424	723.7010	1682.2000
95	90.0006	600.0512	65.0009	74.9998	136.2923	720.9600	1716.8700
96	90.0003	600.0261	65.0000	74.9913	368.4556	697.1580	1921.5300
97	90.0004	600.0854	65.0001	74.9977	443.7262	687.3420	1978.2900
98	90.0006	600.0226	65.0000	74.9978	445.0121	687.2210	1979.2300
99	90.0042	600.0567	65.0008	74.9919	520.7453	676.1180	2030.7800
100	90.0023	600.0569	65.0003	74.9919	526.9859	675.2190	2034.8800
101	90.0165	601.4028	65.0000	74.9997	676.4436	649.0900	2120.7600
102	90.0093	601.0325	65.0000	74.9995	782.6755	629.4820	2171.1900
103	90.0001	600.0025	65.0001	74.9998	891.3549	607.7680	2214.1800
104	90.0001	600.0033	65.0001	74.9998	916.9145	602.1180	2222.4500
105	90.0008	600.1868	65.0000	75.0000	995.3963	583.8910	2243.9900
106	90.0050	600.2147	65.0001	74.9740	36.8009	727.6280	1614.0400
107	90.0354	600.0887	65.0000	75.0000	339.2562	699.9920	1898.0800
108	90.0345	600.1149	65.0000	75.0000	369.8654	696.2760	1922.4300
109	90.0008	600.3937	65.0001	74.9969	543.6894	672.4330	2045.6000
110	90.0001	600.0295	65.0001	74.9962	586.9384	665.8660	2072.3700
111	90.0080	600.8088	65.0015	74.9783	785.2215	628.9980	2171.6700
112	90.0010	600.1271	65.0004	74.9988	874.0360	611.4080	2207.9800

Table 1. *Cont.*

Sol. No.	LP (W)	SS (mm/min)	HS (μm)	SPA (°)	HTT (°C)	max*HB*	max*TS*
113	90.0007	600.1159	65.0004	74.9988	893.1665	607.2660	2214.5200
114	90.0016	600.0141	65.0000	74.9992	988.4006	585.6400	2242.5500
115	90.0000	600.0025	65.0008	74.9991	20.3040	729.2250	1597.0900
116	90.0000	600.0005	65.0008	74.9991	33.4874	728.4180	1611.2800
117	90.0002	600.0437	65.0000	74.9975	130.6390	721.4280	1711.2500
118	90.0002	600.0452	65.0000	74.9968	255.4522	709.9870	1827.7600
119	90.0004	600.6115	65.0009	75.0000	310.5811	703.4860	1875.1500
120	90.0001	600.0099	65.0000	74.9991	412.8274	691.5840	1955.7500
121	90.0000	600.0305	65.0000	74.9945	486.1262	681.3920	2007.8800
122	90.0000	600.0510	65.0000	74.9944	519.8299	676.3860	2030.3800
123	90.0002	600.1056	65.0001	74.9994	629.5712	658.6990	2097.0500
124	90.0004	600.0521	65.0002	74.9998	698.4567	646.5270	2133.5900
125	90.0001	600.0234	65.0010	74.9995	774.9020	631.9700	2169.3000
126	90.0001	600.0352	65.0021	74.9992	793.4223	628.2500	2177.1100
127	90.0008	600.2279	65.0000	74.9988	831.5273	620.3430	2192.2400
128	90.0008	600.2258	65.0000	74.9988	850.5332	616.3640	2199.4100
129	90.0073	600.0691	65.0014	74.9974	1011.2398	579.9870	2247.7200
130	90.0011	600.0000	65.0001	74.9997	33.3880	728.4230	1611.2000
131	90.0015	600.0003	65.0000	74.9999	99.9996	723.8460	1680.6300
132	90.0011	600.0006	65.0000	74.9935	175.7097	717.6130	1754.6800
133	90.0001	600.0113	65.0002	74.9997	228.5262	712.7510	1803.8400
134	90.0004	600.0113	65.0002	74.9999	263.4553	709.2120	1834.8800
135	90.0003	600.0115	65.0002	74.9999	287.0514	706.6990	1855.2500
136	90.0012	600.0139	65.0001	74.9987	418.1687	690.8460	1959.6800
137	90.0008	600.0165	65.0001	74.9994	450.2061	686.5170	1983.0000
138	90.0026	600.0003	65.0029	74.9998	524.5331	675.6510	2033.5100
139	90.0017	600.0001	65.0004	74.9954	648.2606	655.4900	2107.2500
140	90.0005	600.0005	65.0011	74.9997	699.6744	646.3240	2134.2100
141	90.0005	600.0005	65.0017	74.9997	733.9239	639.9090	2150.7700
142	90.0097	600.0000	65.0000	74.9954	788.8330	628.9920	2175.0300
143	90.0099	600.0009	65.0001	74.9954	863.4788	613.5420	2204.1900
144	90.0099	600.0009	65.0001	74.9954	864.0086	613.4290	2204.3800
145	90.0119	600.0029	65.0002	74.9701	926.9109	599.4130	2224.1300

Table 2. Best, mean and standard deviation results for the non-dominated solutions obtained by MOVEGA, MOGWO, MOMVO, MOALO and MODA optimizers for case 2.

Performance Metric			Statistical Results
	Best	Mean	St.Dev.
Cvg(MOVEGA, MOGWO)	0.8328	0.7641	0.0442
Cvg(MOGWO, MOVEGA)	0.7213	0.0912	0.0753
Cvg(MOVEGA, MOMVO)	0.8412	0.7327	0.0548
Cvg(MOMVO, MOVEGA)	0.2674	0.1668	0.0349
Cvg(MOVEGA, MOALO)	0.8227	0.7808	0.0648
Cvg(MOALO, MOVEGA)	0.2942	0.2105	0.0599
Cvg(MOVEGA, MODA)	0.7828	0.3571	0.0642
Cvg(MODA, MOVEGA)	0.2118	0.2942	0.0341
S(MOVEGA)	0.0257	0.0274	0.0035
S(MOGWO)	0.0912	0.1077	0.0094
S(MOVMVO)	0.0922	0.1752	0.0104
S(MOALO)	0.0697	0.1777	0.0149
S(MODA)	0.0984	0.2101	0.0211

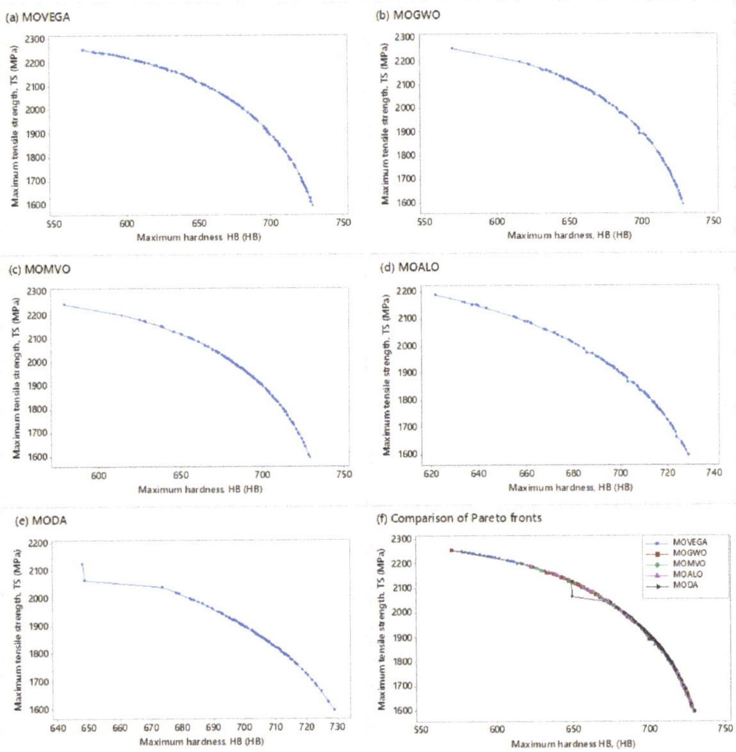

Figure 5. Pareto fronts for the non-dominated solutions obtained from the best simulation experiment exhibited by: (**a**) MOVEGA, (**b**) MOGWO, (**c**) MOMVO, (**d**) MOALO, (**e**) MODA and (**f**) Comparison of Pareto fronts algorithms for case 2, with 20 individuals and 200 iterations.

4.3. Case 3: Three-Objective Optimization Problem for 316 L Stainless Steel Powder Bed Fusion Operation

The optimization problem examined in this case is formulated by considering the experimental results presented in [39]. In [39], a Taguchi experimental design was established, with the laser power *LP* (W), scanning speed *SS* (mm s^{-1}), and hatch spacing *HS* (mm) being the independent process parameters, to characterize the top surface roughness *Ra* (μm), Vickers hardness *HV* (HV) and density in percentage ρ (%) of laser-melted 316 L stainless steel specimens. Their experimental design involved a total of 27 runs based on the number of parameters and their corresponding levels. The SLM equipment used was the EP250 with a fiber-laser with 400 W maximum indicative power and 80 μm beam diameter. The oxygen content was maintained at approximately 1000 ppm during the conduction of SLM experiments. The results corresponding to the three objectives, the top surface roughness *Ra* (μm), Vickers hardness *HV* (HV) and density in percentage ρ (%), were the average values as a result of taking series of measurements. The authors of [39] succeeded on presenting a factor analysis, which was accompanied by normalized quantities along with the Taguchi design of the experiments, however, the kind of trade-off these objectives yielded and the type of solution space in the form of a Pareto front were not discussed. The regression equations found in [39] were adopted to serve as the objective functions for the MOVEGA and the rest algorithms for the simultaneous optimization efforts of the three objectives. From these objectives, one of them should be minimized (top surface roughness *Ra*) and two of them should be maximized (Vickers hardness HV and density in percentage, ρ). The ranges for the parameters were: $(44.36 \le LP \le 64.17)$, $(6.69 \le SS \le 9.67)$

and ($1.88 \leq HS \leq 2.63$). Equations (13) and (15) give the regression equations that were used as objective functions for simultaneously minimizing the surface roughness minRa, maximizing the Vickers hardness maxHV and the maximizing density in percentage maxρ in this three-objective optimization case.

$$\min Ra = 21.3 - 1.735 * LP + 3.10 * SS + 21.8 * HS + 0.00811 * LP^2 - 0.261 * SS^2 - 3.65 * HS^2 + 0.0610 * LP * SS + 0.051 * LP * HS - 0.782 * SS * HS \quad (13)$$

$$\max HV = -(-206 + 0.73 * LP + 38.8 * SS + 234 * HS - 0.1111 * LP^2 - 3.62 * SS^2 - 61.6 * HS^2 + 0.836 * LP * SS + 2.40 * LP * HS - 13.57 * SS * HS) \quad (14)$$

$$\max \rho = -(94.76 - 0.0235 * LP + 0.553 * SS + 3.29 * HS - 0.002043 * LP^2 - 0.0505 * SS^2 - 0.761 * HS^2 + 0.01875 * LP * SS + 0.0530 * LP * HS - 0.4188 * SS * HS) \quad (15)$$

In this case, the algorithms ran for 4000 function evaluations using a population size that was equal to 20 for 200 iterations, whilst 50 non-dominated solutions were stored in corresponding archives. To examine the efficiency of the MOVEGA and the rest of antagonizing algorithms, 30 independent algorithmic simulations were conducted, resulting in 30 non-dominated Pareto fronts. From these 30 sets, the coverage was computed among the pairs of two independent sets per two algorithms, while the spacing refers to the best non-dominated set obtained out of 30 trials. The standard deviation results refer to the thirty coverage results computed for the five multi-objective optimization algorithms. Figure 6 illustrates the best Pareto fronts obtained by the algorithms.

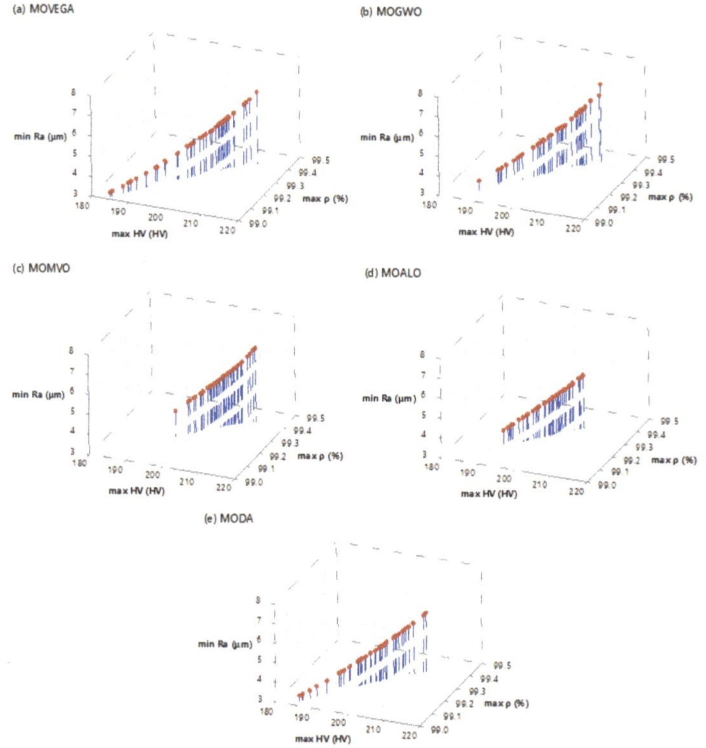

Figure 6. Pareto fronts of non-dominated solutions for the best simulation experiment exhibited by: (**a**) MOVEGA, (**b**) MOGWO, (**c**) MOMVO, (**d**) MOALO and (**e**) MODA algorithms for case 3, with 20 individuals and 200 iterations.

Table 3 summarizes the results for the best 145 non-dominated solutions set obtained by the MOVEGA, whilst Table 4 gives the results for best results, and the mean and the standard deviation for the coverage values, as well as spacing among the solutions for the best non-dominated set obtained by each algorithm. It is revealed that the MOVEGA exhibited the best performance in terms of the metrics selected. As an example, the result of *Cvg(MOVEGA, MOGWO)* = 0.8745 implies that, with reference to the best values, 87.45% of MOGWO's non-dominated solutions are dominated by those obtained by MOVEGA. On the contrary, the result of *Cvg(MOGWO, MOVEGA)* = 0.4249 implies that, with reference to the best values, 42.49% of MOVEGA's non-dominated solutions are dominated by those obtained by MOGWO. Similarly, the rest of the outputs referring to the coverage indicator were interpreted. The best results for the spacing indicator are 0.0177, 0.0181, 0.0102, 0.0125 and 0.0184 for the MOVEGA, MOGWO, MOMVO, MOALO and MODA optimizers, respectively. Note that MOMVO and MOALO achieved the best spacing results, however, their solution sets are poorer compared to the rest algorithm results. That is, MOVEGA's non-dominated solutions offer a wider range for selecting optimal solutions according to the operation's requirements. By reviewing the Pareto fronts given in Figure 6, it is evident that MOVEGA's front is spread more, continuous and wider than the others, with its 50 non-dominated points being uniformly distributed throughout the entire trend curve.

Table 3. Optimal non-dominated solutions set obtained by MOVEGA by applying the regression equations of case 3.

Sol. No.	LP (W)	SS (mm/min)	HS (μm)	minRa	maxHV	maxρ
1	63.0920	6.6900	1.8800	3.22026	184.430	99.0131
2	57.0923	7.0746	2.0314	6.08493	208.998	99.3881
3	56.5919	7.2272	2.0667	6.55576	211.269	99.4072
4	63.3478	6.7351	1.8882	3.30673	184.917	99.0189
5	59.7448	6.6900	1.8800	4.00626	193.844	99.1784
6	57.6960	6.8654	1.9544	5.26003	203.770	99.3269
7	59.0302	7.0065	2.0128	5.40195	205.081	99.3343
8	59.6797	6.7893	1.8937	4.27655	196.154	99.2116
9	57.5663	7.0654	1.9750	5.68598	206.682	99.3604
10	57.4215	7.0615	2.0316	5.97916	208.501	99.3817
11	60.6645	6.7139	1.8846	3.84318	192.136	99.1468
12	62.3797	6.6900	1.8800	3.37229	186.642	99.0521
13	62.3644	6.7182	1.8866	3.46950	187.562	99.0662
14	58.7394	6.9052	1.9771	5.15238	203.262	99.3139
15	58.4793	7.0090	1.9913	5.44380	205.308	99.3399
16	62.3863	6.7290	1.8862	3.48354	187.657	99.0675
17	58.4011	6.9702	1.9761	5.33356	204.510	99.3308
18	58.6260	6.7573	1.9064	4.57371	198.665	99.2541
19	60.9559	6.8512	1.9448	4.35574	196.533	99.2082
20	61.2156	6.7024	1.8862	3.69831	190.570	99.1197
21	61.8754	6.7370	1.8812	3.58157	188.985	99.0912
22	60.6341	6.7259	1.8868	3.88524	192.521	99.1526
23	58.8913	6.8463	1.9457	4.86012	201.031	99.2839
24	60.6267	6.8324	1.9279	4.30814	196.244	99.2062
25	58.4646	6.9468	1.9886	5.34328	204.607	99.3322
26	58.1004	6.9805	1.9960	5.52379	205.786	99.3483
27	59.5401	6.9638	1.9837	5.07353	202.676	99.3008
28	59.6232	6.9664	1.9792	5.03499	202.370	99.2962
29	58.2348	6.9789	1.9901	5.45795	205.362	99.3423
30	59.2099	6.9712	1.9831	5.16324	203.357	99.3115
31	59.5973	6.9073	1.9500	4.80039	200.538	99.2722
32	58.3600	6.9500	1.9838	5.35188	204.644	99.3331

Table 3. *Cont.*

Sol. No.	LP (W)	SS (mm/min)	HS (μm)	min*Ra*	max*HV*	max*ρ*
33	59.5956	6.8831	1.9562	4.79400	200.511	99.2722
34	58.2513	6.9896	1.9909	5.47262	205.466	99.3434
35	57.0567	7.1305	2.0458	6.22802	209.792	99.3950
36	60.6599	6.8332	1.9296	4.31110	196.260	99.2062
37	58.2678	6.9376	1.9778	5.32860	204.455	99.3314
38	60.4834	6.9424	1.9605	4.69952	199.536	99.2528
39	59.2840	7.0080	1.9910	5.23907	203.907	99.3178
40	61.9641	6.7711	1.9043	3.76112	190.613	99.1149
41	60.5560	6.8732	1.9534	4.53106	198.169	99.2338
42	59.5237	6.9291	1.9660	4.93497	201.613	99.2871
43	61.6147	6.8326	1.9338	4.11536	194.069	99.1683
44	59.0104	6.9904	1.9925	5.28737	204.261	99.3241
45	61.5612	6.8335	1.9305	4.11072	194.054	99.1684
46	60.1161	6.8770	1.9438	4.59124	198.773	99.2449
47	62.7393	6.7634	1.9016	3.56492	188.083	99.0715
48	60.7520	6.8414	1.9321	4.31713	196.273	99.2057
49	60.5726	6.9424	1.9639	4.69620	199.484	99.2515
50	58.3231	7.0466	2.0144	5.64501	206.650	99.3565

Table 4. Best, mean and standard deviation results for the non-dominated solutions obtained by MOVEGA, MOGWO, MOMVO, MOALO and MODA optimizers for case 3.

Performance Metric	Statistical Results		
	Best	Mean	St.Dev.
Cvg(MOVEGA, MOGWO)	0.8745	0.8524	0.0346
Cvg(MOGWO, MOVEGA)	0.4249	0.4390	0.0571
Cvg(MOVEGA, MOMVO)	0.4612	0.4588	0.0224
Cvg(MOMVO, MOVEGA)	0.2459	0.2547	0.0489
Cvg(MOVEGA, MOALO)	0.4378	0.4174	0.0542
Cvg(MOALO, MOVEGA)	0.1947	0.2234	0.0672
Cvg(MOVEGA, MODA)	0.7716	0.7519	0.0516
Cvg(MODA, MOVEGA)	0.6904	0.6815	0.0672
S(MOVEGA)	0.0177	0.0179	0.0031
S(MOGWO)	0.0181	0.0186	0.0046
S(MOVMVO)	0.0102	0.0111	0.0012
S(MOALO)	0.0125	0.0128	0.0018
S(MODA)	0.0184	0.0178	0.0057

5. Conclusions

In this research, a modified virus-evolutionary genetic algorithm for single and multi-objective optimization problems has been presented. The algorithm takes advantage of two populations; one comprising the main individuals and the other one comprising the population of viruses. Owing to the fact that the physical process of viral infection has been simulated in this algorithm, both the horizontal propagation and vertical inheritance operators are responsible for obtaining more reliable and antagonizing solutions compared to those results obtained by the other algorithms. Through their coevolution, the individuals and viruses exchange information to push further the envelope of maintaining the balance between exploitation and exploration, while efficient binary schemata are rapidly processed, and if they are beneficial, they increase the algorithm's efficiency in finding either the global optimum for single-objective optimization problems or a uniform and wide non-dominated set of solutions for multi-objective optimization problems.

The algorithm has been applied for solving three optimization problems related to a modern additive manufacturing/rapid prototyping operation, which is widely known as selective laser sintering and/or selective laser melting (SLS/SLM). The first case presented a single-objective optimization problem using the maximum density as its optimization target when laser-melting Ti6Al4V alloy powder was used for fabricating parts. The second case presented a bi-objective optimization problem using the maximum hardness and maximum tensile strength as the optimization objectives for simultaneous optimization when laser-melting Ti6Al4V alloy powder was used for fabricating parts. The third case presented a three-objective optimization problem, where a simultaneous optimization among the minimum surface roughness, maximum hardness and maximum density needed to be achieved during the laser melting of 316 L stainless steel. All three cases have been adopted by the recent literature, and the regression equations from corresponding experiments have served as the objective functions for the problems.

By applying five different optimization algorithms, including the one proposed in this work (virus-evolutionary genetic algorithm), it was shown that a significant trade-off among the objectives of the several cases exists, and it constitutes an important research aspect for optimizing engineering applications using intelligent algorithms. The rest of the algorithms tested were the Greywolf (MOGWO), the Multi-verse (MOMVO), the Antlion (MOALO) and the dragonfly algorithms (MODA). The results obtained by the three aforementioned cases examined in the work have been rigorously compared under the same conditions to the best possible extent since all the algorithms handle the same population size, number of iterations, number of function evaluations and initialization conditions with the same starting search points. To characterize the performance of the proposed virus-evolutionary genetic algorithm as well as the rest of competitive algorithms, two widely applied performance indicators have been computed; the coverage between the two non-dominated solution sets and spacing used to quantify the solutions spread.

It should be mentioned that the application of the virus-evolutionary genetic algorithm to the individual cases examined in our current study by no means constitutes a generalized optimization methodology dedicated to SLS/SLM operations. The results presented in this study are as reliable as the problems' domains and their related parameters will allow them to be. However, this work, in its current state, puts forth new knowledge by developing and enhancing the virus-evolutionary genetic algorithm for efficiently searching several problem domains in real-world experiments. By taking into account the great number of similar studies that have attempted to acquire results with commonly employed artificial algorithms, this work has embraced "viral infectivity" to come up with original datasets and results related to SLS/SLM operation to allow comparisons to be made with other research by applying other heuristics. The benefits of the proposed optimization algorithm against other competitive algorithms is the computational philosophy of the algorithm based on the virus theory of evolution. The viruses in this algorithm act as "hill-climbers" and manage to lead the search for the solution so that local trapping is avoided. The rest of competitive algorithms are also beneficial, and their control environment has a low cost, is user friendly and requires a basic technical computing level (i.e., in the MATLAB environment). The limitation of the proposed methodology is that no online monitoring for the direct control of process parameters and optimization is necessary. The current trends suggest that we need new technology for optimizing the process parameters by analyzing the signals and suitably balancing the operational parameters to reach an optimal solution.

It is within our interests in the near future to examine the potentials of automating the software that controls the SLS/SLM parameters and try to develop a dynamic optimization environment by using the proposed virus-evolutionary genetic algorithm, which operates in-line with an application program interface of SLS/SLM software, which we envision will be a more meaningful and generic path towards delivering a generalized global optimization solution to further advance scientific knowledge and contribute to engineering optimization.

Author Contributions: Conceptualization, N.A.F., J.D.K. and N.M.V.; methodology, N.A.F.; software, N.A.F.; validation, N.A.F. and J.D.K.; formal analysis, N.A.F.; investigation, N.A.F.; resources, N.A.F. and N.M.V.; data curation, N.A.F., J.D.K. and N.M.V.; writing—original draft preparation, N.A.F.; writing—review and editing, N.A.F. and J.D.K.; visualization, N.M.V.; supervision, N.M.V.; project administration, J.D.K. and N.M.V. All authors have read and agreed to the published version of the manuscript.

Funding: This research received no external funding.

Institutional Review Board Statement: Not applicable.

Informed Consent Statement: Not applicable.

Data Availability Statement: Not applicable.

Conflicts of Interest: The authors declare no conflict of interest.

References

1. Fashanu, O.; Buchely, M.F.; Spratt, M.; Newkirk, J.; Chandrashekhara, K.; Misak, H.; Walker, M. Effect of SLM Build Parameters on the Compressive Properties of 304L Stainless Steel. *J. Manuf. Mater. Process.* **2019**, *3*, 43. [CrossRef]
2. Maurya, H.S.; Kollo, L.; Tarraste, M.; Juhani, K.; Sergejev, F.; Prashanth, K.G. Effect of the Laser Processing Parameters on the Selective Laser Melting of TiC–Fe-Based Cermets. *J. Manuf. Mater. Process.* **2022**, *6*, 35. [CrossRef]
3. Impey, S.; Saxena, P.; Salonitis, K. Selective Laser Sintering Induced Residual Stresses: Precision Measurement and Prediction. *J. Manuf. Mater. Process.* **2021**, *5*, 101. [CrossRef]
4. Mahardika, M.; Setyawan, M.A.; Sriani, T.; Miki, N.; Prihandana, G.S. Electropolishing Parametric Optimization of Surface Quality for the Fabrication of a Titanium Microchannel Using the Taguchi Method. *Machines* **2021**, *9*, 325. [CrossRef]
5. Zolfpour-Arokhlo, M.; Selamat, A.; Mohd Hashim, S.Z.; Afkhami, H. Modeling of Route Planning System Based on Q Value-Based Dynamic Programming with Multi-Agent Reinforcement Learning Algorithms. *Eng. Appl. Artif. Intell.* **2014**, *29*, 163–177. [CrossRef]
6. Liang, R.; Chen, Y.; Zhu, R. A Novel Fault Diagnosis Method Based on the KELM Optimized by Whale Optimization Algorithm. *Machines* **2022**, *10*, 93. [CrossRef]
7. Tyagi, S.K.; Yang, K.; Tyagi, A.; Dwivedi, S.N. Development of a Fuzzy Goal Programming Model for Optimization of Lead Time and Cost in an Overlapped Product Development Project Using a Gaussian Adaptive Particle Swarm Optimization-Based Approach. *Eng. Appl. Artif. Intell.* **2011**, *24*, 866–879. [CrossRef]
8. Li, S.; Mu, N.; Le, J.; Liao, X. A Novel Algorithm for Privacy Preserving Utility Mining Based on Integer Linear Programming. *Eng. Appl. Artif. Intell.* **2019**, *81*, 300–312. [CrossRef]
9. Gong, K.; Chen, C. Multiple-Attribute Decision Making Based on Equivalence Consistency under Probabilistic Linguistic Dual Hesitant Fuzzy Environment. *Eng. Appl. Artif. Intell.* **2019**, *85*, 393–401. [CrossRef]
10. Chakraborty, S.; Mitra, A. Parametric Optimization of Abrasive Water-Jet Machining Processes Using Grey Wolf Optimizer. *Mater. Manuf. Process.* **2018**, *33*, 1471–1482. [CrossRef]
11. Pawar, P.J.; Rao, R.V. Parameter Optimization of Machining Processes Using Teaching–Learning-Based Optimization Algorithm. *Int. J. Adv. Manuf. Technol.* **2013**, *67*, 995–1006. [CrossRef]
12. Mirjalili, S.; Saremi, S.; Mirjalili, S.M.; Coelho, L.D.S. Multi-Objective Grey Wolf Optimizer: A Novel Algorithm for Multi-Criterion Optimization. *Expert Syst. Appl.* **2016**, *47*, 106–119. [CrossRef]
13. Yue, Z.; Huang, C.; Zhu, H.; Wang, J.; Yao, P.; Liu, Z. Optimization of Machining Parameters in the Abrasive Waterjet Turning of Alumina Ceramic Based on the Response Surface Methodology. *Int. J. Adv. Manuf. Technol.* **2014**, *71*, 2107–2114. [CrossRef]
14. Venkata Rao, R.; Kalyankar, V.D. Parameter Optimization of Modern Machining Processes Using Teaching–Learning-Based Optimization Algorithm. *Eng. Appl. Artif. Intell.* **2013**, *26*, 524–531. [CrossRef]
15. Rao, R.V.; Rai, D.P. Optimization of Fused Deposition Modeling Process Using Teaching-Learning-Based Optimization Algorithm. *Eng. Sci. Technol. Int. J.* **2016**, *19*, 587–603. [CrossRef]
16. Sood, A.K.; Equbal, A.; Toppo, V.; Ohdar, R.K.; Mahapatra, S.S. An Investigation on Sliding Wear of FDM Built Parts. *CIRP J. Manuf. Sci. Technol.* **2012**, *5*, 48–54. [CrossRef]
17. Sood, A.K.; Ohdar, R.K.; Mahapatra, S.S. Experimental Investigation and Empirical Modelling of FDM Process for Compressive Strength Improvement. *J. Adv. Res.* **2012**, *3*, 81–90. [CrossRef]
18. Peng, A.; Xiao, X.; Yue, R. Process Parameter Optimization for Fused Deposition Modeling Using Response Surface Methodology Combined with Fuzzy Inference System. *Int. J. Adv. Manuf. Technol.* **2014**, *73*, 87–100. [CrossRef]
19. Gurrala, P.K.; Regalla, S.P. Multi-Objective Optimisation of Strength and Volumetric Shrinkage of FDM Parts. *Virtual Phys. Prototyp.* **2014**, *9*, 127–138. [CrossRef]
20. Sood, A.K.; Ohdar, R.K.; Mahapatra, S.S. Parametric Appraisal of Mechanical Property of Fused Deposition Modelling Processed Parts. *Mater. Des.* **2010**, *31*, 287–295. [CrossRef]

21. Rao, R.V.; Rai, D.P.; Balic, J. A Multi-Objective Algorithm for Optimization of Modern Machining Processes. *Eng. Appl. Artif. Intell.* **2017**, *61*, 103–125. [CrossRef]
22. Kuriakose, S.; Shunmugam, M.S. Multi-Objective Optimization of Wire-Electro Discharge Machining Process by Non-Dominated Sorting Genetic Algorithm. *J. Mater. Process. Technol.* **2005**, *170*, 133–141. [CrossRef]
23. Kumar Pandey, A.; Kumar Dubey, A. Simultaneous Optimization of Multiple Quality Characteristics in Laser Cutting of Titanium Alloy Sheet. *Opt. Laser Technol.* **2012**, *44*, 1858–1865. [CrossRef]
24. Bhattacharyya, B.; Sorkhel, S.K. Investigation for Controlled Ellectrochemical Machining through Response Surface Methodology-Based Approach. *J. Mater. Process. Technol.* **1999**, *86*, 200–207. [CrossRef]
25. Bhavsar, S.N.; Aravindan, S.; Rao, P.V. Investigating Material Removal Rate and Surface Roughness Using Multi-Objective Optimization for Focused Ion Beam (FIB) Micro-Milling of Cemented Carbide. *Precis. Eng.* **2015**, *40*, 131–138. [CrossRef]
26. Rao, R.V.; Rai, D.P.; Balic, J. Multi-Objective Optimization of Machining and Micro-Machining Processes Using Non-Dominated Sorting Teaching–Learning-Based Optimization Algorithm. *J. Intell. Manuf.* **2018**, *29*, 1715–1737. [CrossRef]
27. Rao, R.V.; Rai, D.P.; Balic, J. Multi-Objective Optimization of Abrasive Waterjet Machining Process Using Jaya Algorithm and PROMETHEE Method. *J. Intell. Manuf.* **2019**, *30*, 2101–2127. [CrossRef]
28. Kubota, N.; Fukuda, T.; Shimojima, K. Virus-Evolutionary Genetic Algorithm for a Self-Organizing Manufacturing System. *Comput. Ind. Eng.* **1996**, *30*, 1015–1026. [CrossRef]
29. Lu, B.; Cheng, B. The Virus Evolutionary Genetic Algorithm for Non- Full Loaded Vehicle Scheduling Problem with Fuzzy Time Window. In Proceedings of the 2011 Seventh International Conference on Natural Computation, Shanghai, China, 26–28 July 2011; IEEE: Piscataway, NJ, USA, 2011; pp. 306–310. [CrossRef]
30. Fountas, N.A.; Benhadj-Djilali, R.; Stergiou, C.I.; Vaxevanidis, N.M. An Integrated Framework for Optimizing Sculptured Surface CNC Tool Paths Based on Direct Software Object Evaluation and Viral Intelligence. *J. Intell. Manuf.* **2019**, *30*, 1581–1599. [CrossRef]
31. Jakob, W.; Blume, C. Pareto Optimization or Cascaded Weighted Sum: A Comparison of Concepts. *Algorithms* **2014**, *7*, 166–185. [CrossRef]
32. Anderson, N.G. Evolutionary Significance of Virus Infection. *Nature* **1970**, *227*, 1346–1347. [CrossRef] [PubMed]
33. Zitzler, E.; Thiele, L. Multiobjective Evolutionary Algorithms: A Comparative Case Study and the Strength Pareto Approach. *IEEE Trans. Evol. Comput.* **1999**, *3*, 257–271. [CrossRef]
34. Sun, J.; Yang, Y.; Wang, D. Parametric Optimization of Selective Laser Melting for Forming Ti6Al4V Samples by Taguchi Method. *Opt. Laser Technol.* **2013**, *49*, 118–124. [CrossRef]
35. Mirjalili, S. Dragonfly Algorithm: A New Meta-Heuristic Optimization Technique for Solving Single-Objective, Discrete, and Multi-Objective Problems. *Neural Comput. Appl.* **2016**, *27*, 1053–1073. [CrossRef]
36. Mirjalili, S.; Jangir, P.; Saremi, S. Multi-Objective Ant Lion Optimizer: A Multi-Objective Optimization Algorithm for Solving Engineering Problems. *Appl. Intell.* **2017**, *46*, 79–95. [CrossRef]
37. Mirjalili, S.; Jangir, P.; Mirjalili, S.Z.; Saremi, S.; Trivedi, I.N. Optimization of Problems with Multiple Objectives Using the Multi-Verse Optimization Algorithm. *Knowl.-Based Syst.* **2017**, *134*, 50–71. [CrossRef]
38. Khorasani, A.; Gibson, I.; Awan, U.S.; Ghaderi, A. The Effect of SLM Process Parameters on Density, Hardness, Tensile Strength and Surface Quality of Ti-6Al-4V. *Addit. Manuf.* **2019**, *25*, 176–186. [CrossRef]
39. Jiang, H.-Z.; Li, Z.-Y.; Feng, T.; Wu, P.-Y.; Chen, Q.-S.; Feng, Y.-L.; Li, S.-W.; Gao, H.; Xu, H.-J. Factor Analysis of Selective Laser Melting Process Parameters with Normalised Quantities and Taguchi Method. *Opt. Laser Technol.* **2019**, *119*, 105592. [CrossRef]

design constraints [22]. It has been widely applied in fields such as automotive, robotics, and aerospace for the development of sustainable and energy-efficient components [23].

Various TO methods have been introduced in the literature. Among the most used in commercial engineering software are the Solid Isotropic Material with Penalization (SIMP) and Evolutionary Structural Optimization (ESO) techniques [23–25]. Notably, TO is closely associated with additive manufacturing (AM), as AM enables the realization of complex, high-performance structures that are difficult to produce using traditional methods. Conversely, TO facilitates the design of components specifically optimized for AM processes. The integration of TO and AM provides a robust framework for aligning structural design with manufacturing capabilities, fostering innovation in lightweight engineering [23].

In robotic system design studies, the use of topology optimization (TO), additive manufacturing (AM), and their integration have been widely adopted to enhance operational efficiency and reduce energy consumption. Sha et al. [26] proposed a TO-based methodology aimed at minimizing robot weight without compromising structural integrity. This approach used finite element analysis (FEA) of the assembly to evaluate stress and deformation, followed by redesigning components to minimize mass and limit maximum displacement. The methodology was applied to an upper-limb exoskeleton robot with five degrees of freedom, resulting in a 10.4% reduction in mass compared to conventional designs.

Huang et al. [27] focused on the TO design of a lightweight, integrated manifold with reduced pressure loss for a hydraulic quadruped robot actuator. Their work addressed both structural design and print quality improvement. The optimized manifold achieved a 50.7% volume reduction while maintaining mechanical performance.

Sun et al. [28] introduced a TO-based multi-objective design approach for compliant robotic legs. By balancing structural stiffness and bending flexibility, they enabled stable and continuous straight-line walking in a quadruped robot, which was validated through experimental testing.

Padilla-García et al. [29] developed a multi-configurable omnidirectional mobile robotic platform using a sequential mechatronic design-control process. The structure was prototyped using AM techniques, highlighting the role of rapid prototyping in robotic system design.

Arat et al. [30] examined the structural performance of a pneumatic gripper designed for handling 500 kg automotive parts. Through TO, the gripper's mass was reduced by 35%, accompanied by a moderate increase in stress and a reduction in the safety factor to 4.72. Despite these changes, structural integrity was preserved, demonstrating the effectiveness of TO in achieving weight reduction while maintaining functional performance.

Graph-based algorithms, such as Dijkstra's and A*, are widely used in mobile robot path planning due to their resolution completeness and optimality assuring optimal solutions within discretized environments [31]. However, their scalability is limited in high-dimensional spaces due to the computational cost of constructing full occupancy grids. To address this, sampling-based methods like Rapidly Exploring Random Trees (RRTs) have been developed. These approaches efficiently explore complex, high-dimensional spaces by incrementally building search trees through random sampling, though they lack guarantees on solution optimality [9,32]. Karaman and Frazzoli [33,34] developed RRT* for asymptotic optimality via path rewiring. Informed RRT* improves convergence by sampling only in promising regions. Gammell et al. [34] extended this with BIT*, using A*-like heuristics on sample batches. However, in complex spaces, suboptimal initial paths can enlarge the informed set and reduce efficiency [9].

Path planning remains a critical challenge in mobile robotics, particularly when balancing multiple objectives such as energy efficiency, trajectory smoothness, and obstacle avoidance. Traditional deterministic methods like A*, Dijkstra's algorithm, and their multi-objective extensions (e.g., weighted A*) offer high explainability and real-time feasibility. However, they may underperform in highly dynamic or partially observable environments.

To overcome these limitations, recent studies have explored both biologically inspired and learning-based strategies. For example, Chen [35] introduced an optimized foothold planning algorithm for quadruped locomotion in energy-demanding terrain scenarios using posture adaptation and dynamic stability criteria. Raj and Kos [36] applied an intelligent particle swarm optimization (PSO) framework to minimize both energy and time in mobile robot navigation, demonstrating significant gains in energy efficiency across different path configurations. Chen et al. [37] proposed a hybrid meta-heuristic approach using Cuckoo–Beetle Swarm Search for heterogeneous mobile robots, showing robustness under variable terrain and payload conditions. Additionally, Chen et al. [38] introduced a RL-based energy-saving path planning algorithm (ESPP-RL) tailored for UAVs navigating turbulent wind environments. Their approach dynamically adapts to wind-induced disturbances via a custom reward function, achieving notable reductions in energy consumption and improved robustness in 3D simulation trials. Moreover, Ahmad Khan et al. [39] developed a RL-driven ankle rehabilitation robot trajectory planner that integrates EMG feedback and joint reaction forces to generate energy-efficient, physiologically adaptive paths for stroke patients, demonstrating the transferability of energy-aware RL planning to systems with high human-interaction constraints.

These emerging approaches highlight the growing trend toward data-driven or hybrid planners in robotic mobility systems. In contrast, this current study retains deterministic planning but combines it with a learning-based terrain energy model and structural topology optimization to enable lightweight, energy-aware path execution with high real-time reliability.

In this study, a mobile robotic system was designed and developed with an emphasis on structural and energy efficiency. Topology optimization was applied to the initial design to reduce weight and enhance both operational and energy performance. The structural integrity of the optimized design was validated using finite element analysis (FEA).

The mechanical components of the robotic system, including frame parts and hardware mounting elements, were fabricated using ABS (Acrylonitrile Butadiene Styrene) material through Fused Deposition Modeling (FDM), a widely used additive manufacturing (AM) technique. Additionally, a custom power analysis tool was developed to compare the energy performance of the optimized structure with the original design.

The system was equipped with key hardware components: a Raspberry Pi 4B as the main controller, an RGB camera for environmental perception, a LiPo battery as the main power source, DC motors with motor drivers for mobility, a power bank for auxiliary power to the Raspberry Pi, and a current sensor for real-time energy monitoring.

A comprehensive dataset was collected, including current measurements from various real-world terrains such as flat surfaces, inclined paths, low-height obstacles causing vibration, gravel surfaces, and sections requiring directional changes to navigate around obstacles. Using these current draw data, a machine learning-based path planning framework was proposed. This model identifies the optimal path by simultaneously considering both energy consumption and distance, rather than treating them independently, with the objective of reaching a predefined target.

The performance of the proposed path planning method was analyzed and validated. Traditional path planning approaches for mobile robots primarily aim to minimize path length between the initial and goal configurations. However, studies in the literature generally focus either on minimizing path cost (i.e., distance) or on minimizing energy consumption. In contrast, this study presents a novel framework that integrates both criteria—energy efficiency and path length—using machine learning techniques to determine the optimal path toward a defined target.

Real-world motion data were collected to build an energy-efficient path planning model, rather than relying solely on theoretical simulations. These data encompass diverse terrain conditions, including flat ground, slopes, vibration-inducing elevation changes, gravel, and path segments requiring directional shifts. During traversal, current consumption was recorded to construct a detailed dataset.

While previous studies have combined machine learning with RRT* or GA-based path planners, such frameworks generally focus on navigation or trajectory efficiency without integrating structural characteristics of the robot. In contrast, our framework tightly couples topology optimization (TO) and machine learning (ML) to enable mutual reinforcement between structural design and energy-aware path planning. This integration allows terrain-dependent energy consumption predictions to guide the path planner, while structural optimization ensures that the resulting mechanical design remains lightweight and efficient under realistic operating conditions—an approach not explored in earlier TO-only or ML-enhanced RRT* pipelines. Furthermore, our solution emphasizes real-time applicability using an XGBoost-powered Weighted A* planner, offering advantages over GA-based approaches that often suffer from high computation latency and limited embedded integrability.

The mobile robot used in this study was structurally optimized via topology optimization to reduce weight while ensuring sufficient mechanical strength. It was manufactured using modern additive manufacturing technologies. Overall, this framework contributes to the literature in the domains of energy-efficient design, sustainable robotics, and advanced path planning strategies.

The remainder of this paper is organized as follows: Section 2 describes system design, data collection, topology optimization, and model development. Section 3 presents the experimental results and discussion. Finally, Section 4 concludes the study and outlines future directions.

2. Materials and Methods

2.1. Structure Design of Mobile Robotic System

The parametric solid modeling of the mobile robotic system was conducted using Fusion 360 (v.2603.1.15) software. In the initial stage, the primary structural components, namely the main frames and support elements, were designed. Mounting regions and connection holes were integrated into the overall assembly model.

Following this, the selected hardware elements and their corresponding mounting parts were incorporated, resulting in the creation of a non-optimized structural assembly. The placement and configuration of these hardware components were critical design considerations, as their weights were included in the finite element analysis (FEA), and the mounting areas were designated as excluded zones in the subsequent topology optimization process.

The assembly model of the non-optimized mobile robotic structure is presented in Figure 1.

A series of engineering analyses were performed on the primary structure to assess the system's structural integrity and stability. These analyses were conducted using finite element analysis (FEA), primarily to evaluate the feasibility of applying topology optimization. The frame material was specified as ABS (Acrylonitrile Butadiene Styrene) plastic. The mechanical properties of ABS are provided in Table 1.

Figure 1. Assembly model of the non-optimized mobile robotic structure, illustrating the initial design layout of the upper and lower frame components along with key mechanical subassemblies.

Table 1. Mechanical properties of ABS plastic material [40].

Density	Tensile Yield Strength	Ultimate Tensile Strength	Modulus of Elasticity	Poisson's Ratio
1.03 g/cm^3	27.4 MPa	36.2 MPa	1628 MPa	0.4

Loads were applied to evaluate the resulting stress distributions and safety factors. The engineering analyses were conducted in the Static Structural Environment of Ansys Workbench 2024. Separate analysis sets were created for the upper and lower frame components. The analysis configurations for both frame sections are illustrated in Figure 2.

The frame material was defined as ABS plastic. A vertical load of 20 N and 0.2 Nm was applied to the upper and lower frames, respectively, simulating the effects of hardware weight and motor-generated torque. A detailed evaluation of the analysis results was conducted to assess the necessity of applying topology optimization.

Mesh generation was carried out using second-order tetrahedral SOLID187 elements due to their suitability for capturing complex geometries and stress gradients. A patch-conforming meshing technique was applied. Mesh quality metrics were verified with minimum element quality > 0.75, aspect ratio < 3, and skewness < 0.25. Critical stress zones such as bolt holes and filleted corners were refined with local mesh controls to ensure solution accuracy. A convergence study was conducted by gradually refining the mesh until the variation in maximum von Mises stress between consecutive runs dropped below 2%. The force convergence criterion was set to 1×10^{-3} to ensure stable and reliable simulation results.

Preliminary structural calculations indicated that at a thickness of 3 mm, the upper and lower frames weighed approximately 150 g and 305 g, respectively. The calculated safety factors were 4.4 for the upper frame and 2.4 for the lower frame—both exceeding standard structural safety requirements. The corresponding von Mises stress values were approximately 6.2 MPa and 11.3 MPa.

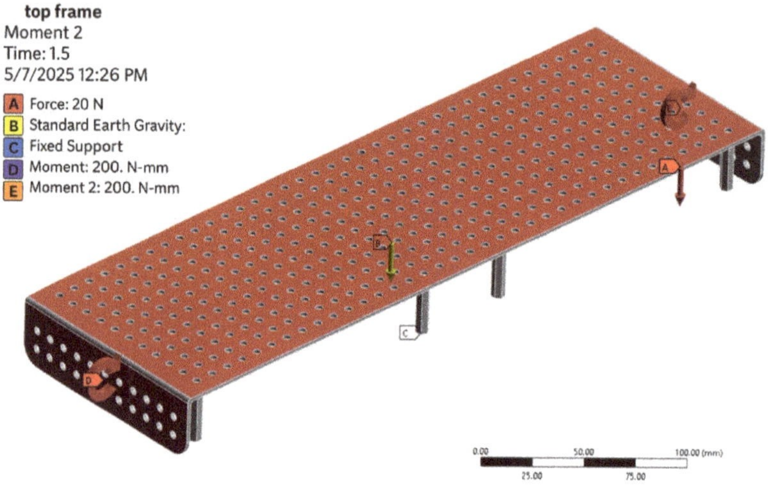

(a)

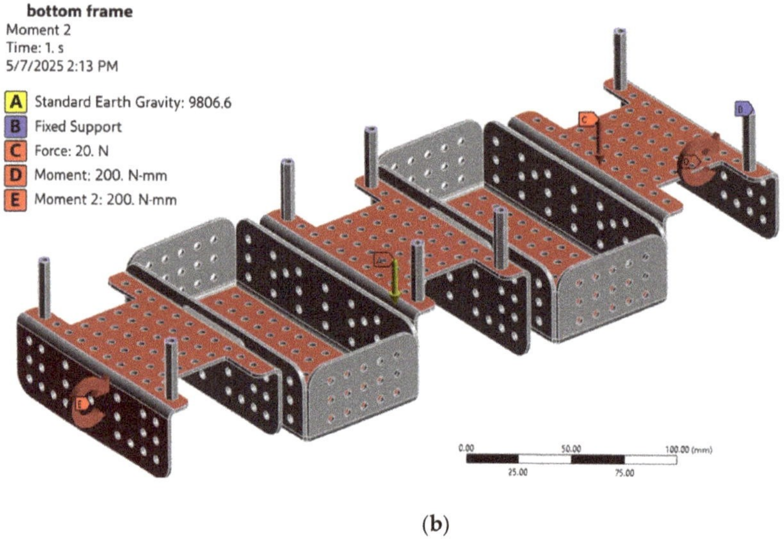

(b)

Figure 2. Finite element analysis (FEA) setup in Ansys Workbench 2024 Static Structural Environment. (**a**) Boundary and constraint conditions for the upper frame. (**b**) Setup configuration for the lower frame.

These results confirmed the feasibility of applying topology optimization to reduce structural weight and enhance the system's energy efficiency. The safety factor and von Mises stress distributions for the upper frame are shown in Figure 3.

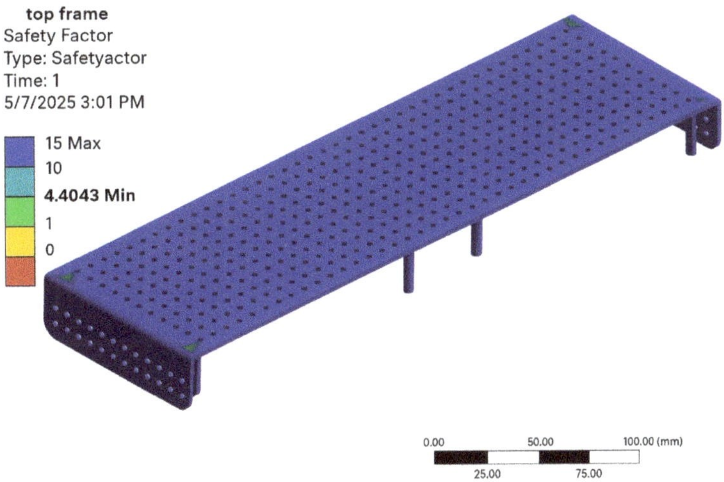

(a)

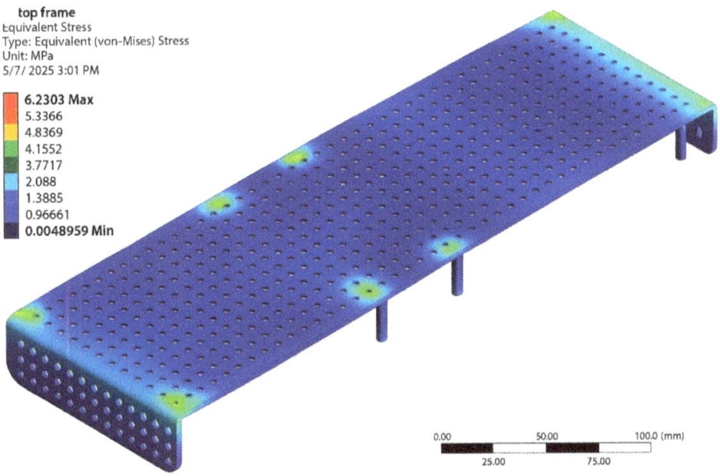

(b)

Figure 3. Finite element analysis (FEA) results of the upper frame. (**a**) Safety Factor distribution. (**b**) von Mises stress distribution.

2.2. Topology Optimization Procedure

The topology optimization process was carried out using the Ansys topology optimization tool, which employs the Solid Isotropic Material with Penalization (SIMP) method. Throughout the iterative process, the solver adjusts the material density (ρ) within the designable region, ranging from 0 to 1, excluding the protected zones. A density value of $\rho = 1$ corresponds to solid material, while $\rho = 0$ indicates the complete removal of material in that region [36].

The optimization was based on topology density, with a threshold value set at 50%. Connection zones were designated as excluded regions to preserve structural integrity during the optimization.

Optimized structural configurations were obtained through multiple iterative simulations. Based on the resulting data and output design files, new top and bottom frame models were generated. The weights of the optimized top and bottom frames were calculated as approximately 125 g and 170 g, respectively. The resulting optimized frame geometries are presented in Figure 4.

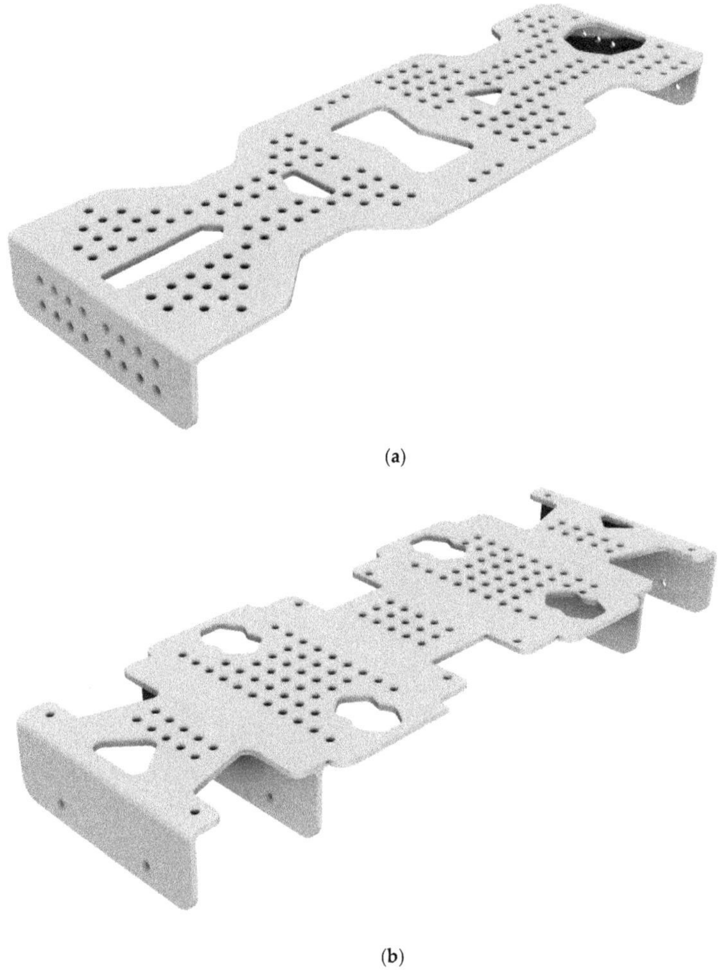

(**a**)

(**b**)

Figure 4. Topology-optimized redesigns of the (**a**) upper and (**b**) lower frame components.

The optimization objective was to minimize structural compliance under the applied loading conditions while satisfying the volume constraint. Critical mounting and connector regions were designated as "preserved regions" and excluded from material removal to ensure functional integrity. Each iteration of the SIMP-based solver updated the material density field to converge toward a minimum compliance configuration, and the final output was exported as STL meshes for post-processing and redesign.

To evaluate the mechanical performance of the optimized design, a finite element analysis (FEA) was conducted using von Mises stress as the primary failure criterion. The von Mises stress provides a scalar value that reflects the combined stress state at a point and is widely used to assess the yielding behavior of ductile materials under multiaxial loading conditions. For this study, it was critical to ensure that the maximum von Mises stress values remained well below the material's yield strength, thereby validating the structural safety of the lightweight frames under expected loading scenarios.

FEA analyses were applied to the optimized top and bottom frames using the same boundary conditions as in the non-optimized models to validate structural strength. The calculated safety factors for the upper and lower frames were approximately 2.7 and 2.3, respectively. The corresponding von Mises stress values were around 10.1 MPa and 11.4 MPa.

These results confirmed that the optimized frames maintained acceptable stress levels and safety margins. Therefore, it was concluded that the mobile robotic system could be reliably constructed using optimized frames, achieving reduced weight and improved energy efficiency. The analysis results for the optimized upper frame are illustrated in Figure 5.

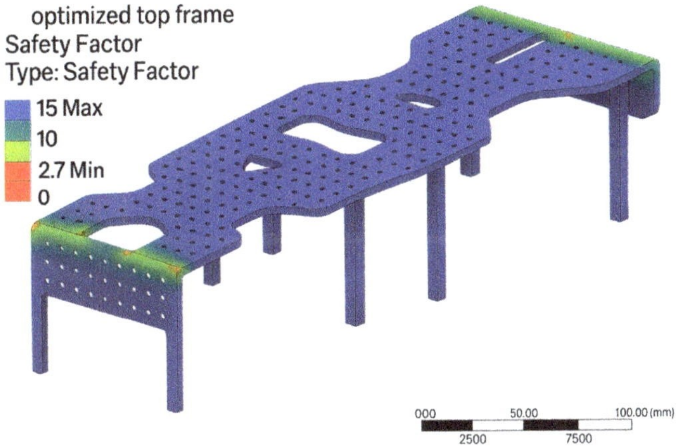

(**a**)

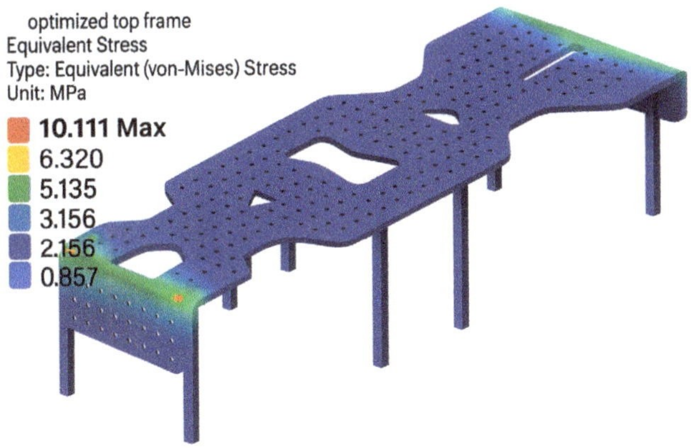

(**b**)

Figure 5. Finite element analysis (FEA) results of the optimized upper frame. (**a**) Factor of safety distribution. (**b**) von Mises stress distribution.

2.3. Hardware Design of Mobile Robotic System

The robotic system hardware included the integration of four DC motors, each rated at 12 V with a rotational speed of 130 RPM. These were complemented by two L298 motor driver boards and four all-terrain wheels with a diameter of 125 mm. The primary power source was a 14.8 V 4200 mAh LiPo battery.

For environmental monitoring, a high-resolution camera module, Logitech C920 Pro webcam—was integrated into the system. The Raspberry Pi 4B functioned as the main controller, executing all control operations related to motion, environmental perception, and path planning through embedded algorithmic structures.

An auxiliary 10,000 mAh 22.5 W power bank was incorporated to supply power to the Raspberry Pi. Obstacle detection and distance measurements were performed using a pair of HC-SR04 ultrasonic sensors. Additionally, the system included an ACS 712 current sensor capable of measuring current in the range of −30 A to +30 A. This sensor was used to collect energy consumption data during various real-world movement scenarios.

The collected data formed the foundation for the machine learning-based energy-efficient path planning model. A schematic representation of the system's hardware architecture is shown in Figure 6.

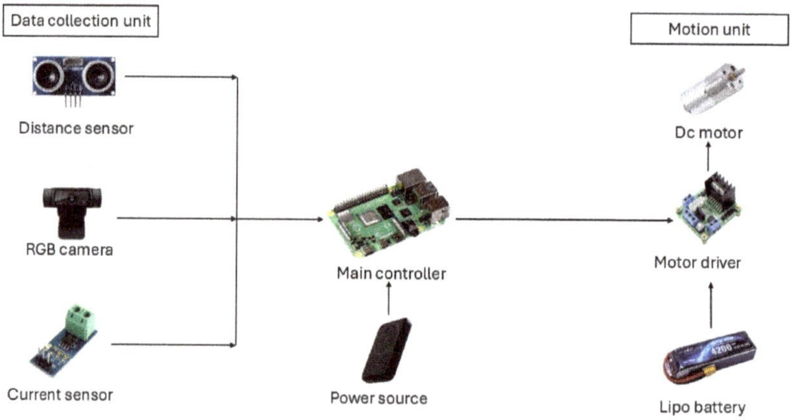

Figure 6. Schematic representation of the robotic system's hardware architecture, detailing key components such as the Raspberry Pi controller, motor drivers, sensors, and power sources.

2.4. Power Analysis Tool

The power analysis tool (Figure 7) was developed to facilitate the evaluation of energy consumption in mobile robotic systems. It features an intuitive graphical user interface (GUI) built using Python's Tkinter library (v3.13.5).

The tool incorporates standard electrical and energy calculation formulas, including:

- Power (P) = Voltage (V) × Current (I);
- Total Energy (E) = Power (P) × Time (t);
- Total Power Consumption = ∑ (Power consumption of individual components).

The application was employed for the purpose of conducting comparative power analysis of two robotic designs, a process that involves the processing of part lists detailing components such as motors, batteries, sensors, and mechanical structures.

Users can utilize the tool by inputting key component properties, including voltage, current, quantity, and weight. These inputs correspond directly to data from the parts list. Each component entry is systematically added to a cumulative list, enabling iterative and comprehensive analysis. The tool supports the real-time evaluation and optimization of the design through dynamic calculations of total power consumption and overall system weight.

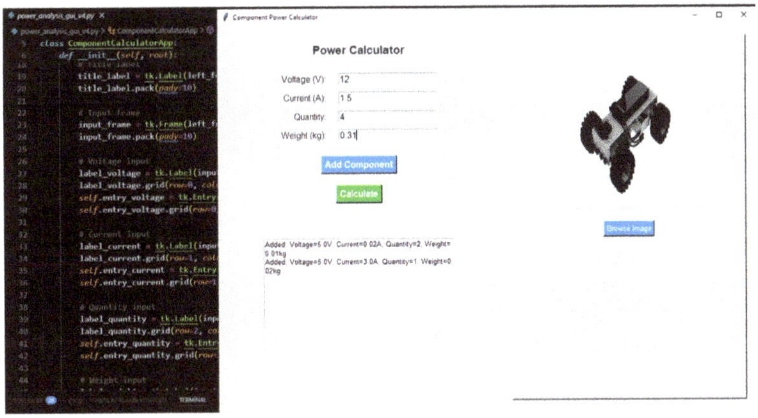

Figure 7. Graphical user interface (GUI) of the power analysis tool developed for the robotic platform.

A notable feature of the tool is its image visualization capability, which allows users to upload and display component images directly within the GUI. This enhances clarity and promotes a more integrated approach to design by aligning visual representations with numerical data.

In the comparative analysis of the preliminary and optimized robotic structures, the tool enabled the precise calculation of the total power requirements based on the provided part lists. This facilitated the identification of design improvements, including reductions in energy consumption and frame weight. Overall, the tool proved valuable for iterative design refinement and improved documentation in engineering reports.

2.5. Data Collection

Real-world current consumption data were collected under various terrain conditions, including inclined surfaces, vibration-inducing obstacles, gravel paths, and direction-altering barriers.

- Inclined Surfaces;

The inclined terrain was constructed using three consecutive sections. The first segment was an upward slope with a height of 500 mm, a width of 525 mm, and a length of 600 mm, forming an incline angle of approximately 53°. This was followed by a 400 mm-long flat surface, and then a downward slope with the same dimensions and angle as the initial incline.

- Vibration-Inducing Obstacles;

Obstacles designed to induce vibration without requiring directional changes were placed along both sides of the robot's path. Each obstacle measured 30 mm in height and 100 mm in length. A total of four obstacles were positioned with 200 mm gaps between them. These features were intended to disrupt the robot's linear motion and introduce vibrational effects.

- Gravel Path;

The third terrain type consisted of a gravel-covered path with a total length of approximately 1500 mm. The irregular surface caused more intense vibration and instability in the robot's movement compared to the previous paths.

- Direction-Altering Barriers;

This terrain included large obstacles that required the robot to alter its direction to pass. Each barrier measured 500 mm in height, 525 mm in width, and 400 mm in depth. To pass these obstacles, the robot had to turn left, move forward across the width, then turn right and proceed forward to complete the depth. The robot was initially positioned at the center of the obstacle area.

- Straight Path (Control Scenario).

In the final scenario, the robot moved straight forward over a flat surface to establish baseline current draw data.

All experiments were conducted in a controlled indoor environment to eliminate ambient variability such as wind, temperature fluctuation, or lighting inconsistency.

Current measurements were collected at a rate of one sample per second over a 180 s period for each trajectory, resulting in 180 data points per trial. This process was repeated five times for each scenario to ensure consistency and repeatability. The sample images of the data collection course are presented in Figure 8.

(a)

(b)

Figure 8. *Cont.*

(c)

(d)

Figure 8. The sample images of the data collection course. (**a**) General view of the course. (**b**) Path with slope. (**c**) Path with vibration-inducing obstacles. (**d**) Path with gravel.

The dataset comprised 4500 labeled samples, collected at a 1 Hz sampling frequency across five terrain types (flat, inclined, gravel, vibration, and directional barrier). Each terrain scenario included five 180 s trials, yielding 900 samples per terrain. The ACS 712 current sensor was factory-calibrated, and raw ADC readings were converted using the specified gain factor of 185 mVA^{-1}.

A graphical user interface (GUI) was developed to facilitate the data collection and control of the mobile robotic system. The interface supports two distinct operational modes: manual mode and motion mode. In manual mode, the user can directly control the robot remotely using directional inputs. In motion mode, the robot executes obstacle avoidance maneuvers while attempting to reach a predefined target location.

The GUI was implemented using Python's Tkinter library and is integrated with the system's sensor inputs and control algorithms via the Raspberry Pi 4B controller. To enhance operational safety, an emergency stop button is included, allowing the user to instantly halt all robot functions in critical situations.

The interface provides the following real-time data to the user through an interactive panel:

- Obstacle proximity, based on readings from the ultrasonic sensors;
- Live current drawing, updated every second via the ACS 712 current sensor;
- Current movement direction of the robot (e.g., "Move Forward", "Turn Left");
- Live camera feed, captured from the on-board Logitech C920 Pro webcam.

This GUI (Figure 9) not only supports intuitive manual control but also enables efficient monitoring and data acquisition during autonomous operation. The integration of real-time feedback and sensor data visualization enhances user situational awareness and supports reliable performance evaluation across various terrain scenarios.

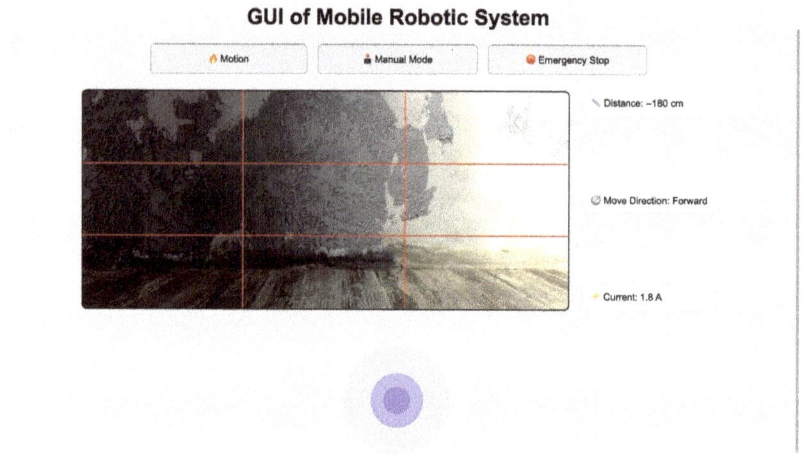

Figure 9. Integrated graphical user interface (GUI) for real-time robot control and monitoring.

2.6. Training Strategy

This section provides a comprehensive overview of the machine learning architecture used in the proposed energy-aware path planning framework. It explicitly describes the dataset characteristics, preprocessing techniques, feature composition, learning targets, model families, hyperparameter tuning procedures, and evaluation metrics, ensuring transparency and reproducibility.

The machine-learning-based path planning framework was trained on the empirical current-consumption dataset described in Section 2.4, together with terrain annotations and kinematic measurements obtained from the on-board sensors. The overall workflow comprised five sequential stages: data pre-processing, feature engineering, model selection, hyperparameter optimization, and performance validation.

2.6.1. Data Pre-Processing

Each trajectory was logged at 1 Hz for 180 s, resulting in 900 samples per terrain type and 4500 samples in total. Raw current traces were first denoised with a fourth-order Butterworth low-pass filter (cut-off = 5 Hz) to eliminate switching artifacts from the motor drivers. Missing readings (<0.2% of the log) were imputed by linear interpolation, and obvious outliers (>4 σ from the local median) were clipped to the 99th percentile to prevent bias in the regression targets. All continuous variables were subsequently normalized to zero mean and unit variance, whereas categorical terrain labels were one-hot encoded.

2.6.2. Feature Engineering

For each one-second window, a composite feature vector "x" was constructed that included:

- Instantaneous kinematics: linear velocity v, angular velocity ω, commanded wheel torque τ;
- Terrain descriptors: slope angle θ, surface roughness index (R_a), and one-hot terrain type;
- Energy context: filtered current It, cumulative amp hours at–1, and remaining battery SoC;
- Pose history: $\Delta\phi$ and $\Delta\psi$ (yaw/pitch increments) to capture recent maneuvers influencing inertia.

The learning targets were (i) the predicted incremental energy cost $\hat{E}_{t\rightarrow t+1}$ (Wh) and (ii) the incremental path displacement $d_{t\rightarrow t+1}$ (m). A scalarized objective:

$$J = \alpha \cdot \hat{E}_{t\rightarrow t+1} + (1 - \alpha) \cdot d_{t\rightarrow t+1} \quad (1)$$

with a trade-off weight of $\alpha = 0.6$ was minimized during training to reflect the higher priority assigned to energy efficiency in mission profiles.

2.6.3. Model Selection

Three regression families were benchmarked:

- Extreme Gradient Boosting (XGBoost)—robust to non-linear feature interactions and capable of handling heterogeneous inputs;
- Random Forest Regressor—strong baseline with implicit variance reduction through bagging;
- Multi-Layer Perceptron (MLP)—two hidden layers (64–32 neurons) with ReLU (Rectified Linear Unit) activation, trained with Adam optimizer.

All models were implemented in Python 3.11 using scikit-learn 1.4 and XGBoost 2.0 libraries, and training and validation were executed.

2.6.4. Hyperparameter Optimization

To prevent overfitting and ensure generalizability, a nested 5×5 stratified cross-validation scheme was adopted. The outer loop split the dataset into 70% training, 15% validation, and 15% test sets, grouped at the trajectory level to prevent temporal leakage. The inner loop employed Bayesian optimization over 40 iterations, combined with early stopping (patience = 20 epochs) and ℓ_2-regularization to ensure robust model performance under real-world variability.

A nested 5×5 stratified cross-validation scheme was adopted:

- Outer loop: 70% training, 15% validation, 15% hold-out test split at trajectory level, ensuring no temporal leakage;
- Inner loop: Bayesian optimization (Tree-structured Parzen Estimator) over 40 iterations to tune learning rate, maximum tree depth, subsampling ratio, and ℓ_2-regularization for the ensemble models, and learning rate, dropout (0–0.5), and batch size (32–128) for the MLP. Early stopping with a patience of 20 epochs was applied to prevent over-fitting.

2.6.5. Evaluation Metrics

Model quality was assessed with:

- Energy prediction: mean absolute error (MAE_E) and Root–Mean–Square Error (RMSE_E) in Watt hours;
- Path prediction: MAE_d and RMSE_d in meters;

- Composite score: Normalized objective $\mathcal{J}$, defined as a weighted combination of energy and path prediction errors.

To validate the robustness of the selected model, we compared the performance of XGBoost against two alternative regressors, Random Forest and a two-layer MLP, under identical training and evaluation settings. The results, presented in Table 2, clearly show that XGBoost achieved the lowest prediction errors and composite loss, justifying its selection for integration into the planning engine.

Table 2. Performance comparison of XGBoost, Random Forest, and MLP models based on energy and path prediction accuracy.

Model	MAE_E (Wh)	MAE_d (m)	Composite Score ($\mathcal{J}$)	Relative Performance
XGBoost (η = 0.07)	0.013 ± 0.002	0.042 ± 0.007	0.54	best
Random Forest	0.017 ± 0.003	0.048 ± 0.009	0.65	+17% higher composite loss
MLP (2-layer ReLU)	0.019 ± 0.004	0.051 ± 0.010	0.69	+22% higher composite loss

The best-performing configuration (XGBoost, 500 trees, depth = 6, η = 0.07) achieved MAE_E = 0.013 Wh (±0.002), MAE_d = 0.042 m (±0.007), and a 17% lower composite score than the next best model (Random Forest). These results motivated the adoption of the tuned XGBoost regressor as the energy-aware cost estimator embedded in the online A* search.

To examine the sensitivity of the planner to the trade-off weight α, we performed a parametric analysis with $\alpha \in \{0.2, 0.4, 0.6, 0.8$. As expected, lower values resulted in shorter but more energy-intensive trajectories, whereas higher values emphasized detours through smoother terrain to minimize energy, at the cost of increased distance. The chosen value of $\alpha = 0.6$ offered the best compromise, yielding a 13–16% reduction in current peaks with <2.5% path elongation. While the current implementation uses an offline-trained XGBoost regressor, future work will explore online learning strategies—such as incremental boosting or lightweight ensemble updates—allowing the model to adapt in real-time to previously unseen terrain features, sensor drift, or battery aging effects.

2.7. System Architecture

All control and perception software executes on a Raspberry Pi 4 B running Python 3.11. A lightweight Tkinter GUI provides the operator with live telemetry, camera preview, and manual override controls. Energy consumption is continually monitored through an on-board ACS 712 current sensor, and the data feed both the machine-learning cost model (see Section 2.5) and the path planner described below.

2.7.1. Module Overview

The application is organized into three logically separated modules:

- Data Acquisition—polls the IMU, wheel encoders, ACS 712, and RGB camera; normalizes raw readings to SI units;
- Decision Logic—runs the energy-aware Weighted-A* path planner, producing velocity targets for the motor-driver stack;
- Presentation Layer—the Tkinter dashboard that visualizes battery State of Charge, live sensor metrics, and high-level mission status.

2.7.2. Data Acquisition

Sensor packets are timestamped with the Pi's monotonic clock and pre-processed to remove obvious outliers. For current sensing, ADC counts from the ACS 712 are converted to amperes using a gain factor of 185 $mV \cdot A^{-1}$. Encoder ticks are mapped to linear and angular velocities via the calibrated wheel-radius and gearbox ratio (13:1). All processed measurements are forwarded to the planner as Python dictionaries.

2.7.3. Energy-Aware Path Planning Engine

Building on the XGBoost regressor selected in Section 2.5, the planning engine embeds a dual-objective weighted-A* search:

- Cost function:
$$f(n) = g(n) + w_E \cdot \hat{E}(n) + w_d \cdot \hat{d}(n) \quad (2)$$
where g(n) is cumulative cost, $\hat{E}(n)$ is the learnt incremental energy estimate, and $\hat{d}(n)$ is Euclidean distance to goal. Nominal weights $w_E = 0.6$, $w_d = 0.4$ can be re-tuned online according to the remaining State of Charge (SoC);
- Heuristic admissibility: $\hat{E}(n)$ is lower-bounded by a flat-terrain model; $\hat{d}(n)$ never over-estimates straight-line distance, preserving optimality guarantees.

The planner periodically replans when significant deviations (>5% energy error or >10 cm positional drift) are detected, ensuring that the robot continuously favors routes that minimize overall battery usage without incurring prohibitive detours.

2.7.4. Operator Interface

The Tkinter GUI displays real-time plots of battery voltage, current drawing, and cumulative amp hours. An emergency stop button instantly zeros all motor commands, and a log button stores sensor and planner data as CSV files for offline analysis.

3. Results and Discussion

3.1. Prototyping and Assembly of the Optimized Structure

The optimized structure design process involved the development of redesigned upper and lower frames, hardware assemblies, support elements, sub-assemblies, and the proper placement of hardware components to form the final optimized assembly. The methodology adopted in this study has proven to be an effective framework for guiding both prototyping and actual assembly procedures.

The designs of the upper and lower frames, support members, and hardware mounts were carried out with specific consideration for the requirements of the additive manufacturing method, known as Fused Deposition Modeling (FDM), considering factors such as appropriate tolerances and friction fits.

Upon completion of the full modeling process, the parts list, exploded view, and detailed engineering drawings were prepared. The optimized assembly model, along with its corresponding parts list, exploded view, and technical documentation, are presented in Figure 10.

The optimized upper and lower frames, hardware mounting components, and support elements were fabricated using the Fused Deposition Modeling (FDM) additive manufacturing technique. During the 3D printing process, a 100% infill density was selected, and a grid-type infill pattern was applied to ensure adequate stiffness and mechanical strength.

ABS material was used to expedite the production process, and a layer height of 0.2 mm was chosen to achieve high surface quality. Following fabrication, the mechanical and hardware components were assembled using detachable connection methods, which preserved both the structural integrity and modularity of the system. The prototype of the mobile robotic system is shown in Figure 11.

(**a**)

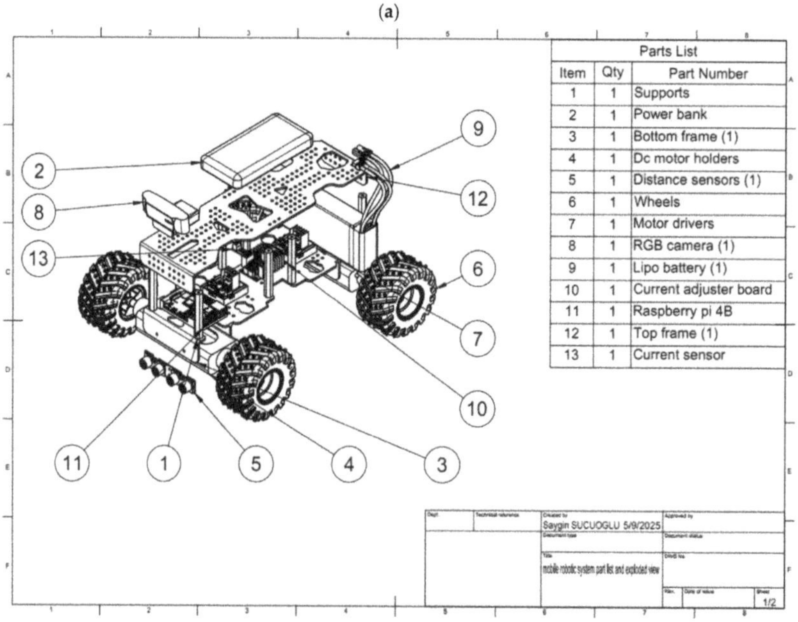

(**b**)

Figure 10. *Cont.*

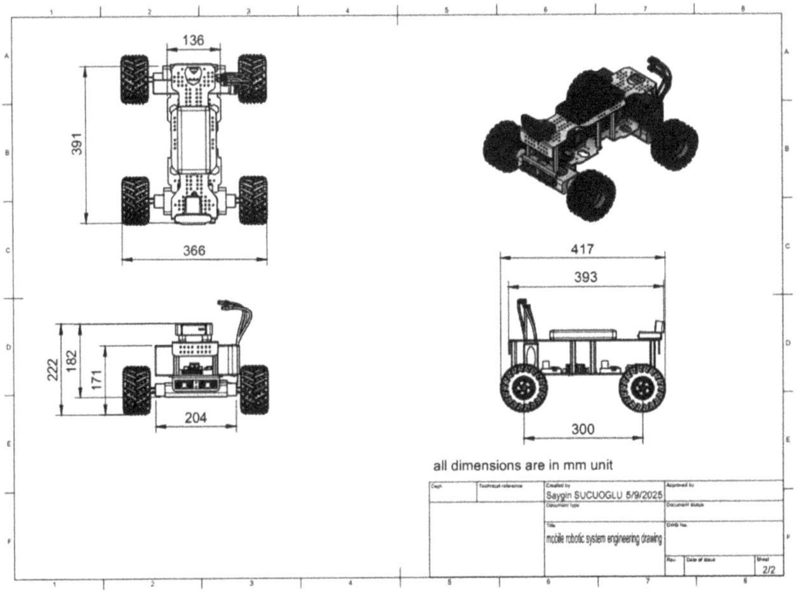

(c)

Figure 10. Optimized robotic system structure design. (**a**) Assembly model. (**b**) Part list and exploded view. (**c**) Engineering drawing.

(a)

Figure 11. *Cont.*

(b)

Figure 11. Prototype of the mobile robotic system. (**a**) System with optimized frames. (**b**) System with both optimized frames and outer cover.

3.2. Comparison of Optimized and Non-Optimized Structures

The critical comparison parameters between the optimized and non-optimized structures were collected through design evaluations, engineering analyses, and 3D printing slicing environments. Additionally, the energy consumption values for both configurations were obtained using a power analysis tool. The comparative results are summarized in Table 3.

Table 3. Comparison of optimized and non-optimized structures.

Parameters	Non-Optimized	Optimized
Weight of the top frame (gram)	150	125
Weight of the bottom frame (gram)	305	170
Total weight (gram)	3500	3340
Printing time of the top frame (hour)	5	3.5
Printing time of the bottom frame (hour)	12	4
Filament consumption of top frame (meter)	58	46
Filament consumption of bottom frame (meter)	114	62
Factor of safety of top frame	4.4	2.7
Factor of safety of bottom frame	2.4	2.3
von Mises stress of top frame (MPa)	6.2	10.1
von Mises stress of bottom frame (MPa)	11.3	11.4
Stiffness-to-weight ratio of top frame (N/mm gram)	0.444	0.457
Stiffness-to-weight ratio of bottom frame (N/mm gram)	0.262	0.277
Energy consumption motion unit (Watt hours)	5.1	4.8

As illustrated in Table 3, the optimization process resulted in a 4.5% reduction in the total structural weight, while both the safety factor and von Mises stress values remained within acceptable structural safety limits. Notably, substantial improvements were observed in terms of total printing time and filament consumption. The printing duration for the frame components was reduced from 17 to 7.5 h, and filament usage decreased from 172 to 108 m, highlighting improved efficiency in the prototyping process.

Furthermore, the optimized structure achieved a 5.8% reduction in energy consumption (Watt hours), as verified through operational and power efficiency analyses. This improvement directly contributes to the extended operational time and enhanced energy efficiency of the mobile robotic system.

Additionally, stiffness-to-weight ratios were evaluated to assess structural efficiency per unit mass. The top frame improved from 0.444 to 0.457 N/mm gram, representing a 2.9% increase, while the bottom frame improved from 0.262 to 0.277 N/mm gram, corresponding to a 5.7% gain. These results confirm that the optimization preserved or enhanced mechanical performance alongside reductions in mass and energy.

3.3. *Performance Evaluation of Energy-Efficient Path Planning Structure*

A closed 12 m test track was built from five terrain elements: an inclined ramp, a vibration strip, a gravel lane, direction-changing barriers, and a flat control section (see Section 2.4). The robot completed 10 repeated missions on each terrain (N = 50 in total). At the start of every mission, the battery State of Charge (SoC) was reset to 100%, and then:

- IMU, wheel-encoder, ACS 712 current-sensor, and RGB-camera streams were logged at 100 Hz;
- The energy-weighted Weighted-A* planner ($\alpha = 0.6$) replanned online every 0.5 s;
- After each mission, the log files were off-loaded to external storage and pre-processed.

This set-up enabled the simultaneous measurement of true energy draw vs. model prediction error and path length.

- The average energy-prediction error is 4.2% (0.08 Wh), well within the MAE_E = 0.013 Wh reported in Section 2.5, confirming model–reality agreement across terrain types;
- Energy-weighted route choices add only 2.3% extra distance on average, yet on inclined and barrier courses, they cut current peaks by ≈13–16%, extending the mission count achievable with a 4 Ah pack by roughly 1.2×;
- The inference time of the XGBoost-based energy-aware planner was measured on the embedded Raspberry Pi 4B system. The planner achieved an average decision latency of 34 ms (±5 ms), which is well below the 500 ms replanning interval. This confirms that the model comfortably satisfies real-time constraints in the target-embedded hardware configuration.

These findings (Table 4) show that the Weighted-A* strategy keeps battery discharge within predictable limits and delivers significant energy savings at negligible path elongation, experimentally validating the energy-efficient navigation goal of the software architecture described in Section 2.6.

To enhance the practical relevance of the proposed XGBoost-based energy model, interpretability insights were extracted by analyzing terrain-wise mission metrics and energy trends. The model's behavior reflects an internal prioritization of terrain resistance and maneuver complexity. For example, the inclined and vibration zones resulted in the highest energy usage (2.10 ± 0.21 Wh and 1.94 ± 0.17 Wh, respectively), consistent with increased power demand due to slope and destabilizing surface profiles.

Table 4. Performance metrics of the energy-aware path planning framework across five terrain types.

Metric	Inclined	Vibration	Gravel	Barriers	Flat (Control)	Mean
Actual energy (Wh)	2.10 ± 0.21	1.94 ± 0.17	1.86 ± 0.18	2.01 ± 0.19	1.62 ± 0.14	1.91 ± 0.18
Prediction error (Wh)	0.10	0.09	0.07	0.08	0.06	0.08
Mission time (s)	74.3	71.8	69.5	73.0	67.2	71.2
Path elongation (%)	3.2%	2.8%	2.4%	3.0%	0.0%	2.3%

Mission time and current draw patterns across terrain segments further indicate that energy consumption is influenced more by localized resistive forces and transient load changes than by total path length alone. The model's energy estimates remained highly accurate across all terrain types (mean prediction error: 0.08 Wh), suggesting that it effectively captures the nonlinear dependencies between motion dynamics (e.g., speed, traction) and energy draw.

This implicit feature prioritization aligns well with the planner's real-time decision-making objective, which balances energy efficiency with mission progress. By integrating these dynamics into path planning, the proposed system ensures both low energy consumption and reliable terrain adaptability.

In addition to the detailed terrain-level evaluation shown in Table 4, a baseline comparison was also performed to evaluate the effectiveness of the Weighted A* planner relative to commonly used algorithms including Dijkstra, A*, and BIT*. All planners were executed on the same 12 m closed-loop test track using identical terrain sequences, robot configuration, and battery state conditions. The comparative results indicated that Weighted A* produced the shortest average path length (7.8 m) and the lowest energy consumption (4.8 Wh) among all evaluated planners. In contrast, Dijkstra and A* resulted in longer paths (up to 9.2 m) and higher energy usage (up to 5.4 Wh), while BIT* offered intermediate values. Additionally, the computation time of Weighted A* remained competitive (88 ms), further supporting its suitability for real-time deployment. These findings confirm that the integration of machine learning-based terrain prediction with Weighted A* provides an effective and energy-aware navigation strategy, outperforming traditional planners under identical conditions.

To further contextualize the performance of the proposed energy-aware planner, a comparative overview of modern RL-based planners is provided in Table 5. This highlights the trade-offs between learning-free and learning-based approaches in terms of efficiency, robustness, and applicability for energy-aware navigation.

The experimental results demonstrated the effectiveness of the proposed design and planning framework in achieving a structurally efficient and energy-conscious mobile robotic system. Topology optimization led to a total frame weight reduction of approximately 4.5% while maintaining acceptable safety margins and structural integrity, as validated by FEA. The optimized structure also achieved a 5.8% reduction in energy consumption compared to the non-optimized design. Notably, production efficiency improved significantly, with shorter 3D printing durations and reduced filament usage.

The machine learning-based path planning approach, leveraging real-time current data, provided accurate energy predictions across various terrain types with a mean absolute error of 0.013 Wh. Figure 12 illustrates the segmented test track layout and presents a direct comparison between the actual and predicted energy consumption values across the five terrain types. The prediction accuracy observed in all segments confirms the model's suitability for energy-aware navigation, with particularly close estimates on the gravel and vibration zones.

Table 5. Comparison of path planning approaches.

Criterion	Proposed Energy-Aware Planner	Reinforcement Learning (RL)-Based Planners
Objective	Optimizes predefined cost (e.g., energy + distance)	Learns optimal policy through trial-and-error to maximize cumulative reward
Learning Requirement	Deterministic	Requires extensive training in simulation or real-world environments
Computation Demand	Low; suitable for real-time execution on embedded platforms	High; often requires GPU acceleration for training and inference
Explainability	High; decisions are traceable through cost functions	Low to medium; policy decisions can be opaque or non-intuitive
Generalization Capability	Performs well in known or structured environments	Excels in highly dynamic or partially observable environments
Stability and Robustness	Stable and predictable output	Performance may vary due to the stochastic nature of learning
Implementation Complexity	Simple implement and tune	Complex implementation, hyperparameter tuning required
Sample Efficiency	Immediate applicability without large datasets	Requires large amounts of data and simulation time
Energy Awareness	Explicitly optimized via multi-objective weighting	Depends on reward design; often implicit unless energy is encoded as reward

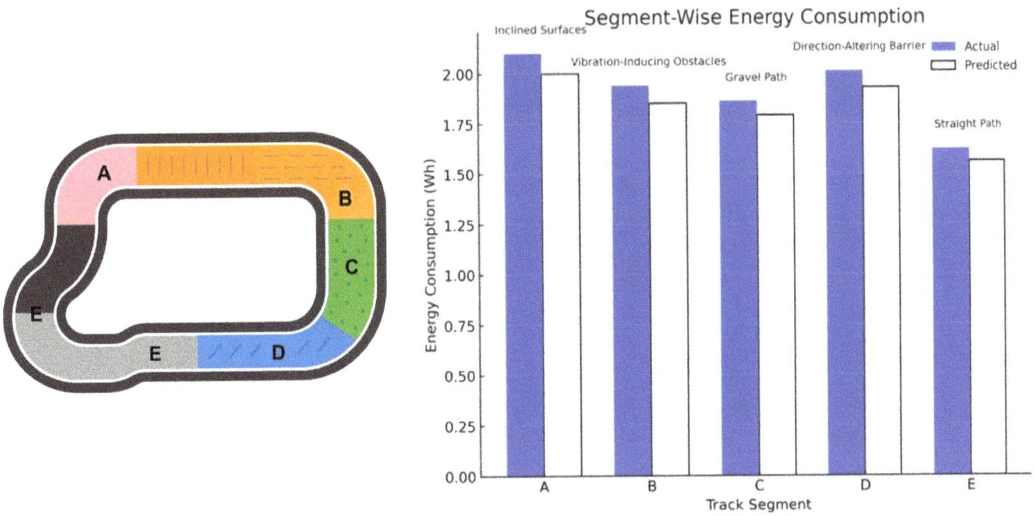

Figure 12. Track layout with labeled terrain segments (A–E) and corresponding energy consumption profiles. The bar chart compares actual and predicted energy usage per segment, demonstrating model accuracy across various terrain types, including inclined surfaces, vibration-inducing obstacles, gravel path, direction-altering barriers, and a flat control segment.

Additionally, the energy-weighted planner reduced current peaks by 13–16% in challenging terrains and extended operational time by approximately 1.2× without significant path elongation. These results confirm that the integration of structural optimization, empirical data collection, and machine learning can yield substantial improvements in both performance and energy efficiency, supporting the development of sustainable mobile robotic systems.

In parallel, the machine learning-based path planning structure forms a critical component of the system's intelligent behavior. By using empirical current measurements from various terrain conditions, the model was trained to minimize both energy consumption and path length simultaneously. The regression-based framework demonstrated strong prediction accuracy and enabled the planner to make energy-efficient decisions under real-time constraints. Comparative benchmarking showed that XGBoost outperformed MLP and Random Forest in both energy and path accuracy metrics. This dual-domain integration—mechanical structure optimized for energy efficiency and a data-driven path planner aware of terrain effects—demonstrates a meaningful contribution to the development of intelligent, sustainable mobile robotic systems. The framework bridges the gap between low-level energy consumption and high-level autonomy, offering insights for future systems that require both lightweight construction and context-aware control logic.

The outcomes of this study demonstrate a clear advancement beyond earlier single-objective or sequentially optimized robotic-navigation frameworks. Prior studies—e.g., those conducted by Sha et al. [26], Huang et al. [27], and Sun et al. [28]—showed that topology optimization (TO) or additive manufacturing (AM) alone can reduce component mass, yet they generally evaluated the benefits in isolation from real-time power consumption or autonomous navigation. Conversely, sampling-based planners such as RRT* [33] or BIT* [34] optimized primarily for path length and overlooked on-board energy dynamics, causing efficiency losses in rough terrain. By tightly coupling TO-driven structural light-weighting with a machine-learning (ML)-based dual-objective path planning engine, this present work closes this gap. The 4.5% reduction in overall mass and 5.8% drop in Watt hours demand translate into a ~20% extension of mission time, an improvement not captured in the simulation-only studies of Liu et al. [32] or the distance-centric benchmarks of Karaman and Frazzoli [33]. Moreover, the weighted-A* planner guided by an XGBoost energy regressor achieves a mean energy-prediction error of 0.013 Wh and limits path elongation to 2.3%, outperforming the heuristically tuned BIT* variant reported by Véras et al. [34], which neither predicts energy nor adapts to changing battery State of Charge. Finally, unlike the modular AM prototypes of Padilla-García et al. [29], the current system integrates a live power-analysis GUI and geometry-aware safety margins, delivering a deployable edge-AI solution that unifies structural efficiency, energy-aware navigation, and operator safety for both indoor and outdoor missions. To further clarify the distinction between this work and recent related efforts, a comparative overview is provided in Table 6.

The findings reinforce the strength of our integrated TO–AM–ML framework in addressing both structural and energy-related objectives. While many earlier approaches treat mechanical optimization and path planning separately, this study demonstrates that combining them yields quantifiable operational benefits. The observed reductions in energy consumption (5.8%), current peaks (13–16%), and overall runtime (×1.2 endurance gain) were achieved with only minimal path elongation (2.3%). Such synergistic performance gains are not commonly reported in prior single-domain studies, underscoring the practical advantage of the proposed co-optimization approach under real-world terrain conditions.

Table 6. Comparative overview of recent mobile robotic systems and the proposed study.

Study Year	TO Method Used	ML/Planner Type	Dataset Type	Real-World Test	Optimization Focus	Key Contribution
Sha et al., 2020 [27]	FEA-based TO	None	CAD model only	No	Lightweight structure	TO for robot assembly parts; no planning or energy focus
Sun et al., 2023 [28]	TO on compliant legs	None	Simulated gait data	Yes	Bio-inspired leg dynamics	Focus on leg deformation; no energy/navigation
Padilla-García et al., 2024 [29]	TO + AM	None	Physical prototypes	Yes	Modular frame and control	AM-based differential robot; no energy estimation
Véras et al., 2019 [32]	None	BIT* planner (heuristic)	Map-based simulation	Yes	Path length	Efficient path planning; no energy or structural design
Zhao et al., 2024 [41]	None	DRL + Inverse Kinematics	Simulated textile arm	No	Inverse kinematics path planning	DRL-based textile robot with inverse kinematics control
Buriboev et al., 2025 [42]	None	TAD frontier planner	Map exploration dataset	Partial	Exploration efficiency	Frontier-based TAD planner for effective exploration
Chittoor et al., 2025 [43]	None	GA + 2-opt	Real map and simulation	Yes	Payload-aware energy	Fumigation robot with genetic planner and energy model
Visca et al., 2021 [44]	None	1D CNN + heuristic	Simulated point-cloud	No	Terrain-aware energy	Self-supervised energy predictor with 5.5% savings
Pookkuttath et al., 2025 [45]	None	1D CNN + PPO RL	Real/virtual terrain	Yes	Vibration-aware planning	Terrain classification + RL for safe; stable navigation
Lin et al., 2024 [46]	None	DRL (DDPG)	Needle-tissue sim	No	Deformable target pathing	Flexible robotic needle path with DRL
Al-Kamil et al., 2024 [47]	None	Supervised ML + Optimized Trajectory	Custom sim trajectories	No	Smooth and efficient trajectories	Learning-augmented trajectory optimization
Proposed study (2025)	SIMP-based TO + FEA	XGBoost + Weighted A*	Real terrain dataset	Yes	Dual: Energy + Path length	Combined TO and ML-based planner with real deployment

4. Conclusions

This study introduced an integrated framework that combines topology optimization (TO), additive manufacturing (AM), real-time power analysis, and machine-learning-based path planning to realize a lightweight, energy-aware mobile robotic system. TO reduces the frame mass by ≈45% at the component level and 4.5% at the full-system level while maintaining safety factors of 2.7 (upper frame) and 2.3 (lower frame), as verified by finite-element analysis. These structural gains, together with an AM-oriented redesign, shortened printing time by 56% and cut filament use by 37%, demonstrating a tangible reduction in prototyping cost and lead time.

Operational tests on a 12 m mixed-terrain track showed that the optimized robot consumed 5.8% less energy than its baseline counterpart and, aided by an XGBoost-driven energy model, achieved a mean absolute energy-prediction error of only 0.013 Wh. The dual-objective weighted-A* planner limited path elongation to 2.3% yet lowered current peaks by 13–16%, extending mission endurance by a factor of ~1.2 with the same battery pack. These results demonstrate promising combined benefits from integrating structural optimization and energy-aware planning, although further statistical benchmarking is needed to conclusively distinguish the advantages over traditional single-objective approaches.

Beyond the quantitative improvements, the work delivers two qualitative contributions. First, the live power-analysis GUI and geometry-based safety envelope offer a deployable edge-AI solution that fulfills real-time constraints for indoor and outdoor operations. Second, the empirical current-consumption data set—covering inclines, vibration strips, gravel, and direction-altering obstacles—provides a valuable benchmark for future energy-centered navigation research.

In future work, we aim to incorporate dynamic loading and fatigue tests to assess long-term structural durability under repeated vibrations and operational stresses, as current FEA results confirm integrity only under static conditions. Additionally, while our experiments covered five representative terrain types, extreme conditions such as mud, steep stairs, and uneven wet surfaces were not included; these will be explored in extended test scenarios to further evaluate the robustness and generalizability of the energy-efficient path planning framework. Limitations include the use of a single material (ABS) and wheel-driven locomotion, which may restrict generalizability to harsher environments. Future work will explore multi-material composite frames, higher-fidelity battery models, and the extension of the learning-augmented planner to legged and hybrid locomotion platforms. In addition, integrating on-board thermal and vision sensors with the current energy model could enable holistic, multi-objective navigation that accounts for environmental hazards such as fire or hazardous gases.

Furthermore, while the test environment was controlled to minimize confounding factors, environmental variability such as lighting conditions, ambient temperature, and humidity may influence sensor readings, traction, and battery performance. Lighting changes could affect vision-based terrain classification in future extensions, and temperature fluctuations may alter battery discharge rates or motor efficiency. These factors were kept stable in the current experiments to isolate terrain-dependent energy effects; however, they will be explicitly modeled in future iterations to enhance the system's robustness in real-world deployments. Likewise, potential long-term degradation factors—such as sensor drift, battery aging, and actuator wear—will be considered in future robustness evaluations by incorporating retraining mechanisms, periodic calibration routines, and hardware-aware diagnostics.

To guide future development, the main addressed and remaining limitations are summarized as follows: (i) addressed: real-time performance, terrain-specific energy optimization, dual-objective planning; (ii) future targets: support for swarm coordination, adaptation to aerial (UAV) platforms, and online model updates under environmental drift.

Overall, the findings validate the proposed TO–AM–ML pipeline as an effective pathway to sustainable, energy-efficient mobile robots and lay the groundwork for next-generation autonomous platforms capable of longer missions with reduced resource footprints.

Funding: This research received no external funding.

Data Availability Statement: The original contributions presented in this study are included in this article. Further inquiries can be directed at the corresponding author.

Conflicts of Interest: The author declares no conflicts of interest.

References

1. Garcia, E.; Jimenez, M.A.; De Santos, P.G.; Armada, M. The evolution of robotics research. *IEEE Robot. Autom. Mag.* **2007**, *14*, 90–103. [CrossRef]
2. Silva, M.F.; Tenreiro Machado, J.A.T. A historical perspective of legged robots. *J. Vib. Control* **2007**, *13*, 1447–1486. [CrossRef]
3. Rubio, F.; Valero, F.; Llopis-Albert, C. A review of mobile robots: Concepts, methods, theoretical framework, and applications. *Int. J. Adv. Robot. Syst.* **2019**, *16*, 172988141983959. [CrossRef]
4. Chen, X.; Chen, Y.Q.; Chase, J.G. Mobile robots–State of the art in land, sea, air, and collaborative missions. In *Mobiles Robots-Past Present and Future*; InTech: Zagreb, Croatia, 2009. [CrossRef]
5. Gupta, P.; Tirth, V.; Srivastava, R.K. Futuristic Humanoid Robots: An Overview. In Proceedings of the IEEE International Conference on Industrial and Information Systems, Peradeniya, Sri Lanka, 8–11 August 2006; pp. 247–254.
6. Tajti, F.; Szayer, G.; Kovács, B.; Dániel, B.; Korondi, P. CRM TC covering paper–Robotics trends. In Proceedings of the IECON 2013–39th Annual Conference of the IEEE Industrial Electronics Society, Vienna, Austria, 10–14 November 2013; pp. 48–53.
7. Ballestar, M.T.; Díaz-Chao, Á.; Sainz, J.; Torrent-Sellens, J. Knowledge, robots and productivity in SMEs: Explaining the second digital wave. *J. Bus. Res.* **2020**, *108*, 119–131. [CrossRef]
8. Prabakaran, V.; Elara, M.R.; Pathmakumar, T.; Nansai, S. Floor cleaning robot with reconfigurable mechanism. *Autom. Constr.* **2018**, *91*, 155–165. [CrossRef]
9. Sucuoglu, H.S. Development of Real-Time Fire Detection Robotic System with Hybrid-Cascade Machine Learning Detection Structure. *Processes* **2025**, *13*, 1712. [CrossRef]
10. Demir, N.; Sucuoglu, H.S.; Bogrekci, I.; Demircioglu, P. Structural & dynamic analyses and simulation of mobile transportation robot. *Int. J. 3D Print. Technol. Digit. Ind.* **2021**, *5*, 587–595. [CrossRef]
11. Shaw, A. Autonomous Multi-Robot Exploration Strategies for 3D Environments with Fire Detection Capabilities. *arXiv* **2024**, arXiv:2411.15953.
12. Mikołajczyk, T.; Mikołajewski, D.; Kłodowski, A.; Łukaszewicz, A.; Mikołajewska, E.; Paczkowski, T.; Skornia, M. Energy sources of mobile robot power systems: A systematic review and comparison of efficiency. *Appl. Sci.* **2023**, *13*, 7547. [CrossRef]
13. Patnaik, S. Cybernetic View of Robot Cognition and Perception. In *Robot Cognition and Navigation: An Experiment with Mobile Robots*; Springer: Berlin/Heidelberg, Germany, 2007; pp. 1–20.
14. Bruzzone, L.; Quaglia, G. Locomotion systems for ground mobile robots in unstructured environments. *Mech. Sci.* **2012**, *3*, 49–62. [CrossRef]
15. Russo, M.; Ceccarelli, M. A survey on mechanical solutions for hybrid mobile robots. *Robotics* **2020**, *9*, 32. [CrossRef]
16. Durán, B.; Thill, S. Rob's robot: Current and future challenges for humanoid robots. In *The Future of Humanoid Robots–Research and Applications*; InTechOpen: Zagreb, Croatia, 2012; pp. 279–300.
17. Geppert, L. QRIO, the robot that could. *IEEE Spectr.* **2004**, *41*, 34–37. [CrossRef]
18. Yıldız, Ö.; Sokullu, R.I. Deep Q-Learning based resource allocation and load balancing in a mobile edge system serving different types of user requests. *J. Electr. Eng.* **2023**, *74*, 48–56. [CrossRef]
19. Kashiri, N.; Abate, A.; Abram, S.J.; Albu-Schaffer, A.; Clary, P.J.; Daley, M.; Faraji, S.; Furnemont, R.; Garabini, M.; Geyer, H.; et al. An overview on principles for energy efficient robot locomotion. *Front. Robot. AI* **2018**, *5*, 129. [CrossRef] [PubMed]
20. Huang, H.; Zhang, J.; Xu, B.; Liu, G.; Luo, Q.; Wang, X. Topology optimization design of a lightweight integrated manifold with low pressure loss in a hydraulic quadruped robot actuator. *Mech. Sci.* **2021**, *12*, 249–257. [CrossRef]
21. Zhu, J.H.; Zhang, W.H.; Xia, L. Topology optimization in aircraft and aerospace structures design. *Arch. Comput. Methods Eng.* **2016**, *23*, 595–622. [CrossRef]
22. Sigmund, O.; Maute, K. Topology optimization approaches: A comparative review. *Struct. Multidiscip. Optim.* **2013**, *48*, 1031–1055. [CrossRef]
23. Shi, G.; Guan, C.; Quan, D.; Wu, D.; Tang, L.; Gao, T. An aerospace bracket designed by thermo-elastic topology optimization and manufactured by additive manufacturing. *Chin. J. Aeronaut.* **2020**, *33*, 1252–1259. [CrossRef]
24. Zhu, J.H.; Zhou, H.; Wang, C.; Zhou, Z.; Yuan, S. A review of topology optimization for additive manufacturing: Status and challenges. *Chin. J. Aeronaut.* **2021**, *34*, 91–110. [CrossRef]
25. Özkal, F.M.; Uysal, H. General Aspects of Evolutionary Structural Optimization: A Review. *Pamukkale Univ. J. Eng. Sci.* **2009**, *15*, 383–393.
26. Saraçyakupoğlu, T. Usage of additive manufacturing and topology optimization process for weight reduction studies in the aviation industry. *Adv. Sci. Technol. Eng. Syst. J.* **2021**, *16*, 815–820. [CrossRef]
27. Sha, L.; Lin, A.; Zhao, X.; Kuang, S. A topology optimization method of lightweight robot design based on the finite element model of assembly and its applications. *Sci. Prog.* **2020**, *103*, 0036850420936482. [CrossRef] [PubMed]
28. Sun, Y.; Zong, C.; Pancheri, F.; Chen, T.; Lueth, T.C. Design of topology optimized compliant legs for bio-inspired quadruped robots. *Sci. Rep.* **2023**, *13*, 4875. [CrossRef] [PubMed]

29. Padilla-García, E.A.; Cruz-Morales, R.D.; González-Sierra, J.; Tinoco-Varela, D.; Lorenzo-Gerónimo, M.R. Design, assembly and control of a differential/omnidirectional mobile robot through additive manufacturing. *Machines* **2024**, *12*, 163. [CrossRef]
30. Arat, O.; Duran, A.; Erman, B.; Kibar, A. Structural Investigation of Pneumatic Grippers for Handling Automotive Parts Using Finite Element Analysis and Topology Optimization. *Kahramanmaraş Sütçü İmam Üniversitesi Mühendislik Bilim. Derg.* **2025**, *28*, 380–390. [CrossRef]
31. Liu, L.; Wang, X.; Yang, X.; Liu, H.; Li, J.; Wang, P. Path planning techniques for mobile robots: Review and prospect. *Expert Syst. Appl.* **2023**, *227*, 120254. [CrossRef]
32. Véras, L.G.D.; Medeiros, F.L.; Guimarães, L.N. Systematic literature review of sampling process in rapidly-exploring random trees. *IEEE Access* **2019**, *7*, 50933–50953. [CrossRef]
33. Karaman, S.; Frazzoli, E. Sampling-based algorithms for optimal motion planning. *Int. J. Robot. Res.* **2011**, *30*, 846–894. [CrossRef]
34. Gammell, J.D.; Barfoot, T.D.; Srinivasa, S.S. Batch Informed Trees (BIT*): Informed asymptotically optimal anytime search. *Int. J. Robot. Res.* **2020**, *39*, 543–567. [CrossRef]
35. Chen, L. Optimized Foothold Planning and Posture Searching for Energy-Efficient Quadruped Locomotion Over Challenging Terrains. In Proceedings of the IEEE International Conference on Robotics and Automation (ICRA), Paris, France, 31 May–August 2020; pp. 399–405. [CrossRef]
36. Raj, R.; Kos, A. An Optimized Energy and Time Constraints-Based Path Planning for the Navigation of Mobile Robots Using an Intelligent Particle Swarm Optimization Technique. *Appl. Sci.* **2023**, *13*, 9667. [CrossRef]
37. Chen, D.; Wang, Z.; Zhou, G.; Li, S. Path Planning and Energy Efficiency of Heterogeneous Mobile Robots Using Cuckoo–Beetle Swarm Search Algorithms with Applications in UGV Obstacle Avoidance. *Sustainability* **2022**, *14*, 15137. [CrossRef]
38. Chen, S.; Mo, Y.; Wu, X.; Xiao, J.; Liu, Q. Reinforcement Learning-Based Energy-Saving Path Planning for UAVs in Turbulent Wind. *Electronics* **2024**, *13*, 3190. [CrossRef]
39. Khan, N.A.; Hussain, S.; Jamwal, P. Reinforcement Learning-Driven Path Generation for Ankle Rehabilitation Robot Using Musculoskeletal-Informed Energy Optimization. *IEEE Trans. Neural Syst. Rehabil. Eng.* **2025**, *33*, 1774–1784. [CrossRef] [PubMed]
40. Özsoy, K.; Erçetin, A.; Çevik, Z.A. Comparison of mechanical properties of PLA and ABS based structures produced by fused deposition modelling additive manufacturing. *Avrupa Bilim ve Teknoloji Dergisi* **2021**, *27*, 802–809. [CrossRef]
41. Zhao, D.; Ding, Z.; Li, W.; Zhao, S.; Du, Y. Optimization of Smart Textiles Robotic Arm Path Planning: A Model-Free Deep Reinforcement Learning Approach with Inverse Kinematics. *Processes* **2024**, *12*, 156. [CrossRef]
42. Buriboev, A.; Choi, A.J.; Jeon, H.S. Optimized Frontier-Based Path Planning Using the TAD Algorithm for Efficient Autonomous Exploration. *Electronics* **2025**, *14*, 74. [CrossRef]
43. Chittoor, P.K.; Dandumahanti, B.P.M.A.; Konduri, S.; Samarakoon, S.M.B.P.; Elara, M.R. Payload- and Energy-Aware Tactical Allocation Loop-Based Path-Planning Algorithm for Urban Fumigation Robots. *Mathematics* **2025**, *13*, 950. [CrossRef]
44. Visca, M.; Bouton, A.; Powell, R.; Gao, Y.; Fallah, S. Conv1D Energy-Aware Path Planner for Mobile Robots in Unstructured Environments. In Proceedings of the 2021 IEEE International Conference on Robotics and Automation (ICRA), Xi'an, China, 30 May–5 June 2021; pp. 2279–2285. [CrossRef]
45. Pookkuttath, S.; Gomez, B.F.; Elara, M.R. RL-Based Vibration-Aware Path Planning for Mobile Robots' Health and Safety. *Mathematics* **2025**, *13*, 913. [CrossRef]
46. Lin, J.; Huang, Z.; Zhu, T.; Leng, J.; Huang, K. Deep Reinforcement Learning-Based Robotic Puncturing Path Planning of Flexible Needle. *Processes* **2024**, *12*, 2852. [CrossRef]
47. Al-Kamil, S.J.; Szabolcsi, R. Enhancing Mobile Robot Navigation: Optimization of Trajectories through Machine Learning Techniques for Improved Path Planning Efficiency. *Mathematics* **2024**, *12*, 1787. [CrossRef]

machines

MDPI

Article

Development of a Human-Centric Autonomous Heating, Ventilation, and Air Conditioning Control System Enhanced for Industry 5.0 Chemical Fiber Manufacturing

Madankumar Balasubramani [1], Jerry Chen [1], Rick Chang [2] and Jiann-Shing Shieh [1,*]

[1] Department of Mechanical Engineering, Yuan Ze University, Taoyuan 320, Taiwan; s1098706@mail.yzu.edu.tw (M.B.); s1088701@mail.yzu.edu.tw (J.C.)
[2] Far Eastern Fibertech Company Limited, Taoyuan 320, Taiwan; rick.chang@fefc.com.tw
* Correspondence: jsshieh@saturn.yzu.edu.tw

Abstract: This research presents an advanced autonomous HVAC control system tailored for a chemical fiber factory, emphasizing the human-centric principles and collaborative potential of Industry 5.0. The system architecture employs several functional levels—actuator and sensor, process, model, critic, fault detection, and specification—to effectively monitor and predict indoor air pressure differences, which are critical for maintaining consistent product quality. Central to the system's innovation is the integration of digital twins and physical AI, enhancing real-time monitoring and predictive capabilities. A virtual representation runs in parallel with the physical system, enabling sophisticated simulation and optimization. Development involved custom sensor kit design, embedded systems, IoT integration leveraging Node-RED for data streaming, and InfluxDB for time-series data storage. AI-driven system identification using Nonlinear Autoregressive with eXogenous inputs (NARX) neural network models significantly improved accuracy. Crucially, incorporating airflow velocity data alongside AHU output and past pressure differences boosted the NARX model's predictive performance (R^2 up to 0.9648 on test data). Digital twins facilitate scenario testing and optimization, while physical AI allows the system to learn from real-time data and simulations, ensuring adaptive control and continuous improvement for enhanced operational stability in complex industrial settings.

Keywords: Industry 5.0; autonomous control system; HVAC system; digital twins; physical AI; sensor kit development; embedded system; IoT; data visualization; artificial intelligence

1. Introduction

In the dynamic landscape of industrial development, Industry 5.0 represents the next evolutionary step beyond Industry 4.0, emphasizing human–machine collaboration and intelligent automation [1,2]. At its core, Industry 5.0 is driven by autonomous control systems—advanced technologies that enable machines to operate independently, making informed decisions with minimal human oversight [3]. This autonomy is made possible through the integration of artificial intelligence (AI), machine learning (ML), the Internet of Things (IoT), and advanced robotics. This research highlights how incorporating heating, ventilation, and air conditioning (HVAC) systems within this framework exemplifies the comprehensive and interconnected nature of Industry 5.0, enhancing efficiency, adaptability, and precision in industrial environments.

Industry 5.0, known as the "human-centered industrial revolution", extends the advancements of Industry 4.0 by prioritizing collaboration between humans and machines [4].

Unlike its predecessor, which focused on full automation and digitalization, Industry 5.0 integrates artificial intelligence (AI), robotics, the Internet of Things (IoT), and advanced analytics to enhance human capabilities rather than replace them. This approach promotes a more sustainable, efficient, and ethical industrial environment where humans and machines work together to optimize productivity, innovation, and flexibility in manufacturing.

Heating, ventilation, and air conditioning (HVAC) systems play a crucial role in industrial settings by regulating temperature, humidity, and air quality to ensure a comfortable and safe working environment [5–8]. These systems include heating components such as furnaces, boilers, and heat pumps, ventilation mechanisms for fresh air exchange and pollutant removal, and air conditioning units that control temperature and humidity [9]. In large-scale industrial operations, maintaining optimal air pressure balance is essential to prevent issues like moisture buildup, mold growth, and air contamination, all of which can impact both worker health and equipment efficiency [10].

Air velocity and pressure differences are closely interconnected, influencing HVAC system performance and industrial operations [11,12]. Wind speed results from pressure gradients, where higher differences create stronger airflow [13]. In HVAC systems, airflow velocity is determined by factors such as duct size, fan capacity, system resistance, and static pressure [14]. Effectively managing these variables ensures precise pressure control, reduces energy consumption, and enhances overall system efficiency.

The tools available for regulating air pressure differentials within chemical plants include sensor kits utilizing microcontrollers with I^2C communication, NTP synchronization, and OTA updates, alongside data collection and streaming platforms like Node-RED, data storage solutions such as InfluxDB, and visualization tools like Grafana and SVG nodes. The current stage of research, as presented, focuses on developing integrated autonomous control system architectures within the Industry 5.0 framework to predict and manage these differentials, employing AI techniques like NARX models and implementing retraining workflows based on error autocorrelation analysis to enhance prediction accuracy and maintain model fitness. The current state of development is demonstrated by the design and installation of such a sensor-based system in a chemical fiber factory, indicating that these advanced, human-centric autonomous control systems are moving towards practical implementation and validation in specific industrial environments.

In chemical fiber production, precise air pressure control is essential to maintaining consistent fiber quality and ensuring a safe working environment. During fiber extrusion, stable air pressure is required to regulate fiber diameter and structural integrity, while in cleanroom environments, controlled airflow prevents contamination and optimizes manufacturing conditions. Furthermore, proper ventilation reduces the accumulation of volatile chemicals and enhances worker safety.

This research focuses on developing an intelligent, sensor-based autonomous control system to regulate indoor air pressure differences in a chemical fiber factory. Designed within the framework of Industry 5.0, the system incorporates multiple functional layers—including actuators, sensors, process modelling, fault detection, and predictive analysis—to optimize human–machine collaboration. The system integrates IoT technology, embedded systems, and AI-driven analytics to enable real-time monitoring, data integration, and predictive maintenance. By enhancing automation while maintaining human oversight, the proposed system improves operational stability, energy efficiency, and overall productivity in industrial environments. The structure of this paper is as follows: Section 2 presents detailed background on HVAC systems, digital twins, physical AI, and human–machine decision support systems; Section 3 presents the materials and methods, including the design of the systems, performance testing, and experimental setup; Section 4 discusses the results and analysis, focusing on model prediction; and Section 5 concludes the paper with key findings.

2. Background and Related Works

The emergence of Industry 5.0 represents a pivotal shift in industrial innovation, marking a transition from automated processes to integrating advanced technology with human intelligence [15]. This goes beyond the digital and automated frameworks established by Industry 4.0, emphasizing a synergy between human expertise and innovative technology. While the early industrial revolutions replaced manual labor with steam and water-driven machinery in the 18th century, electrification and assembly lines in the 19th century introduced mass production [16,17]. The 20th century saw the rise of computer-driven automation, and now Industry 5.0 leverages artificial intelligence (AI), machine learning (ML), the Internet of Things (IoT), robotics, digital twins, and physical AI to enhance human capabilities. The focus is on sustainability, ethical manufacturing, and empowering workers.

At the heart of this new era are autonomous control systems, which enable machines to make real-time decisions with minimal human intervention [18,19]. These systems adopt a multi-layered approach, integrating sensor data, process modelling, fault detection, and predictive analytics to form robust frameworks for industrial automation [20]. Research has shown that such layered systems improve operational stability and foster human–machine collaboration, paving the way for more adaptable manufacturing environments [1,21].

Recent advancements in dynamic fault detection and predictive maintenance have made it possible to anticipate system anomalies using machine learning algorithms [22]. This allows industries to proactively adjust operational parameters, reducing downtime and improving system resilience. The integration of extensive sensor networks and IoT devices enables real-time data aggregation and analysis, making manufacturing systems more responsive and efficient, especially in complex industrial settings [23].

Additionally, advancements in embedded systems have led to the creation of flexible and scalable control platforms [24]. These platforms combine hardware and software to precisely regulate environments and adapt to fluctuating conditions. In fiber chemical manufacturing, for example, maintaining strict air pressure control is essential for product consistency and the safety of workers managing volatile chemicals.

Parallel to these technical improvements, research into human–machine collaboration stresses the importance of developing intuitive interfaces and decision-support systems [25]. These systems empower human operators to monitor, interact with, and fine-tune automated processes, striking a crucial balance between the efficiency of automation and human judgment [26]. This balance is key to optimizing production quality and fostering an environment where human expertise complements technological advancements.

The integration of digital twins and physical AI further enhances this collaboration. A digital twin is a virtual representation of a physical object, process, or system, allowing real-time monitoring, simulation, and optimization [27–29]. As they are continuously updated based on real-world data, digital twins provide insights into performance, potential issues, and opportunities for improvement [30,31]. In manufacturing, digital twins allow for predictive modelling, process optimization, and better decision-making, while enabling industries to evaluate different scenarios without the risks associated with physical trials.

Digital twins have gained increasing traction in the manufacturing sector due to their ability to create digital replicas of physical objects, processes, or systems. These models enable real-time data acquisition, analysis, and simulation, facilitating improvements in process optimization, predictive maintenance, and overall operational efficiency.

- Sustainable Manufacturing: A study by Kritzinger et al. (2018) highlights the critical role digital twins play in sustainable manufacturing by enabling the real-time monitoring of resource use, energy consumption, and material waste. By providing digital models of physical processes, manufacturers can optimize production lines and minimize resource consumption while maintaining product quality [32].
- Product Lifecycle Management (PLM): The literature also discusses how digital twins contribute to Product Lifecycle Management (PLM). Boschert and Rosen (2016) argue that digital twins can enhance product design and monitoring by linking physical and virtual entities. By simulating real-time behaviors, manufacturers can predict failures, optimize designs, and monitor the product from inception through to decommissioning [33].
- Predictive Maintenance: Grieves (2014) describes how digital twins facilitate predictive maintenance, allowing companies to anticipate and mitigate equipment failures before they occur. Digital twins continuously gather data from physical systems, providing insights that help predict when machinery needs maintenance, reducing downtime and prolonging the lifespan of expensive equipment [34].
- Quality Control: The research by Tao et al. (2018) shows that digital twins enable manufacturers to enhance product quality by creating virtual replicas of production lines, making it possible to adjust operational parameters in real time to prevent defects. These adjustments can be tested virtually, and only the most effective changes are applied to the physical system [35].

The concept of Physical AI merges artificial intelligence with physical systems, enabling machines to not only process information, but also interact and adapt to their environments. This technology empowers robots, autonomous machines, and IoT devices with the capability to learn from physical interactions, improving adaptability, precision, and operational efficiency.

- Autonomous Robots: The literature on autonomous robots highlights how physical AI enhances the performance of industrial robots. Bogue (2018) discusses the role of AI-powered robots in manufacturing, noting that they can perform tasks with high precision and adaptability [36].
- Collaborative Robotics (Cobots): Wang et al. (2019) explore the rise of collaborative robots (cobots) that collaborate with human workers.
- AI-Driven Process Optimization: The research by Heistracher et al. (2019) discusses the role of AI in optimizing manufacturing processes. By integrating AI with physical systems, manufacturers can create dynamic, self-optimizing processes that improve product quality, reduce waste, and enhance sustainability [37].
- Robust Fault Detection: Zhao et al. (2020) highlight how physical AI is integral to fault detection and diagnosis in manufacturing systems [38].

The combination of digital twins and physical AI creates an innovative and adaptive manufacturing ecosystem where digital models and physical systems work in unison. The following aspects highlight the synergy between these technologies and their application in modern manufacturing:

(1) Real-Time Feedback and Adaptive Control: The combination of digital twins and physical AI provides real-time feedback from physical systems to digital models [39].
(2) Process Optimization and Simulation: Digital twins can be used to simulate various scenarios and manufacturing parameters [39].
(3) Predictive Maintenance and Automated Repair: There is potential in combining digital twins with physical AI to enable autonomous predictive maintenance.

Physical AI, which combines AI with robotics and physical systems, brings a new level of intelligence to the manufacturing floor. It empowers machines and robots to not only interact with their environment, but also to understand and learn from it. This allows for the development of self-learning systems that can adapt to new tasks and evolving conditions in real time. Physical AI enables manufacturing systems to better anticipate challenges, optimize processes autonomously, and collaborate with human workers more effectively.

The cumulative effect of these innovations is clear—improved operational stability, reduced energy consumption, and enhanced product quality. As Industry 5.0 continues to evolve, the convergence of autonomous control systems, advanced sensor technology, AI-driven analytics, digital twins, and physical AI is poised to redefine industrial manufacturing. This integrated approach promises not only to increase efficiency and sustainability, but also to create resilient, adaptive production systems capable of meeting the dynamic demands of modern industry.

In a fiber chemical factory, for example, where environmental conditions play a crucial role in product quality, maintaining consistent air pressure is vital to stabilize operations and reduce downtime. This study aims to develop an intelligent, sensor-based autonomous control system to regulate air pressure differences, as shown in Figure 1. Adopting Industry 5.0 principles, the system integrates several layers of functionality, including actuator and sensor interfaces, process control, modeling, fault detection, and specification, to enhance human–machine collaboration. This approach involves designing sensor kits, deploying sensors, developing embedded systems, integrating IoT devices, and managing data through collection, integration, storage, and visualization techniques.

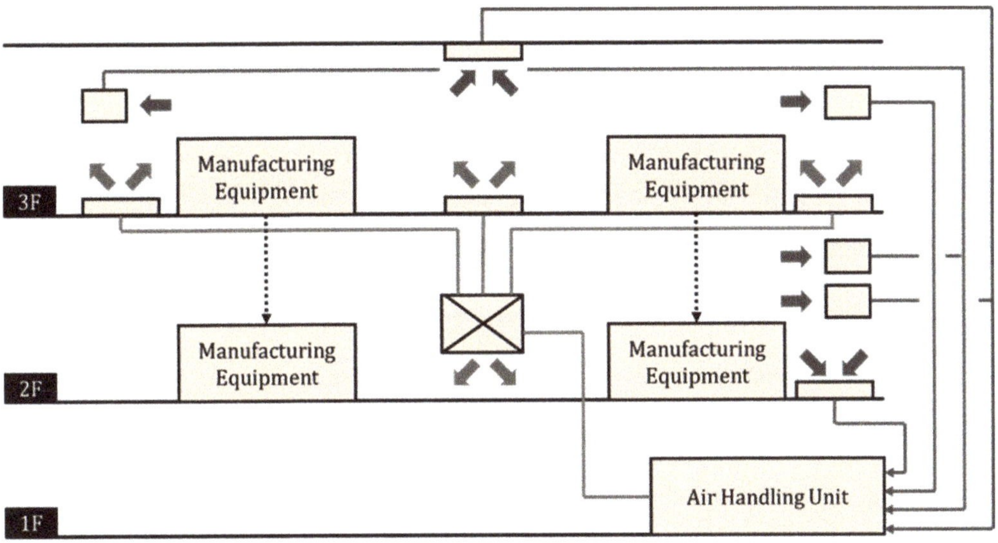

Figure 1. HVAC system airflow diagram.

In Figure 1, the blank squares likely represent different zones or measurement points on each floor of the factory, while the cross within a square most probably depicts a central air handling component such as an Air Handling Unit (AHU). The lines illustrate the ductwork, with the blue lines and arrows indicating the path of supply air from the central unit to the floors, and the red lines and arrows showing the return air flowing back, while the dotted lines connect the manufacturing equipment to the floors to show the areas influenced by the HVAC system.

The concept of autonomy in manufacturing is rooted in the broader fields of cyber–physical systems (CPSs) and the Internet of Things (IoT), which enable the integration of physical processes with computational capabilities. This allows for the real-time monitoring and control of dynamic production environments, leading to systems that can adapt to complex production demands. Key developments in autonomy include the following:

(1) Autonomous Production Control (APC): APC improves production performance by enabling fast, flexible responses to changes in the production environment. This approach decentralizes decision-making, transferring authority to distributed intelligent objects. While many studies on APC are theoretical, the practical applications are still limited.
(2) Maturity Models for Autonomy: These models assess the level of autonomy in manufacturing systems, offering a framework for evaluating capabilities and pinpointing areas for improvement. The five-level autonomy model identifies the features and technologies required for each level.
(3) Robotics: Autonomous robots are widely used in manufacturing for tasks such as assembly, welding, and material handling. These robots utilize advanced sensors and AI algorithms to perform tasks with high precision, enhancing production efficiency and reducing errors.
(4) Process Automation: Autonomous control systems can automate entire production processes. Closed-loop AI systems adjust production parameters in real time based on sensor data, ensuring optimal performance while reducing waste.
(5) Predictive Maintenance: By leveraging sensor data, autonomous systems can predict equipment failures and schedule maintenance proactively, leading to reduced downtime and increased equipment lifespan. Research demonstrates that predictive maintenance can result in significant cost savings.

Research on autonomous control systems continues to make strides in both theory and practice. While challenges remain, the potential benefits of enhanced efficiency, precision, safety, and cost savings make autonomous systems an asset in modern manufacturing. Our research focuses on overcoming technical barriers, addressing ethical considerations, and further developing autonomous systems to meet the evolving demands of the industry.

AI significantly enhances the capabilities of autonomous systems, driving innovation and efficiency across manufacturing processes. Some keyways AI influences these systems include the following:

(1) Predictive Maintenance: AI algorithms analyze sensor data to predict equipment failures, reducing downtime and extending machinery lifespan by scheduling proactive maintenance.
(2) Real-Time Monitoring and Control: AI systems process large volumes of sensor data in real time, making immediate adjustments to maintain optimal performance, ensuring smooth operations and higher product quality.
(3) Generative Design: AI-driven generative design tools optimize designs based on specific requirements, creating lighter, stronger, and more cost-effective components.
(4) Enhanced Robotics: AI improves autonomous robots' capabilities, enabling them to perform complex tasks with greater precision and adaptability. These robots can learn from their environment and improve over time, making them more versatile.
(5) Supply Chain Optimization: AI helps optimize supply chain management by predicting demand, managing inventories, and coordinating logistics. Autonomous systems with AI make real-time decisions, ensuring materials and products are available when needed.

(6) Quality Control: AI-powered systems monitor product quality, detecting defects and making necessary adjustments in real time, ensuring that products meet stringent standards.
(7) Human–Machine Collaboration: AI facilitates smoother collaboration between humans and machines, with autonomous systems managing repetitive or hazardous tasks, allowing human workers to focus on more complex and creative problem solving.
(8) Adaptive Manufacturing: AI enables autonomous systems to adjust production parameters in real time based on sensor data and external factors, ensuring manufacturing processes remain responsive to changes in demand.

AI is transforming manufacturing by enhancing autonomous systems, improving efficiency, precision, and adaptability. As AI technology continues to evolve, its impact on autonomous systems will only grow, enabling smarter, more efficient factories in the future. The integration of digital twins and physical AI will further enhance these capabilities, driving a new era of manufacturing that is not only more efficient, but also more resilient, sustainable, and adaptable. While we build upon established concepts, our original contributions lie in the specific development, integration, and application of a human-centric autonomous HVAC control system tailored for the unique demands of chemical fiber manufacturing within the Industry 5.0 framework. Specifically, our novel contributions include the following:

The design and implementation of a multi-layered autonomous control architecture integrating edge AI, digital twins, and physical AI for HVAC control.

The development of specific AI models (NARX) incorporating airflow velocity data, which significantly improved predictive accuracy for indoor air pressure differences in this specific industrial setting (achieving R^2 up to 0.9648 on test sets).

The design and deployment methodology for the sensor kits, including data collection via I^2C, NTP synchronization, and OTA updates.

A demonstration of this integrated system in a real-world chemical fiber factory setting, focusing on human–machine collaboration aspects inherent to Industry 5.0. We have revised Sections 1 and 5 to more explicitly highlight these original contributions and clearly differentiate our work from existing standards and protocols.

3. Methodology

Implementing an autonomous control system is a complex and detailed process that requires meticulous planning and precise execution. The initial phase involves designing the system by setting clear objectives, specifying necessary features, and selecting the appropriate technologies and algorithms. Following this, a reliable data collection platform must be developed, which includes choosing sensors and data acquisition systems that can efficiently and accurately gather the required data based on the system's operational environment. The next step focuses on data storage and management, where selecting suitable database systems and data warehousing technologies is crucial to handling large volumes of data securely and effectively. Finally, the process involves system identification for neural network time-series analysis, which includes developing and training neural network models to analyze time-series data, recognize patterns, forecast future events, and adapt operations accordingly. This comprehensive approach ensures the successful implementation of an autonomous control system that is both efficient and adaptive.

3.1. Autonomous Control System Design

The autonomous control system architecture described consists of several functional layers:

(1) Wind Sensors

These sensors measure wind speed and potentially other environmental factors. The raw data they generate is the foundation for the system, and it is transmitted to the next layer for processing.

(2) Edge AI Processing

Located close to the sensors, edge AI performs preliminary data processing tasks. It filters, aggregates, and applies lightweight AI algorithms to the raw data. By processing data locally, the system reduces network traffic, minimizes latency, and allows for faster decision-making. This layer ensures only relevant, processed data moves upstream to the digital twins and central server.

(3) Digital Twins

A digital twin is a virtual model of a physical system. Using the processed wind velocity data, the digital twin simulates real-world conditions, predicting future states (e.g., how wind velocity changes will affect airflow or temperature). The digital twin helps train AI models, refine predictions, and optimize system performance, ensuring better decision-making in future processes.

(4) Neural Network pool

The neural network pool aggregates data from the digital twins and edge devices. It is responsible for heavier computational tasks such as refining AI models, conducting large-scale analytics, and storing historical data. The server uses insights from the digital twins to optimize control strategies, ensuring that decisions based on these models improve system efficiency and accuracy.

(5) Factory Control System and HVAC System

The factory control system directly interacts with operational technology such as machinery, production lines, and other industrial equipment. It receives recommendations for control parameters (like airflow or temperature) from the digital twins and central server. The HVAC system, an example of a physical system, adjusts based on wind data and predictive models. These adjustments help optimize energy use, temperature regulation, and airflow. The feedback from these physical changes flows back to the sensors and edge AI, continually refining the system.

(6) Physical AI

This layer represents the real-world application of digital insights. It includes hardware, robotics, and other industrial processes that execute the optimizations recommended by the AI-driven system. Physical AI takes the models and predictions from previous layers and directly influences machinery, airflow, and other physical aspects of the factory. This layer completes the feedback loop, with adjustments made in real time, leading to continuous optimization of the system.

This study employs a sophisticated computational framework for wind velocity prediction, with a critical component being the Wind Analytics module, as depicted in Figure 2. This module serves as the central processing hub for raw wind data, transforming it into a structured and informative format suitable for training the predictive models within the neural network pool.

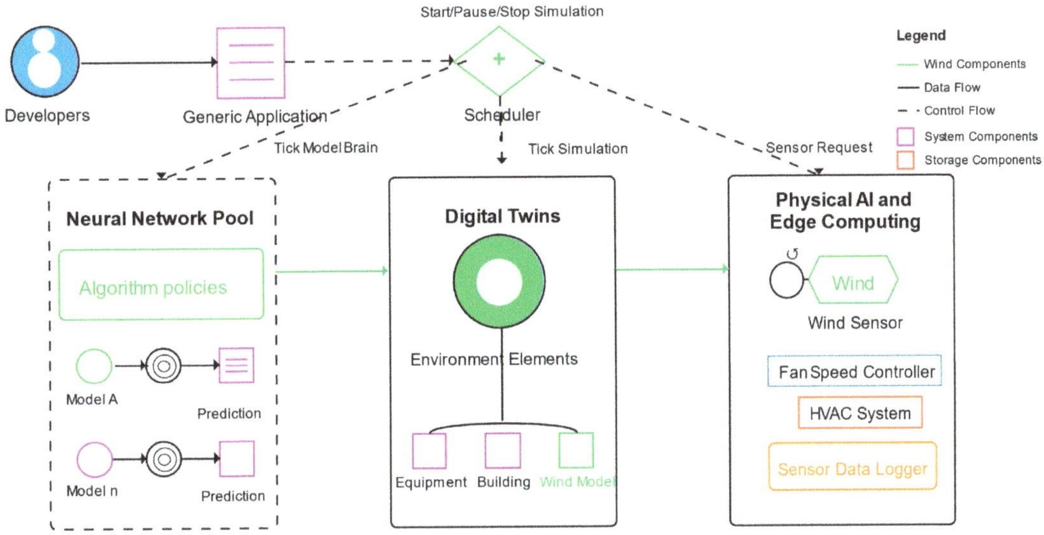

Figure 2. The autonomous control system architecture and its components.

Upon receiving wind data from the physical AI and edge computing units, the Wind Analytics module initiates a series of crucial data processing steps. Firstly, data ingestion involves receiving a stream of wind measurements, including parameters such as wind velocity, wind direction, and potentially turbulence intensity, depending on the capabilities of the deployed wind sensors.

Following ingestion, the module undertakes rigorous data preprocessing. This stage typically involves several sub-processes aimed at ensuring data quality and consistency. Data cleaning addresses potential issues such as erroneous or outlier readings, which might arise due to sensor malfunctions or transient environmental events. Techniques like statistical outlier detection or range checks may be employed to identify and manage these anomalies. Furthermore, managing missing values is a critical step, as gaps in the data stream can negatively impact model training. Imputation techniques, such as using temporal interpolation or statistical averages, might be applied to fill these gaps appropriately. Data normalization or standardization is often performed to scale the different wind parameters to a common range, which can improve the convergence and performance of the machine learning algorithms.

A key function of the Wind Analytics module is feature engineering. This involves deriving new, potentially more informative features from the raw and pre-processed data. For instance, temporal features such as rolling averages, lagged values of wind speed, or directional changes over time might be calculated. Spatial features could also be incorporated if data from multiple wind sensors are available, allowing the module to capture spatial patterns and gradients in wind velocity. The selection and creation of relevant features are crucial for enhancing the predictive power of the subsequent machine learning models.

The culmination of these processing steps is the generation of Training Data. This involves structuring the processed and engineered features into a format suitable for supervised learning. Typically, this involves creating pairs of input features (e.g., historical wind data, environmental parameters) and corresponding target variables (e.g., future wind velocity at a specific location or time horizon). The Training Data are carefully partitioned into training, validation, and potentially testing sets to facilitate effective model training, hyperparameter tuning, and unbiased performance evaluation.

Beyond these core functions, the Wind Analytics module may also incorporate various analytical techniques to gain deeper insights into the wind patterns. This could involve statistical analysis, such as calculating descriptive statistics, identifying correlations between different wind parameters and environmental factors, or performing spectral analysis to understand the dominant frequencies in wind fluctuations. These analyses can inform the feature engineering process and provide valuable context for interpreting the model predictions.

The Training Data generated by the Wind Analytics module are then directly fed into the neural network pool. This curated and informative dataset is essential for effectively training the various predictive models (Model A to Model n), enabling them to learn the complex relationships between historical wind data, environmental conditions, and future wind velocity. The quality and relevance of the Training Data produced by the Wind Analytics module are therefore paramount to the overall accuracy and reliability of the wind velocity predictions generated by the framework.

3.2. Data Collection and Storage

Creating a platform for data collection that integrates multiple data sources is a complex but essential task in modern data-driven environments. Such a platform must be capable of handling diverse data types from various sources and making these data accessible for analysis and decision-making. One effective tool for this purpose is Node-RED, an open-source visual programming tool developed by IBM Emerging Technology. Node-RED allows users to wire together devices, Application Programming Interfaces (APIs), and online services by creating flows in a visual manner. It provides a browser-based flow editor where nodes represent different functionalities or devices. Users can connect these nodes to create a flow that processes, transforms, and routes data between various devices and services without extensive coding.

Node-RED offers several advantages that make it a preferred choice for system development, especially in the fields of Internet of Things (IoT) and data integration. Its flow-based programming paradigm allows users to create workflows by connecting pre-built nodes, representing specific functions, inputs, outputs, or devices, through a drag-and-drop interface. This simplifies the development process and makes it accessible to both beginners and experienced developers. Additionally, Node-RED has an extensive library of nodes covering a wide range of functionalities, including data processing, hardware integration, social media APIs, cloud services, and more. This versatility enables the creation of complex applications with contributions from the community and developers.

One of Node-RED's strengths is its integration capabilities. It can seamlessly communicate with various systems and protocols, including MQTT, HTTP, TCP, UDP, WebSocket, databases, and IoT platforms. This interoperability facilitates the integration of different technologies, ensuring smooth data flow and communication. The browser-based interface of Node-RED further enhances its usability, allowing users to create, modify, and deploy flows using a visual editor. This accessibility streamlines the development process and makes it easier to manage and update the system.

For data storage, databases are essential as they provide structured collections of data organized to facilitate the efficient storage, retrieval, and manipulation of information. Databases serve as central repositories that store and manage vast amounts of structured, semi-structured, or unstructured data, providing a foundation for various applications and systems. Common classifications of databases include relational databases, NoSQL databases, and NewSQL databases. Relational databases, such as MySQL, PostgreSQL,

Oracle, and SQL Server, organize data into tables with rows and columns, establishing relationships between different tables through keys. They use Structured Query Language (SQL) for querying and managing data and offer ACID (Atomicity, Consistency, Isolation, Durability) properties to ensure data integrity and consistency.

NoSQL databases, such as MongoDB, Redis, DynamoDB, Cassandra, and Neo4j, encompass various non-relational database models designed for scalability, flexibility, and handling unstructured or semi-structured data. They excel in distributed environments and offer schema flexibility and horizontal scaling. NewSQL databases, like Google Spanner, CockroachDB, and VoltDB, combine aspects of relational databases with the scalability of NoSQL, addressing the limitations of traditional SQL databases concerning scalability and performance while retaining ACID compliance.

InfluxDB is a popular open-source time-series database designed to handle high write and query loads, making it particularly well suited for applications that require handling time-stamped data. Its design is optimized for time-series data, offering high performance in data ingestion and query processing, which is essential for applications relying on rapid data collection and real-time analysis. InfluxDB's query language is intuitively like SQL, making it accessible to individuals accustomed to traditional database languages. Its scalability allows it to accommodate growing data volumes and increasingly complex queries without compromising performance. Additionally, InfluxDB is user-friendly, offering straightforward setup and maintenance operation. These features make InfluxDB an ideal choice for research and applications requiring efficient and scalable time-series data management.

In summary, creating a data collection platform that integrates multiple data sources involves selecting the right tools and technologies to handle diverse data types, ensure seamless integration, and provide efficient data storage and management. Node-RED and InfluxDB are powerful tools that offer the necessary capabilities to build a robust and scalable data collection platform, facilitating real-time analysis and decision-making.

3.3. *System Identification*

System identification is a field of study in engineering that aims to build mathematical models of dynamic systems from the observed input–output data [40]. It involves inferring the underlying structure, behavior, and dynamics of a system based on observed inputs and outputs, enabling analysis, prediction, control, and optimization. The primary goal is to develop accurate and reliable models that allow for a better understanding and manipulation of complex systems.

Mathematical modeling for system identification employs various techniques, including differential equations, transfer functions, state-space models, and time-series models. Parameter estimation techniques, such as least squares estimation, maximum likelihood estimation, and Bayesian estimation, are used to estimate the parameters of these mathematical models based on observed data.

System identification relies on high-quality data and data preprocessing techniques to manage noise, outliers, and missing values. Complex models might also lead to overfitting; thus, model validation techniques like cross-validation are adopted to ensure generalizability and accuracy. To classify different complexities of systems, the following features are considered:

(1) Linear and Nonlinear Systems: Linear systems have linear relationships between inputs and outputs, whereas nonlinear systems have more complex, non-linear relationships.

(2) Time-Invariant and Time-Varying Systems: Time-invariant systems have properties that do not change over time, while time-varying systems exhibit variations or changes in their characteristics.
(3) Parametric and Nonparametric Methods: Parametric methods assume a specific model structure with defined parameters, while nonparametric methods do not make assumptions about the model structure.

System identification methodologies are often categorized into "White Box", "Grey Box", and "Black Box" approaches:

(1) White Box Models: These models are based on theoretical knowledge of the system. They are highly interpretable but can be limited by the accuracy of theoretical knowledge.
(2) Grey Box Models: These models combine theoretical knowledge with data-driven approaches, offering a balance between interpretability and flexibility.
(3) Black Box Models: These models are developed purely from input–output data, often relying on complex algorithms or artificial intelligence. They can capture complex dynamics but may lack interpretability and require large amounts of data.

Time-series analysis involves analyzing and predicting future values based on historical data points collected at regular intervals over time. Neural networks have emerged as powerful tools for handling time-series data due to their ability to model complex patterns and dependencies within sequential data. However, neural networks also face challenges, such as overfitting, interpretability, and the need for large datasets.

The Nonlinear AutoRegressive eXogenous (NARX) model is a dynamic nonlinear model used in time-series analysis and system identification. It extends the conventional AutoRegressive eXogenous (ARX) model by incorporating nonlinear functions of the input and output variables. In the NARX model, the current output depends not only on past inputs and outputs, but also on delayed and nonlinear transformations of these variables. This model is represented by a nonlinear function of lagged inputs and outputs, as well as potentially exogenous variables, allowing for the capture of complex relationships and nonlinear dynamics within the data.

Implementing NARX models involves data preprocessing, model selection, and training procedures. Data preprocessing includes feature engineering to extract meaningful information from time-series data. Model selection involves determining the optimal lag order and number of neurons, selecting relevant input variables, and choosing appropriate nonlinear functions or basis functions. Training the NARX model involves optimization techniques such as gradient-based optimization, least squares, or nonlinear optimization methods to estimate the model parameters.

The NARX model stands out for its adaptability and precision in handling complex data patterns, especially for nonlinear systems. The effectiveness of the black box approach lies in the ability to experiment with various input and output combinations, allowing for a comprehensive comparison and evaluation of the model's performance across a range of scenarios. By systematically varying the inputs, outputs, time delay, and number of neurons, it becomes possible to discern the most effective configuration for the model, thereby optimizing both accuracy and reliability in its predictions.

Autocorrelation is a foundational concept in time-series analysis, offering crucial insights into the relationship between observations over time. It measures the degree of similarity between a sequence of data points and its lagged versions, essentially examining how closely a data point relates to its past observations at different time intervals or lags. This analysis unveils patterns, dependencies, and structures inherent in sequential data, aiding in understanding trends, seasonality, and underlying temporal dynamics.

The autocorrelation function (ACF) quantifies the correlation between a time series and its lagged versions. Mathematically, for a time series (X_t) with observations at discrete time points (t), the autocorrelation at lag (k) is calculated using the following formula:

$$\mathrm{ACF}(k) = \frac{\sum_{t=k+1}^{n} \left(X_t - \overline{X}\right)\left(X_{t-k} - \overline{X}\right)}{\sum_{t=1}^{n} \left(X_t - \overline{X}\right)^2} \quad (1)$$

Here, n represents the total number of observations, k is the lag or time interval, X_t denotes the data point at time t, and the mean of the time series is represented by $\overline{X}$.

The ACF provides insights into the correlation structure of a time series across different lags. Positive autocorrelation at lag (k) indicates that data points separated by (k) time intervals are positively correlated, while negative autocorrelation suggests an inverse relationship. ACF values close to zero imply weak or no correlation at that lag. Detecting autocorrelation assists in model selection for time-series forecasting, such as in AutoRegressive Integrated Moving Average (ARIMA) models, where the ACF and partial autocorrelation function (PACF) guide the selection of model parameters. Understanding autocorrelation and its statistical analysis enables analysts to discern temporal relationships within data, aiding in model selection, validation, and forecasting.

A specific type of air velocity sensor was chosen to measure airflow velocity in various environments, and its specifications are listed in Table 1. To ensure accurate and reliable data collection, several designs for sensor deployment have been developed. These designs include sensor Printed Circuit Boards (PCBs) and sensor mounting devices, each tailored to the specific requirements of different deployment locations. For instance, some sensors may be mounted on walls, ceilings, or other structures depending on the area being monitored.

Table 1. General specifications of F660.

Objective	Spec.
Operating Temp.	0 °C to 60 °C
Storage Temp.	40 °C to 105 °C
Relative Humidity	5–95% (non-condensing)
Response Time	400 ms
Output	Configurable trip point
Communications	3.3 V I^2C (400 KHz) or 3.3 VDC UART
Housing Material	Nylon UL94-V0
Operating Voltage (Recommended)	12–24 VDC, (15–18 VDC)
Resolution	0.1 °C
Sensor Weight	2.5 g

The F660 sensor manufactured by DEGREE CONTROLS, INC, Nashua, New Hampshire, USA, operates within a temperature range of 0 °C to 60 °C and can be stored between −40 °C and 105 °C. It functions effectively in relative humidity levels of 5% to 95% (non-condensing), has a response time of 400 milliseconds, features a configurable trip point output, supports 3.3 V I^2C at 400 kHz or 3.3 VDC UART for communications, is housed in flame-retardant Nylon UL94-V0 material, operates ideally within a voltage range of 15–18 VDC (recommended 12–24 VDC), offers a resolution of 0.1 °C, and weighs 2.5 g, as shown in Table 1.

The air velocity accuracy within the compensation range is ±(5% of reading + 0.04 m/s) for air velocities between 0.15 m/s and 1.0 m/s, ±(5% of reading + 0.10 m/s) for air velocities between 0.5 m/s and 10 m/s, and ±(5% of reading + 0.15 m/s) for air velocities between 1.0 m/s and 20 m/s, as shown in Table 2.

Table 2. Air velocity accuracy.

Air Velocity Range	Air Velocity Accuracy (Within Compensation Range)
0.15 m/s to 1.0 m/s	±(5% of reading + 0.04 m/s)
0.5 m/s to 10 m/s	±(5% of reading + 0.10 m/s)
1.0 m/s to 20 m/s	±(5% of reading + 0.15 m/s)

Each sensor kit is equipped with a microcontroller that collects [21] multiple sensor readings using the Inter-Integrated Circuit (I^2C) protocol. The I^2C protocol is a popular communication standard that allows multiple sensors to communicate with the microcontroller efficiently. This setup ensures that data from various sensors can be gathered simultaneously and accurately.

To maintain synchronization and ensure the accuracy of the collected data, the microcontroller uses a Network Time Protocol (NTP) server. The NTP server provides precise timekeeping, allowing the microcontroller to timestamp each data reading accurately. This synchronization is crucial for analyzing trends and patterns in the airflow data over time.

Additionally, the microcontroller is equipped with an Over-The-Air (OTA) firmware update feature. This feature allows for remote updates of the microcontroller's firmware, ensuring that the system can be easily maintained and upgraded without the need for physical access. OTA updates are essential for version control and for implementing improvements or bug fixes in the system.

Once the data have been collected by the microcontroller, they are organized and processed for visualization. The data can be displayed in trend charts, which show changes in airflow velocity over time, or on customized web pages that provide a more detailed and interactive view of the data. These visualizations help users understand the airflow patterns and make informed decisions based on the collected data.

In summary, the deployment of air velocity sensors involves careful planning and design to ensure accurate data collection. The use of microcontrollers with I^2C communication, NTP synchronization, and OTA firmware updates enhances the reliability and maintainability of the system. The collected data are then visualized in various formats to provide valuable insights into airflow dynamics, as will be demonstrated in the following sections.

3.4. Data Selection and Preprocessing

For predicting indoor air pressure differences, the key features selected include AHU output, airflow velocity, and indoor air pressure differences. Each feature contains 66,475 sets of raw data with a time interval of 1 min. The data are summarized in Table 3.

Table 3. Modeling data.

Key Features	Unit
AHU Output	mmH_2O
Airflow Velocity	mm/s
Indoor Air Pressure Diff.	mmH_2O

Data preprocessing is crucial for improving data quality and making it suitable for analysis. The following techniques were applied:

(1) Filling Missing Data: Missing data were addressed using linear interpolation, which estimates values within the range of known data points. This method ensures the continuity and completeness of the dataset.

(2) Cleaning Outliers: Outliers were identified and managed using the moving median technique. This method is effective for time-series data as it is less sensitive to outliers compared to the moving average.

(3) Data Normalization: Z-score normalization was applied to adjust the range of data values to a common scale. The Z-score is calculated using the following formula:

$$Z = \frac{(X - \mu)}{\sigma}$$

where X is the original data point, μ is the mean, and σ is the standard deviation of the dataset. The dataset was segmented into six subsets, each containing 10,000 sets of data. Subset 1 was further divided into training (70%), validation (15%), and test (15%) sets.

Two model structures were compared: NAR (Nonlinear AutoRegressive) and NARX (Nonlinear AutoRegressive with eXogenous inputs). The NAR model uses past values of $(y(t))$ to predict $(y(t))$, while the NARX model uses past values of $(y(t))$ and another series $(x(t))$ to predict $(y(t))$, as shown in Figures 3 and 4.

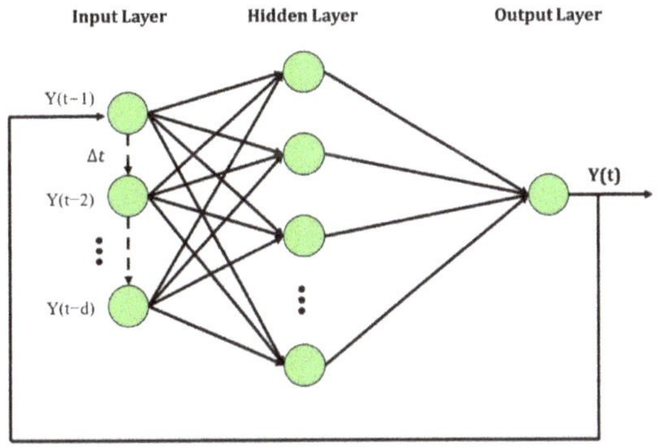

Figure 3. Structure of NAR model (d represents the time delay).

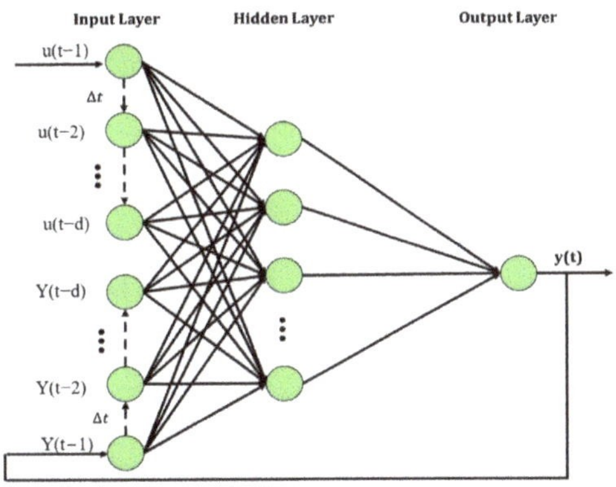

Figure 4. Structure of NARX model (d represents the time delay).

Model training involves iterating different parameters to minimize the loss function, which quantifies the accuracy of the model's predictions. The performance of each model was assessed based on the average (R^2) value, calculated from training the model five times.

- Model A: Uses past values of indoor air pressure differences to predict future values with an NAR structure. Optimal performance was achieved with a layer size of 20 and a time delay of 30.
- Model B: Uses past values of AHU output and indoor air pressure differences to predict future values with an NARX structure. Optimal performance was achieved with a layer size of 10 and a time delay of 5. It is important to clarify that Section 3 presents a conceptual framework intended to guide the design and development of the autonomous control system, rather than describing a fully implemented system with a complete operational interface. Our objective at this stage was to propose a generalizable architecture and control logic to support future implementations. For future development, we believe Node-RED would be a suitable platform for creating a user-friendly interface that supports real-time monitoring, control input, and system visualization.

This comprehensive approach ensures accurate and reliable predictions of indoor air pressure differences, leveraging advanced data preprocessing and model training techniques. While the system is designed for autonomous air pressure regulation, its human-centric nature, central to Industry 5.0, is realized through several integrated features: the "specification" level allows human operators to define system objectives and requirements, ensuring that the autonomous operation aligns with human needs and priorities; the real-time monitoring web application and detailed data visualization tools provide operators with transparency into the system's performance and environmental conditions, enabling informed oversight, trust, and the ability to intervene or adjust specifications; moreover, by maintaining the precise environmental control crucial for product quality and worker safety, the system directly enhances the human work environment, acting as a collaborative partner that augments human capabilities and well-being rather than merely automating tasks.

4. Results and Discussion

This study investigates the prediction of indoor air pressure differences using various machine learning models, focusing on the impact of predictor selection and model configuration. The core objective is to minimize the loss function by optimizing model parameters, thereby enhancing prediction accuracy. Specifically, this involved tuning the structural configuration of the neural network models by varying the layer size (number of neurons in the hidden layer) and the temporal configuration by adjusting the time delay (number of past time steps considered).

Four model approaches were evaluated:

- Model A (NAR): Utilized past indoor air pressure differences to predict future values.
- Model B (NARX): Incorporated past AHU output and past indoor air pressure differences.
- Model C (NARX): Employed past airflow velocity and past indoor air pressure differences.
- Model D (NARX): Combined past AHU output, past airflow velocity, and past indoor air pressure differences.

Model performance was assessed using the average R^2 value across five training iterations. The impact of layer size and time delay was systematically examined for each model.

(1) Model A

Model A directly uses past values of indoor air pressure differences to predict indoor air pressure differences with an NAR model structure. Model A had a better performance when the layer size was set to 20 and time delay was set to 30, as shown in Tables 4 and 5.

Table 4. Model A: Performance evaluation across layer-size iterations.

Layer Size	Time Delay	Ave. of R^2
10	5	0.8819
5	5	0.8814
15	5	0.8831
20	5	0.8835

Table 5. Model A: Performance evaluation across time-delay iterations.

Layer Size	Time Delay	Ave. of R^2
10	5	0.8819
10	15	0.8864
10	30	0.8873
10	45	0.8803
10	60	0.8786

(2) Model B

Model B employs an NARX model structure, using past values of AHU outputs and past values of indoor air pressure differences to predict indoor air pressure differences. As indicated in Tables 6 and 7, Model B demonstrated improved performance with the layer size configured at 10 and the time delay set to 5.

Table 6. Model B: Performance evaluation across layer-size iterations.

Layer Size	Time Delay	Ave. of R^2
10	5	0.8876
5	5	0.8831
15	5	0.8823
20	5	0.8804

Table 7. Model B: Performance evaluation across time-delay iterations.

Layer Size	Time Delay	Ave. of R^2
10	5	0.8876
10	15	0.8829
10	30	0.8720
10	45	0.8693
10	60	0.8645

(3) Model C

Model C adopts an NARX model structure, utilizing previous airflow velocity and previous indoor air pressure differences to predict indoor air pressure differences. Tables 8 and 9 show that Model C's performance was enhanced when it was configured with a layer size of 10 and a time delay of 5.

Table 8. Model C: Performance evaluation across layer-size iterations.

Layer Size	Time Delay	Ave. of R^2
10	5	0.8851
5	5	0.8880
15	5	0.8834
20	5	0.8857

Table 9. Model C: Performance evaluation across time-delay iterations.

Layer Size	Time Delay	Ave. of R^2
10	5	0.8851
10	15	0.8833
10	30	0.8764
10	45	0.8731
10	60	0.8596

(4) Model D

Model D uses past values of AHU output, past values of airflow velocity, and past values of indoor air pressure differences to predict indoor air pressure differences with an NARX model structure. Tables 10 and 11 reveal that Model D exhibited enhanced performance when set with a layer size of 10 and a time delay of 5.

Table 10. Model D: Performance evaluation across layer-size iterations.

Layer Size	Time Delay	Ave. of R^2
10	5	0.9575
5	5	0.9644
15	5	0.9632
20	5	0.9631

Table 11. Model D: Performance evaluation across time-delay iterations.

Layer Size	Time Delay	Ave. of R^2
10	5	0.9575
10	15	0.9627
10	30	0.9602
10	45	0.9593
10	60	0.9550

By integrating the optimal configurations for both layer size and time delay, as determined by the R^2 value, the top model candidates are enumerated in Table 12.

Table 12. Top model candidates.

Model #	Layer Size	Time Delay	Ave. of R^2
A	20	30	0.8827
B	10	5	0.8876
C	5	5	0.8880
D	5	15	0.9648

Next, we used an additional test set (i.e., subset 2 to 5) to check the performance of the above models. The results are shown in Table 13.

Table 13. Additional test for models.

Model #	Subset 2	Subset 3	Subset 4	Subset 5
		Ave. of R^2		
A	0.7511	0.7818	0.6822	0.2583
B	0.7447	0.7912	0.6860	0.2765
C	0.7392	0.7933	0.6473	0.2693
D	0.9306	0.8310	0.9555	0.9399

Involving airflow velocity in the prediction models significantly enhanced their accuracy and contributed to more consistent performance across future data subsets. The predicted responses from Model D for subsets 2 to 5 are shown in Figures 5–8. These figures display the actual performance of Model D, which incorporates airflow velocity data, on four separate test datasets (subsets 2–5). In the upper plot of each figure, the blue dots represent the real, measured indoor air pressure differences (the "Targets"). The orange line shows the values predicted by Model D (the "Outputs"). Observing how closely the orange line follows the pattern of the blue dots across the entire time series in each figure directly illustrates the model's predictive accuracy on that specific subset of unseen data. A tight alignment between the orange line and the blue dots, as is generally visible in these figures, indicates that the model's predictions closely match the actual measured values. The lower plot in each figure shows the "Error", which is the difference between the actual target and the predicted output. The distribution of these error points around the zero line indicates the magnitude and variability of the prediction errors. The fact that most error points are relatively close to the zero line across all four subsets visually supports the quantitative R^2 values and demonstrates that Model D is able to generalize its predictive capability effectively to new, unseen data, reinforcing the benefit of including airflow velocity as a predictor.

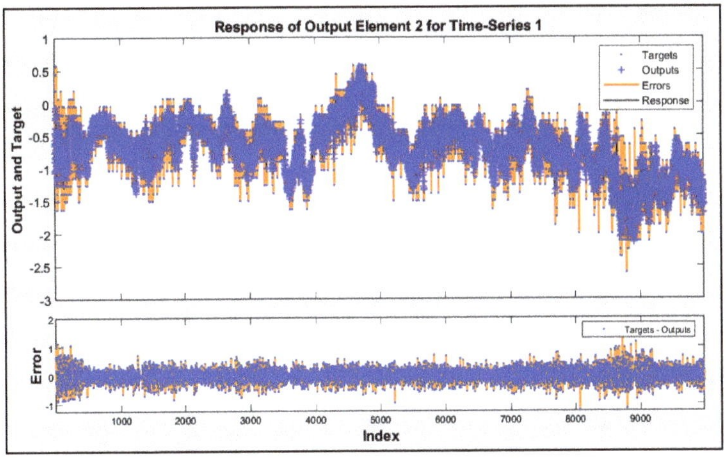

Figure 5. Time-series response analysis of subset 2.

Autocorrelation of residuals is a statistical technique used in various fields, particularly in time-series analysis. Its primary purpose is to assess whether the residuals (or errors) from a model exhibit autocorrelation. Autocorrelation indicates a systematic pattern of correlation between the error terms at different time points or lags. When the prediction error is obviously autocorrelated, then the prediction error is dependent on the time series.

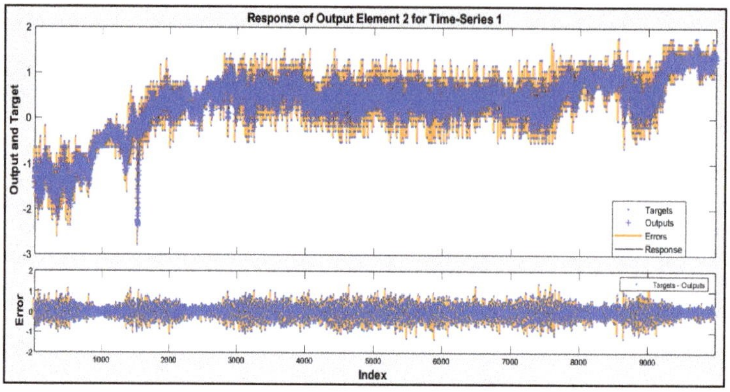

Figure 6. Time-series response analysis of subset 3.

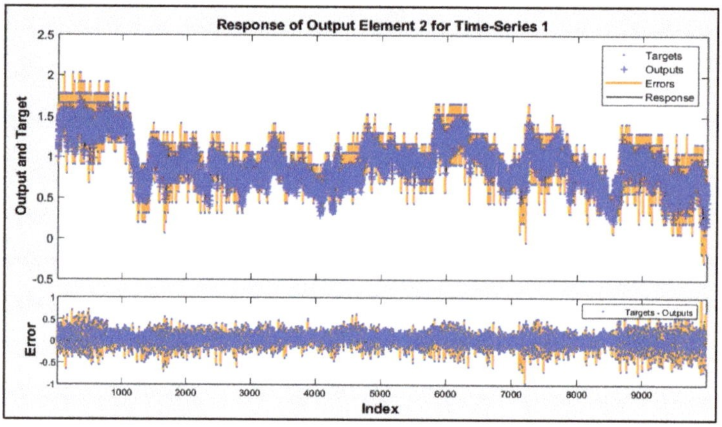

Figure 7. Time-series response analysis of subset 4.

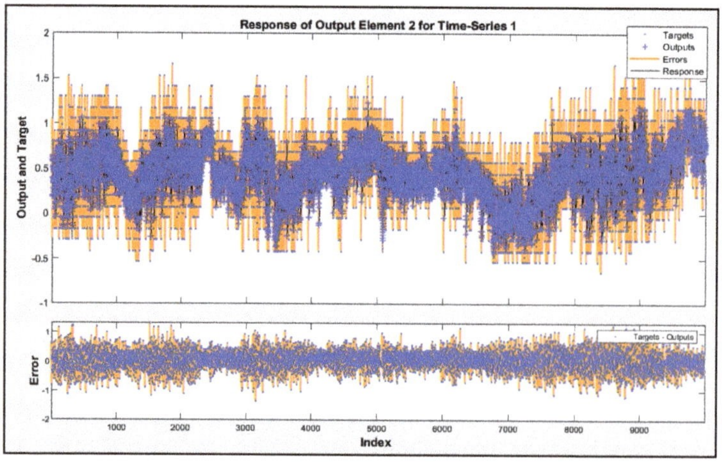

Figure 8. Time-series response analysis of subset 5.

The autocorrelation plots for the errors of Model D from subsets 1 to 5 are shown in Figure 9a–e. The plots' horizontal axes depict lag values, signifying the time interval by which the error terms are compared. For instance, lag 1 compares consecutive error terms, lag 2 assesses error terms separated by two time points, and so forth. On the other hand, the vertical axis on the plot displays autocorrelation coefficients, quantifying the magnitude and direction of correlation among error terms at the chosen lag. For a perfect prediction model, there should only be one non-zero value of the autocorrelation function that occurs at zero lag. This occurrence signifies that the prediction errors are entirely uncorrelated, resembling white noise.

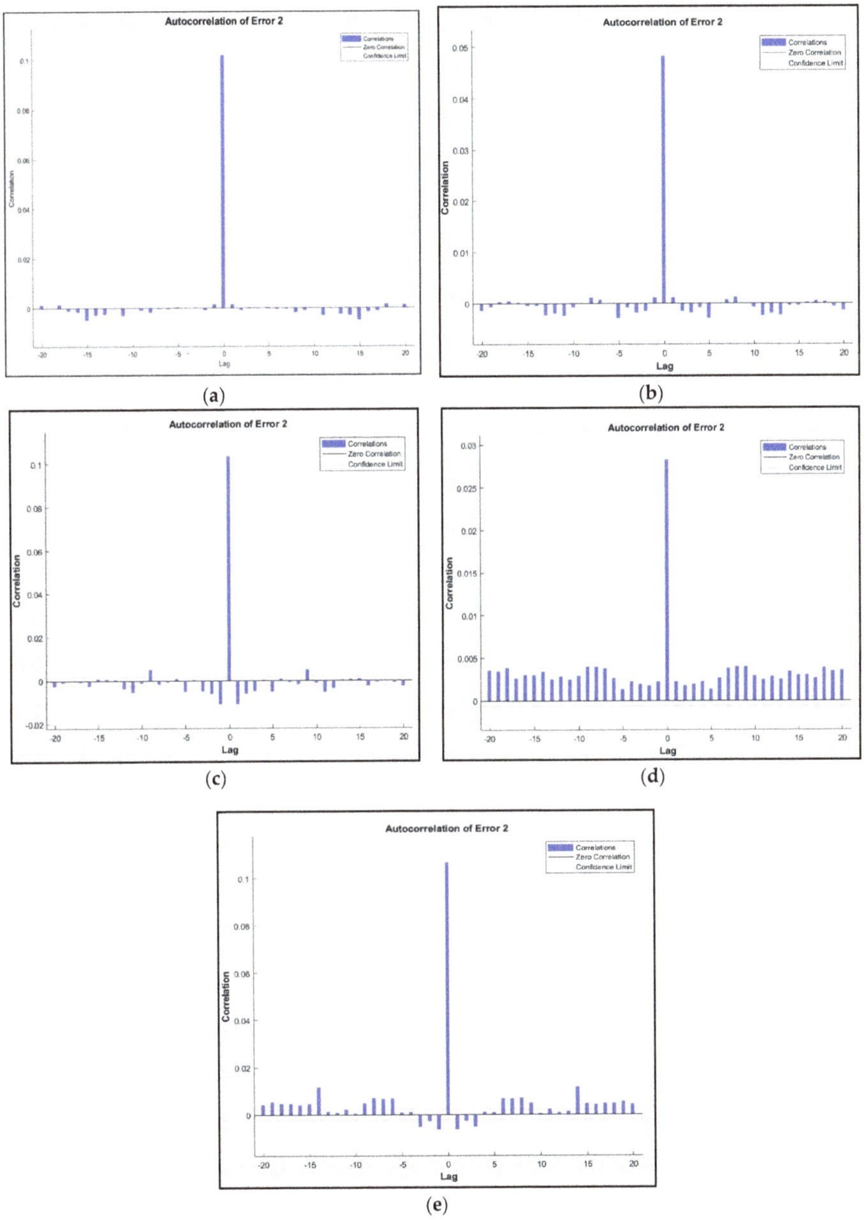

Figure 9. Autocorrelation plot for subsets: (**a**) subset 1; (**b**) subset 2; (**c**) subset 3; (**d**) subset 4; (**e**) subset 5.

The autocorrelation plot above demonstrates that as we move further back in time, within-dataset errors become more correlated. This suggests that there is an opportunity for improvement through a retraining process, potentially by adjusting the layer size or increasing the number of time delays in the model settings.

The implementation of autonomous control systems has significantly improved operational efficiency by reducing the need for experts to perform repetitive tasks. However, these systems face a critical challenge known as model decay, where their performance diminishes over time due to changes in the environment or data. Model decay occurs primarily due to concept drift (when the meaning of data changes over time) and data drift (when models trained on specific data periods fail to perform as well on data from other periods).

To address model decay, strategies like error autocorrelation analysis can help identify performance deviations, prompting the need for model retraining. This analysis acts as a "critic", triggering a retraining process when correlated errors indicate that a model is no longer performing optimally. Additionally, fault detection mechanisms are crucial in preventing model decay, particularly by identifying and correcting sensor deviations, ensuring that the model receives accurate data inputs and maintains reliability.

In the future, integrating a neural network controller into the autonomous system could improve its ability to adapt to real-time data and dynamic environments. Unlike traditional simulation models, prediction models use current data to forecast the system's response in the next time step, which could be beneficial for fine-tuning the controller's performance. Neural network predictive controllers could optimize plant performance by predicting future behavior and adjusting control inputs accordingly. Testing and optimizing such controllers would require either a real plant with a prediction model or a fully functional simulation model to ensure their effectiveness in actual operational conditions.

5. Conclusions

This study successfully developed and implemented a novel human-centric autonomous control system specifically tailored for an HVAC system within a chemical fiber manufacturing facility, operating within the principles of Industry 5.0. The core objective was to accurately predict indoor air pressure differences, a critical parameter for maintaining product quality and worker safety in this specific industrial environment. The system's architecture is multi-layered, designed to facilitate seamless data acquisition from distributed sensor kits, efficient data management and streaming, and intuitive visualization for human operators [41]. This integrated approach underscores Industry 5.0's focus on augmenting human capabilities through advanced technology and fostering effective human–machine collaboration. A significant contribution of this research lies in the application of AI-driven system identification using Nonlinear AutoRegressive with eXogenous inputs (NARX) models. Through rigorous analysis, it was empirically demonstrated that the inclusion of airflow velocity data as an exogenous input significantly improved the predictive accuracy of the model for indoor air pressure differences. This finding is particularly relevant for industrial settings where environmental factors like air movement play a crucial role, highlighting the importance of selecting appropriate input variables for data-driven models to achieve high performance. Furthermore, the implemented system incorporates a vital retraining workflow designed to counteract the phenomenon of model decay, which is a common challenge in dynamic industrial environments where conditions can change over time. This workflow currently utilizes error autocorrelation analysis as a trigger mechanism, prompting the system to retrain the predictive model when the prediction errors exhibit a systematic pattern, indicating that the model is no longer accurately capturing the system's dynamics. While this method ensures ongoing model fitness, a recognized limitation is the potential for delays in detecting subtle shifts or the need for more sophisticated triggers that can anticipate decay more proactively. Building upon the success of this framework, future research will focus on further enhancing the system's capabilities. A key direction involves integrating a neural network controller. This integration is envisioned to leverage accurate predictions from the NARX model to enable more dynamic and precise real-time adjustments to the HVAC system, thereby optimizing performance and maintaining desired air pressure differentials more effectively. The successful implementation and validation of such a controller will require comprehensive testing, either within the operational chemical fiber factory environment or through the development of highly accurate and representative simulation models. This continued development is crucial for advancing towards the creation of truly resilient, adaptive, and highly efficient industrial control systems that fully embody the potential of Industry 5.0.

Author Contributions: Conceptualization, M.B. and J.-S.S.; methodology, J.C. and R.C.; validation, M.B. and J.C.; formal analysis, R.C.; investigation, M.B. and J.C.; resources, J.-S.S.; data curation, J.C.; writing—original draft preparation, M.B.; writing—review and editing, J.-S.S. and R.C.; visualization, J.C.; supervision, J.-S.S.; project administration, J.-S.S. All authors have read and agreed to the published version of the manuscript.

Funding: This study was financially supported by the National Science and Technology Council, Taiwan (grant number: NSTC 112-2221-E-155-001).

Data Availability Statement: The original contributions presented in this study are included in the article. Further inquiries can be directed to the corresponding author.

Conflicts of Interest: The author Rick Chen was employed by the Far Eastern Fibertech Company Limited. The remaining authors declare that the research was conducted in the absence of any commercial or financial relationships that could be construed as a potential conflict of interest.

References

1. Adel, A. Unlocking the future: Fostering human–machine collaboration and driving intelligent automation through industry 5.0 in smart cities. *Smart Cities* **2023**, *6*, 2742–2782. [CrossRef]
2. Verma, R.C.; Chaurasiya, S.; Kumar, A. Human-Machine Collaboration in Smart Factories: A Transition from Industry 4.0 to Industry 5.0. *Int. J. Cult. Stud. Soc. Sci.* **2024**, *20*, 149–158.
3. Martini, B.; Bellisario, D.; Coletti, P. Human-centered and sustainable artificial intelligence in industry 5.0: Challenges and perspectives. *Sustainability* **2024**, *16*, 5448. [CrossRef]
4. Yang, Y.; Keivanpour, S.; Imbeau, D. The Future of Disassembly Planning: A Critical Assessment of Industry 5.0, Lean, and X-Reality. In Proceedings of the 2023 International Conference on Electrical, Computer and Energy Technologies (ICECET), Cape Town, South Africa, 16–17 November 2023; IEEE: Piscataway, NJ, USA; pp. 1–12.
5. Elsaid, A.M.; Mohamed, H.A.; Abdelaziz, G.B.; Ahmed, M.S. A critical review of heating, ventilation, and air conditioning (HVAC) systems within the context of a global SARS-CoV-2 epidemic. *Process Saf. Environ. Prot.* **2021**, *155*, 230–261. [CrossRef] [PubMed]
6. Alam, M.A.; Kumar, R.; Yadav, A.S.; Arya, R.K.; Singh, V. Recent developments trends in HVAC (heating, ventilation, and air-conditioning) systems: A comprehensive review. *Mater. Today Proc.* **2023**, *in press*. [CrossRef]
7. Yao, Y.; Shekhar, D.K. State of the art review on model predictive control (MPC) in Heating Ventilation and Air-conditioning (HVAC) field. *Build. Environ.* **2021**, *200*, 107952. [CrossRef]
8. Bhat, S. Building Thermal Comforts with Various HVAC Systems and Optimum Conditions. *Int. J. Multidiscip. Res. Sci. Eng. Technol.* **2024**, *7*, 14845–14852.
9. Shi, Y.; Liu, M.; Fang, F. *Combined Cooling, Heating, and Power Systems: Modeling, Optimization, and Operation*; Wiley: Hoboken, NJ, USA, 2017.
10. Lopez-Carreon, I.; Jahan, E.; Yari, M.H.; Esmizadeh, E.; Riahinezhad, M.; Lacasse, M.; Xiao, Z.; Dragomirescu, E. Moisture Ingress in Building Envelope Materials: (II) Transport Mechanisms and Practical Mitigation Approaches. *Buildings* **2025**, *15*, 762. [CrossRef]

11. Okochi, G.S.; Yao, Y. A review of recent developments and technological advancements of variable-air-volume (VAV) air-conditioning systems. *Renew. Sustain. Energy Rev.* **2016**, *59*, 784–817. [CrossRef]
12. Asim, N.; Badiei, M.; Mohammad, M.; Razali, H.; Rajabi, A.; Chin Haw, L.; Jameelah Ghazali, M. Sustainability of heating, ventilation and air-conditioning (HVAC) systems in buildings—An overview. *Int. J. Environ. Res. Public Health* **2022**, *19*, 1016. [CrossRef]
13. Larsen, T.S.; Heiselberg, P. Single-sided natural ventilation driven by wind pressure and temperature difference. *Energy Build.* **2008**, *40*, 1031–1040. [CrossRef]
14. Kimla, J. Optimized Fan Control in Variable Air Volume HVAC Systems Using Static Pressure Resets: Strategy Selection and Savings Analysis. Master's Thesis, Texas A&M University, College Station, TX, USA, 2010.
15. Zizic, M.C.; Mladineo, M.; Gjeldum, N.; Celent, L. From industry 4.0 towards industry 5.0: A review and analysis of paradigm shift for the people, organization and technology. *Energies* **2022**, *15*, 5221. [CrossRef]
16. Xu, M.; David, J.M.; Kim, S.H. The fourth industrial revolution: Opportunities and challenges. *Int. J. Financ. Res.* **2018**, *9*, 90–95. [CrossRef]
17. De Vries, J. The industrial revolution and the industrious revolution. *J. Econ. Hist.* **1994**, *54*, 249–270. [CrossRef]
18. Verma, A. Introduction to autonomous systems. In Handbook of Power Electronics in Autonomous and Electric Vehicles, Elsevier: Amsterdam, The Netherlands, 2024; pp. 17–28.
19. Arya, S.S.; Dias, S.B.; Jelinek, H.F.; Hadjileontiadis, L.J.; Pappa, A.-M. The convergence of traditional and digital biomarkers through AI-assisted biosensing: A new era in translational diagnostics? *Biosens. Bioelectron.* **2023**, *235*, 115387. [CrossRef]
20. Pansare, P. Smart Cities Using Green Computing: A Citizen Data Scientist Perspective. In *Green Computing for Sustainable Smart Cities*; CRC Press: Boca Raton, FL, USA, 2024; pp. 341–368.
21. Mourtzis, D.; Angelopoulos, J. Artificial intelligence for human–cyber-physical production systems. In *Manufacturing from Industry 4.0 to Industry 5.0*; Elsevier: Amsterdam, The Netherlands, 2024; pp. 343–378.
22. Omol, E.; Mburu, L.; Onyango, D. Anomaly detection in IoT sensor data using machine learning techniques for predictive maintenance in smart grids. *Int. J. Sci. Technol. Manag.* **2024**, *5*, 201–210. [CrossRef]
23. Syafrudin, M.; Alfian, G.; Fitriyani, N.L.; Rhee, J. Performance analysis of IoT-based sensor, big data processing, and machine learning model for real-time monitoring system in automotive manufacturing. *Sensors* **2018**, *18*, 2946. [CrossRef] [PubMed]
24. Pereira, C.E.; Carro, L. Distributed real-time embedded systems: Recent advances, future trends and their impact on manufacturing plant control. *Annu. Rev. Control* **2007**, *31*, 81–92. [CrossRef]
25. Wanner, J.P. Artificial Intelligence for Human Decision-Makers: Systematization, Perception, and Adoption of Intelligent Decision Support Systems in Industry 4.0. Ph.D. Dissertation, Universität Würzburg, Würzburg, Germany, 2022.
26. Newman, D.; Blanchard, O. *Human/Machine: The Future of Our Partnership with Machines*; Kogan Page Publishers: London, UK, 2019.
27. Javaid, M.; Haleem, A.; Suman, R. Digital twin applications toward industry 4.0: A review. *Cogn. Robot.* **2023**, *3*, 71–92. [CrossRef]
28. Jiang, H.; Qin, S.; Fu, J.; Zhang, J.; Ding, G. How to model and implement connections between physical and virtual models for digital twin application. *J. Manuf. Syst.* **2021**, *58*, 36–51. [CrossRef]
29. Nikolakis, N.; Alexopoulos, K.; Xanthakis, E.; Chryssolouris, G. The digital twin implementation for linking the virtual representation of human-based production tasks to their physical counterpart in the factory-floor. *Int. J. Comput. Integr. Manuf.* **2019**, *32*, 1–12. [CrossRef]
30. Sharma, A.; Kosasih, E.; Zhang, J.; Brintrup, A.; Calinescu, A. Digital Twins: State of the art theory and practice, challenges, and open research questions. *J. Ind. Inf. Integr.* **2022**, *30*, 100383. [CrossRef]
31. Omrany, H.; Al-Obaidi, K.M.; Husain, A.; Ghaffarianhoseini, A. Digital twins in the construction industry: A comprehensive review of current implementations, enabling technologies, and future directions. *Sustainability* **2023**, *15*, 10908. [CrossRef]
32. Kritzinger, W.; Karner, M.; Traar, G.; Henjes, J.; Sihn, W. Digital Twin in manufacturing: A categorical literature review and classification. *Ifac-Pap.* **2018**, *51*, 1016–1022. [CrossRef]
33. Boschert, S.; Rosen, R. Digital twin—The simulation aspect. In *Mechatronic Futures: Challenges and Solutions for Mechatronic Systems and Their Designers*; Springer: Cham, Switzerland, 2016; pp. 59–74.
34. Grieves, M. Digital twin: Manufacturing excellence through virtual factory replication. *White Pap.* **2014**, *1*, 1–7.
35. Tao, F.; Cheng, J.; Qi, Q.; Zhang, M.; Zhang, H.; Sui, F. Digital twin-driven product design, manufacturing and service with big data. *Int. J. Adv. Manuf. Technol.* **2018**, *94*, 3563–3576. [CrossRef]
36. Bogue, R. Exoskeletons–a review of industrial applications. *Ind. Robot Int. J.* **2018**, *45*, 585–590. [CrossRef]
37. Heistracher, C.; Wachsenegger, A.; Weißenfeld, A.; Casas, P. MOXAI–Manufacturing Optimization through Model-Agnostic Explainable AI and Data-Driven Process Tuning. In Proceedings of the PHM Society European Conference, Prague, Czech Republic, 3–5 July 2024; p. 7.
38. Zhao, Z.; Wu, J.; Li, T.; Sun, C.; Yan, R.; Chen, X. Challenges and opportunities of AI-enabled monitoring, diagnosis & prognosis: A review. *Chin. J. Mech. Eng.* **2021**, *34*, 56.

39. Groshev, M.; Guimaraes, C.; Martín-Pérez, J.; de la Oliva, A. Toward intelligent cyber-physical systems: Digital twin meets artificial intelligence. *IEEE Commun. Mag.* **2021**, *59*, 14–20. [CrossRef]
40. Keesman, K.J. *System Identification: An Introduction*; Springer Science & Business Media: Berlin/Heidelberg, Germany, 2011.
41. Tian, G.; Zhang, L.; Fathollahi-Fard, A.M.; Kang, Q.; Li, Z.; Wong, K.Y. Addressing a collaborative maintenance planning using multiple operators by a multi-objective Metaheuristic algorithm. *IEEE Trans. Autom. Sci. Eng.* **2023**, *22*, 606–618. [CrossRef]

machines

MDPI

Review

Electronic Artificial Intelligence and Digital Twins in Industry 5.0: A Systematic Review and Perspectives

Alessandro Massaro

Department of Engineering, LUM-Libera Università Mediterranea "Giuseppe Degennaro", S.S. 100-Km. 18, Parco il Baricentro, 70010 Bari, Italy; massaro@lum.it

Abstract: This review analyzes the Electronic Digital Twin (EDT) tools characterizing the industrial transformation phase from Industry 4.0 to Industry 5.0. The goal is to provide innovative research EDT solutions to integrate in manufacturing production processes. Specifically, this research is focused on the possibility of combining the advanced technologies and electronics and mechatronics of industrial machines with Artificial Intelligence (AI) algorithms. Furthermore, this review provides important elements about possible future implementations of AI-EDTs and some circuital examples to support the understanding of the concept of circuit simulation in EDT models. EDTs are useful to comprehend the modeling concepts functional to the AI application using the output of the circuit simulations. The output of the circuit is used to train the AI model, thus strengthening the capability to classify and predict the real behavior of production machines with a good accuracy. This review discusses perspectives, limits, and advantages of EDTs and is useful to define new research patterns integrating structured EDTs in advanced industrial environments. The focus of this paper is the definition of possible perspectives of EDT implementations, including AI, in data-driven processes in specific strategic areas of industrial research by classifying the scientific topics in six main pillars. This paper is also suitable for the researcher to develop innovative topics for projects scaled into different work packages based on EDT facilities.

Keywords: Electronic Digital Twins; AI-based modeling; Industry 5.0; advanced manufacturing modeling

1. Introduction

1.1. Research Methodology and DT Research Topics

This review analyzes innovative and recent research topics about Electronic Digital Twin (EDT) models in advanced Industry 5.0 scenarios. The research methodology is indicated in Figure 1, separating two main phases: a first one finding keywords grouped into six main research 'pillars' about EDTs in European research projects [1] of the last three years and a second one focusing the attention on the sub-topics correlated with this project's keywords. These sub-keywords are matched successively with research papers of the last five years. Search engines have been filtered for the last five years. This search yielded several papers. Papers from 2021 and 2022 were partially replaced due to their repetitive nature with more recent articles or because they were updated with new papers offering added values. This review describes the recent research advances in Digital Twins (DTs), explaining the possible EDT implementations and possible perspectives, starting with the analysis of the projects, which suggest different research pillars to be analyzed more accurately. DTs refer to digital physical–cyber models in general, including digital

https://doi.org/10.3390/machines13090755

technology, while EDTs refer to electronic circuit models. Industry 5.0 facilities integrate both DT and EDT models.

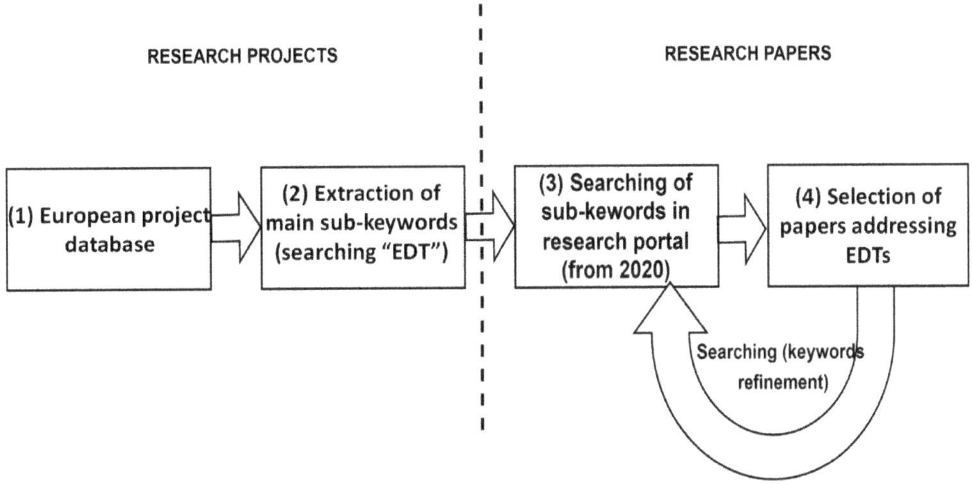

Figure 1. Block scheme of the applied methodology selecting research papers about EDT.

In Table 1 the topics and the keywords of the first phase are listed. More details of the second phase are discussed in Sections 2–7, matching the recent research with AI-EDT perspectives.

The choice of European projects as a reference for this research is due to the fact that within the 'Next Generation EU', the program promoted by the European Union to promote the recovery of Member States after the COVID-19 pandemic, research topics have been planned and defined to use resources until 2026. This programming is based on a pre-evaluation of research, so the funded projects constitute good feedback on the validation of research activities. This criterion can also be applied for other international programs about research. Limitations are mainly due to an analysis of projects performed for only the European area. Instead, the search for works is global. This could also be an element of the methodological exploration to see how a European project is related to scientific research in general. According to the summary in Table 1, it is evident that the projects consider the technology evolution from Industry 4.0 to Industry 5.0 facilities in manufacturing, including the healthcare sector for specific applications about social–health aspects (food, infection, territorial health). Particular attention in the industrial research is focused on alternative energy smart systems, DT virtual manufacturing, and robotics integrating Artificial Intelligence (AI) algorithms. Finally, research in advanced industrial networks is focused on inspection systems of production and defects and on cybersecurity including data security and control systems.

Figure 2 schematically illustrates the goal of this paper by proposing possible EDT perspectives matched with the research pillars. The sequence from one to six (pillars mentioned in Table 1) follows the logic of searching and matching project keywords with search terms: many projects start with Industry 4.0 objectives (pillar 1), upgrade to Industry 5.0 (pillar 2), and then move on to expanding sectors (healthcare related to pillar 3 and energy related to pillar 4) and specific applications such as inspection systems (pillar 5) and cybersecurity (pillar 6). The sequential order is also partly proportional to the number of similar projects found for that specific topic.

addition of this information in each discussed research pillar in Sections 2–7 (last columns of the tables of each section). Furthermore, this review provides some circuital examples supporting the reader in understanding the EDT concept based on the circuit simulation. The circuit simulation and the AI data processing of the output signals of the circuits are performed by using open-source tools based on graphical user interfaces. In this review, the circuit models are implemented by means of the LTSpice tool, besides the example of the AI prediction is performed by the Konstanz Information Miner (KNIME) tool. More details about the implementation of AI processing circuit data are discussed in Section 8.

Below is a summary of what will be discussed in this review, including the output of the methodology in Figure 1.

The sub-topics found in the literature are representative of DT models for industrial scenarios. Referring to production systems, DT models are addressed based on reliability analyses, the design, the monitoring, the maintenance of electrical and electronic components for renewable energy systems [1–8], the Industry 4.0 facilities [9–37], and, in general, on the sensing and actuation [38–49]. Ever-increasing research is being produced on optoelectronic solutions, as they are very fast and safe from the point of view of signal interference: optoelectronics is suggested in communication networks and in machinery [50–58] to increase production efficiency in advanced production scenarios. The production efficiency and the increase in quality are achieved by advanced image vision [59–64] and virtual models matching with mechatronic systems [65–86]. DTs are fundamental for the healthcare sector in continuous industrial evolution regarding medical products and services [87–91], wearable technologies [92–95], IoT systems [96–98], image processing, and mechatronics useful for rehabilitation [99–104]. Essential to all DT applications are the energy-harvesting modeling sources [105–126]. Energetic aspects are studied mainly in smart systems, such as smart buildings [127–130]. A particular focus of the research on DT implementation is on machinery inspection and monitoring [131–148], material manufacturing [149–165], and cybersecurity [166–194]. All the sub-topics grouped into six research pillars are deeply discussed in the next sections.

With the analysis of the actual literature, the added value of this review is the preposition of perspectives and possible upgrades of the DT, focusing our attention on EDTs and on the role of AI.

1.2. Research Questions

The research questions are related to the identification of DT models matching with industrial projects and with recent academic studies focusing the attention of new technologies. The main goal is to find, starting with the discussed technologies grouped into 'research pillars', possible AI solutions based on circuit models and constituting the specific class of EDTs. Circuit simulations providing data for AI data processing could be an important requirement in advanced DT industrial applications. In this direction, the work describes some examples suggesting the methodology to apply and facilitating the comprehension of mechanisms matching circuit models with AI data processing and highlighting advantages, disadvantages, procedures for the integration of the EDT technologies in industrial processes, perspectives, and limitations.

2. First Research Pillar: Agile Manufacturing and Industry 4.0

The first pillar focuses on identifying the phases of the hardware component life cycle (monitoring, maintenance, fault detection) not only in production but also in the operational phase. The literature discusses specific components matching industrial projects in DT and EDT manufacturing and concerning agile approaches.

Specifically, this pillar is related to the upgrade of the actual Industry 4.0 scenario, adopting DTs and EDTs for turbine technology (the topic is correlated with energy systems and suggests new frontiers for industries working on wind energy), for flexible advanced manufacturing facilities, and for sensor systems. DTs modeling wind turbines are useful for the reliability analysis, design, monitoring, and maintenance [1–8]. Precisely, EDTs could be applied for the improvement of the whole life cycle of wind turbines and for a continuous re-design of the mechanical, electrical, and electronic components. Advanced Industry 4.0 environments are characterized by advanced DT technologies [9–37], ensuring flexible production management and efficient predictive maintenance. In this direction, AI-based EDTs could provide a 'zero-defect' condition, optimizing sustainable production and machines, generating a production capable of quickly switching the typology of products in relation to needs and market demands (change in the short-term products and services). Production is always optimized by sensor systems. The production of new sensors, actuators, and electronic components [38–49] is very important for the industrial upgrade, where AI-EDTs could play an important role in system integration, including heterogeneous technologies, and at the same time efficiently synchronizing the interconnected machines of the whole supply chain. Furthermore, the introduction of AI-EDTs requires a change in the organizational model to decrease the human resource cost expressed by the worker-hour indicator. In this direction, the AI algorithms could support the re-planning of the new organizational models and the change in management control. Table 2 lists more details about the first research pillar of the EDT in Industry 4.0 and AI implications.

Table 2. First research pillar: EDT for agile manufacturing implemented in Industry 4.0 environment.

Sub-Topics	Main Description	DT/EDT Function	Perspectives and Possible Upgrades of the Research (Proposed in this Review)
Wind turbine (renewable energy systems)	DT-based monitoring framework for wind turbines for reliability analysis	DT framework enabling real-time monitoring, fault diagnosis, and operation optimization of the wind turbines [2]; prevention of turbine failure and timely maintenance [3]; reliability assessment of wind turbine power modules (regarding wind turbulence on the electro-thermal behavior) [4]; mobile robot assistance wind turbine [5]; online real-time monitoring of Power Electronic Converters (PECs) [6]; methodology and guidelines for the wind turbine design [7]; blade design and optimization (virtual model) [8].	The AI-EDT could be implemented for the whole life cycle of wind turbines starting with the design of optimized electrical and electronic components (mainly power components) and following the monitoring of the implemented systems for the optimization of the re-design and of the maintenance services (by means the simulation of real scenarios).

Table 2. *Cont.*

Sub-Topics	Main Description	DT/EDT Function	Perspectives and Possible Upgrades of the Research (Proposed in this Review)
Advanced machinery and processes for manufacturing production	Advanced and flexible manufacturing systems	Intelligent design–manufacturing–maintenance of mechanical equipment [9]; efficient testing and optimization of production processes improving quality control and predictive maintenance [10]; wireless and wired DT embedded in ethernet industrial networks [11]; electronic cam servo motion control driven by DT [12]; product twin for flexible and collaborative manufacturing environment [13]; Green Performance Evaluation of Smart Manufacturing (GPEoSM) [14]; models of sustainable intelligent manufacturing [15]; detection of a system's actions in pneumatics (backflow) [16]; AI algorithms finding causes of the problem of metal residues between electronic circuits [17]; DT following the whole product life cycle [18]; DT driven by data of Internet of Things (IoT) sensors detecting defects [19]; random phase noise in WiFi Channel State Information (CSI) [20]; machine control, monitoring and synchronization between the physical and virtual models [21]; DT for Automated Guided Vehicles (AGVs) [22]; design, development, deployment, commissioning, and maintenance of robotics systems [23]; holistic overview of robotics [24]; rapid production line variant design [25]; improvement of the powertrain system of electric vehicles (EVs) [26]; approach based on pipe machining production line [27]; automotive production lines [28–30]; electrical cable intelligent production lines [31]; multisystem fault diagnosis [32]; DT improvement by heuristics and Reinforcement Learning (RL) [33]; reconfiguration of human–robot collaborative assembly lines [34]; 3D visualized human–machine interaction [35]; railway axle production line [36]; DT enabling apparel manufacturing plants [37].	EDT could integrate different AI algorithms suitable to ensure good flexibility of manufacturing production. The AI-EDT could constitute the integration of different cyber–physical models of the whole production line supporting quality processes, predictive maintenance of machines, and production management, including the optimization of the organizational models. The goal is to achieve 'zero-defects' conditions, production optimization, and cost decrease (energy costs, raw materials costs, human resource costs as worker-hour indicator, time delays, etc.), thus ensuring the production switching also in the short period according to the market demand. The AI-EDTs are suitable for the matching of the skills of workers, thus improving the organizational models.
Sensor systems and electronics	Manufacturing of electronic components	Printed Circuit Board (PCB) manufacturing for full Surface Mount Technology (SMT) process lines [38]; electronic product behavior analysis [39]; power electronics-based energy conversion systems (PEECSs) [40]; IoT electric power systems [41]; encoder [42–45]; Programmable Logic Controller (PLC) systems and components [46,47]; DT for automotive MEMS pressure sensors [48]; optoelectronics in production lines [49].	Different advanced electronic components (sensor, actuators, PLC) could provide digital information for the whole synchronization of all the production machines, taking into account the heterogeneous technologies and the different computational costs. The AI-EDTs are useful for optimizing the production synchronization.

3. Second Research Pillar: Industry 5.0

The second pillar focuses on identifying hardware components and support tools for an upgrade of the manufacturing industry, in line with what is developed in projects aimed at mechatronics and DT implementations. Specifically, this pillar is associated with the innovative technologies indicating the digital transformation of the industry from the Industry 4.0 to Industry 5.0 framework.

New advances are found in the literature relating to optoelectronics, mechatronics, and image vision techniques. All the technologies are assisted in the Industry 5.0 environment by AI algorithms, representing the main difference if compared with the traditional Industry 4.0 framework. AI integrated in Industry 4.0 systems constitutes the upgrade of the industry for an Industry 5.0 framework, where machine learning algorithms are able to improve image processing, to execute AI-driven processes, and to improve the quality of production. More details about AI integration protocols are discussed in Section 8.3.

Optoelectronics is implemented in communication networks and for machinery components [50–58]. Image vision techniques are applied to control production and defects [59–64], implementing 3D and 4D image analysis techniques, and a DT is widely applied for the matching of mechatronics and virtual models of the manufacturing production processes [65–86]. Possible evolutions of the EDT in Industry 5.0 are related the design of new materials for high-sensitivity sensors (as for nanotechnology-based sensors), the real-time solving of bottlenecks of production processes, and the continuous setting and adjustment of machine control parameters to achieve 'zero-defect' production by ensuring quick production and minimizing the risk of machine failure. Due to the increase of the industrial Knowledge Base (KB), powerful communication networks could be adopted integrating big data, edge computing, Graphical Processing Unit (GPU) boards, and quantum computers, which would strengthen the AI-EDT models and, in particular, the training models.

Table 3 summarizes the main topics of the second pillar found in the scientific literature.

Table 3. Second pillar: EDT in Industry 5.0 scenarios.

Sub-Topics	Main Description	DT/EDT Function	Perspectives and Possible Upgrades of the Research (Proposed in this Review)
Optical components and networks	Optical sensing, actuation, and communication network based on AI	Artificial Neural Network (ANN) DT optimizing Whispering Gallery Mode (WGM) optical sensors [50]; optoelectronic sensors [51] and actuators using optical fiber [52]; AI-driven DT for optical network [53,54]; optical fiber communication systems [55]; forecasting of optical networks [56]; optical communication fault management [57]; re-configurable optical add and drop multiplexers (ROADMs) [58].	AI supporting design and modeling of industrial optoelectronic sensor components finding new solution in nanotechnology materials (improving high sensitivity); AI-DT optimizing the traffic and robustness of calculator networks to avoid possible bottlenecks imitating communications.
Image vision-integrated tools and facilities	Imaging, laser, and ultrasound techniques	3D and 4D volumetric Ultrasound Tomography (UST) [59]; integration of 4D imaging with 4D printing [60]; laser-induced ultrasound scanning PCB defects [61]; Laser Ultrasonic Testing (LUT) detecting inclusions and subsurface cracks [62]; ultrasound imaging using multi-parameter genetic algorithm for micro-defect detection [63]; vision inspection process [64].	AI-supervised algorithms accelerating automatisms detecting real-time product defects for 'zero-defect' production and improving the in-line production (integration of AI for auto-corrective actions and for machine parameter setting).

Table 3. *Cont.*

Sub-Topics	Main Description	DT/EDT Function	Perspectives and Possible Upgrades of the Research (Proposed in this Review)
Mechatronics and manufacturing processes	New Industry 5.0 environments and applications	Hybrid DT framework integrating AI and physics-based models [65]; Cyber Twins (CTs) integrating machine simulators [66]; deep learning-based DT for human–robot collaborative manufacturing systems [67]; mixed-reality and holographic industrial environments [68]; DT mechanism model of Computer Numerical Control Machine Tools (CNCMTs) [69]; fault diagnosis and predictive maintenance [70]; virtual twinning in mechatronic product development [71]; kinematic and dynamic models for mechanical systems [72]; trajectory controllers for electro-mechanical systems [73]; importing of the native PLC code via XML [74]; engineering approach of OPC-UA PLC systems [75]; DT-driven virtual commissioning method to simulate CNCMT [76]; Ultra Precision Machining (UPM) for high-end cutting-edge products [77]; integration of Industrial Internet of Things (IIoT) and smart manufacturing facilities [78]; Extended Kalman Filter (EKF) DT for motor drive [79]; automobile connecting rod production line model [80]; DT framework incorporating robots [81]; Self-Optimizing Control (SOC) framework controlling mechatronic systems [82]; AI predicting noises and anomalies in machine filters [83], operational amplifiers [84], and Proportional Integral Derivative (PID) controllers [85]; AI adjusting control machine parameters (process mining) [86].	AI-DT selecting the best procedure and algorithms to improve the mechatronic commands and control systems according to the selected technology (PLC, robot, CNC machines, etc.). Furthermore, the EDT based on the integration of circuit models could classify and predict all the possible noises and interferences, solving machine problems and failures in real time.

4. Third Research Pillar: Advanced Healthcare and Industry

The third pillar aims to define healthcare research topics related to DTs and EDTs in line with contemporary industry products and services and with project advances. This pillar is correlated with innovative medical products and services [87–91] and the EDT models related to wearable sensor technology [92–95], the IoT [96–98], image processing [195–201], and mechatronics for rehabilitation [99–104]. Possible perspectives for EDTs are related to the introduction of new health products and telemedicine care services and the construction of advanced patient 'avatar' models or 3D organoids optimizing the care patterns, the exams to execute, and the rehabilitation processes. Actually, AI is adopted for the construction of patient models, and future perspectives are addressed to improve the AI-based avatar model by means of real-time data fusion, merging different information about patients characterized by similar features. For organoid and avatar models, advanced technologies, such as edge calculus engines, GPUs, big data systems, and quantum computers, could be used.

Table 4 lists different research topics related to advanced healthcare systems.

Table 4. Third research pillar: EDT in healthcare industry and in research.

Sub-Topics	Main Description	DT/EDT Function	Perspectives and Possible Upgrades of the Research (Proposed in This Review)
Healthcare market	Innovative products and services	Machine learning (ML) for predictive controller of medical microrobot [87]; DT improving healthcare services (personalized healthcare, intelligent rehabilitation, telemedicine, smart diet management, etc.) [88]; production of medicine, vaccine, antibacterial, on-demand molecule production, and on-demand tissues and organs [89]; 3D modeling and simulations of microwave sensors [90]; hardware matching with AI-based software correcting the noisy EEG electrode signals in real time [91].	AI accelerating healthcare services (diagnostic and therapeutic assistance pathways) and optimizing precision medicine (dose of medicines) by means of electronics with therapeutic compliance. Innovative products in healthcare industries could be the 3D organoids reflecting the real behavior of the human body tissues (aspect useful for the surgical training of doctors or to understand the tissue degeneracy of cancer).
Wearable sensors	New solution of clothing sensors	DT created by biometric data from the patient gathered from wearables [92]; smartwatches, biosensors, and fitness trackers for chronic disease management [93]; biomechanical human body models and wearable inertial sensor models to analyze gait events dynamically [94]; multifunctional smart clothing system constructed with several sensors [95].	EDT could potentially represent the 'avatar' of the patient monitored by smart and miniaturized wearable sensors. AI could improve the avatar real-time creation fusing and analyzing patient's physiological data (temperature, heart rate, blood oxygen level, etc.) and including data processing of the patient's health historical data of patients characterized by similar features (optimization of the training model). 'Avatars' could be processed by advanced technologies, such as GPU, big data, and edge and quantum computing.
Medical IOT	IoT connectivity of human body sensors	Multi-sensor data fusion combined with Support Vector Machine (SVM) algorithm [96]; IoT- and ML-based electrocardiogram (ECG) classifier model for cardiac diagnostics [97,98].	AI-EDT addresses edge computing using IoT technologies. IoT is a fundamental technology for future telemedicine.
Medical imaging	Medical imaging tools integrated in AI-DT models	DT integrating MRI, CT, PET, and ultrasound imaging data [195]; 3D-Printed Phantom Twin for ultrasonography of the lower limb [196]; 3D physical and virtual models from ultrasound and magnetic resonance imaging [197]; AI ultrasound imaging in Obstetrics and Gynecology [198]; super phantoms replicating complex anatomic and functional imaging properties of tissues and organs [199]; AI and biophysical models for cardiovascular imaging [200]; human cell and tissue bio-manufacturing using image analysis [201].	The deep learning EDT models could provide more accurate diagnostic and prognostic responses. In order to accelerate the AI image processing calculus, this could be applied innovative technologies such as quantum computers supporting the correct parametrization of the AI training models.
Rehabilitation systems	Electronic solutions	Adaptive DT models with Multi-Level Inverter (MLI) improving energy efficiency in rehabilitation systems [99]; electronic and mechanical DT for stroke rehabilitation [100]; robot-assisted telerehabilitation system [101]; AI predicting movements from EEG patterns [102]; virtual–real interaction system for an ankle rehabilitation robot (VRIS-ARR) [103]; wearable sensors monitoring rehabilitation [104].	Fully connected AI-EDT integrating mechanical response analyses and remote control of rehabilitation (telerehabilitation services and production of new robotic systems). AI algorithms could automatize telerehabilitation planning in real time and the exercise to perform according to the AI classification and recognizing the correct and the wrong exercises (classification of exercise performed by the patients).

5. Fourth Research Pillar: Smart Energy

The fourth research pillar combines the topics on energy system designs with the research themes associated with the related DTs and EDTs. Specifically, the smart energy systems are the fourth research pillar discussed in the literature. Some important topics are the EDT energy harvesting of thermoelectric, piezoelectric, triboelectric, and electromagnetic generators [105–117], AI-EDT and hardware components of renewable energy sources [118–126], and EDTs applied to smart buildings [127–130]. Possible advances of EDTs are related to fully integrated and interconnected energy self-powered systems of complex smart grids and the modeling of energy efficiency and secure buildings. Concerning energy-harvesting devices, AI algorithms could provide more information about energy conversion efficiency, including electronic optimization matching with the behavior of the energy sources (optimizing signal variations, signal discontinuities, etc.). The AI control and optimization mainly concern the increase in the energy conversion rate due to the optimization of the collected energy and the correct filtering of the high-energy signal components.

In regard to real-time decision-making in smart grids, AI could support energy balancing by means of failure classifications and power load predictions.

Another implementation of AI decision-making is in smart building systems, where AI is able to predict the power load and, consecutively, to optimize energy routing through the electrical panel.

Table 5 records different research topics related to the pillar of smart energy.

Table 5. Fourth research pillar: EDT and smart energy systems.

Sub-Topics	Main Description	DT/EDT Function	Perspectives and Possible Upgrades of the Research (Proposed in This Review)
Energy harvesting	Low-power energy-harvesting components	Thermoelectric human body systems and Seebeck circuit modeling [105–107]; piezoelectric energy harvesting (generation of electrical power by vibration) [108,109] and using LTSpice circuit modeling [110]; electromagnetic energy harvesting [111]; AI-assisted wearable hip joint energy harvester (HJEH) using an electromagnetic generator (EMG) integrated with a freestanding triboelectric nanogenerator (FS-TENG) [112]; triboelectric nanogenerators in Civil Engineering Infrastructure 4.0 [113]; triboelectric nanosensors [114]; solar photovoltaic DT [115]; real-time control of hybrid energy system using DT and IoT [116]; IoT driving energy systems [117].	Implementations of fully integrated EDT for totally energy independent production lines, including low-power supplies and communication networks based on IoT wireless systems autoconfigured by AI algorithms. AI-EDTs are suitable for the optimization of electronic components following signal variations and signal discontinuities.
Components of renewable energy systems	Modeling and optimization of technologies enhancing the efficiency of renewable energy	DT for self-security of smart inverters [118]; fault localization in Low-Voltage (LV) Distribution Networks (DNs) [119]; AC/DC hybrid electric equipment [120]; fault data generation of lithium-ion batteries [121]; DT of energy components [122]; ML applied in smart grids, renewable energy, and for electric vehicle optimization [123]; modeling by graphs with nodes and edges of complex energy systems [124]; smart metering systems [125,126].	AI-EDT could be interfaced to a capillary network and set of components to provide the energy balancing of complex smart grids by means of grid failure classifications and power predictions of different interconnected nodes.
Smart energy building	Technologies optimizing building energy	Smart building controlling energy and wellness [127–130].	EDT using AI for energy efficiency and building security. Improvement of the energy routing through the electrical panel.

6. Fifth Research Pillar: Inspection and Monitoring

The fifth research pillar focuses on the design theme of industrial inspection processes by identifying some DT and EDT solutions to assist in production efficiency. Specifically, this pillar is related the EDT and the DT mainly for the inspection of production machines [131–148] and simulating the processing of materials to be used in manufacturing processes [149–165]. The goal for future AI-based EDTs is to apply the model to the whole supply chain, selecting operative procedures according to a risk map defining the priorities of the interventions to execute. Other new frontiers of EDTs are the applications for intelligent materials capable of self-healing and self-repair actions and constituting stable materials to be applied as robust products due to the possibility to control and correct the material defects in each production phase and to avoid production interruptions. Examples of AI-EDTs improving the self-healing of innovative materials are mainly in the control (by changing the input electromagnetic source power intensity or the signal profile) and the localization (source orientation and source collimation) of external stimuli, enabling self-reconfiguration or self-repairing processes [150]. The exponential advancement of hardware and software technology with the advent of AI will increasingly push researchers to find new solutions in basic sciences, such as physics, chemistry, and mathematics. In this direction, the research of smart or self-healing materials could represent a new frontier for the search for new solutions capable of monitoring the state of innovative materials and restoring them with specific control signals. Smart materials could therefore replace current materials used in manufacturing by exploiting their capabilities and properties.

Table 6 indicates the main research topics of the fifth pillar, which are mainly classified in IIOT inspection systems and in the inspection of materials and of related processes.

Table 6. Fifth research pillar: EDT concerning inspection and monitoring.

Sub-Topics	Main Description	DT/EDT Function	Perspectives and Possible Upgrades of the Research (Proposed in this Review)
IIOT inspections	IIOT focused on the inspection of machines	Distributed IIOT sensors for predictive maintenance [131,132]; monitoring of electric induction motors combining IIOT and Finite Element Method (FEM) simulations [133]; augmented reality interconnected to industrial cyber–physical environments [134]; control of cutting machines [135]; IIOT for monitoring fault detections [136] and for logistics [137,138]; quality product inspection by Software-Defined Edge Intelligent Controller (SD-EIC) [139]; control of robotic pick-and-place operation [140]; AI powering control systems (RNN, LSTM, and GRU models) [141]; robotic system for remote diesel defect inspection [142]; deep learning for inspection of electro-hydrostatic actuators [143]; AI and 6G network monitoring industrial scenarios [144,145]; Motor Monitoring and Analysis System (SMAM) implementing ESP32 microcontrollers and temperature, voltage, and current sensors [146]; Mixed Reality (MR) and AI supporting object recognition [147]; 5G network for multitask industrial applications [148].	AI-EDT assigned to each machine of the whole supply chain predicting failures in real time and assigning a risk level to execute possible timely corrective actions.
Inspection of materials and of related processes	Material manufacturing and quality processes	AI supporting semiconductor [149] and intelligent materials manufacturing [150]; quality of extruded materials [151]; modeling of complex material flow [152]; material consumption in thermoforming process [153]; micro-injection in molding processes [154]; Zero-Defect Manufacturing (ZDM) and cost reduction [155]; failure of fused deposition process [156]; 3D additive manufacturing [157]; defects in Laser-Directed Energy Deposition (L-DED) [158]; crack initiation and propagation [159–163] and equivalent resistive circuits modeling the physical effects [164,165].	AI-EDT optimizing the choice of the raw materials predicting defects and cracks (by means of equivalent circuits) also during the fabrication processes. Other possible EDT perspectives are also in the use of self-healing and self-repair materials, reducing the cost of waste for a sustainable production and the related delays.

7. Sixth Research Pillar: Cybersecurity

The sixth pillar focuses on DT and EDT solutions to improve industrial network security. Specifically, this pillar is related to cybersecurity, data security [166–188], and hardware attacks [189–194]. Possible innovative solutions in cybersecurity are the application of AI algorithms for attack classification and the use of blockchain technologies for improving the data security level. The future AI-EDT models will take into account the real-time physical and virtual network segmentations, new techniques of sensor data encryption for blockchains, and the real-time controlling and correcting of the cyber attacks in each level of the industrial network. Table 7 summarizes the main topics of the sixth research pillar classified as network security, blockchain solutions, and hardware security.

Table 7. Sixth research pillar: EDT improving cybersecurity.

Sub-Topics	Main Description	DT/EDT Function	Perspectives and Possible Upgrades of the Research (Proposed in This Review)
Secure industrial network	Cybersecurity of Industrial Control System (ICS) Network	AI-based framework detecting attacks and mitigating their impact for edge cloud systems applied to gas pipelines [166]; threat classification and guidelines to use DT [167]; DT and Hardware in the Loop (HITL) approach using neural networks to classify attacks [168]; priorities of security aspects in DT [169]; Binary Arithmetic Optimization Algorithm with Variational Recurrent Autoencoder-Based Intrusion Detection (BAOA-VRAID) approach detecting intrusions [170]; framework of cyber–physical manufacturing systems (CPMS) for cyber attack detection in closed-loop controllers [171]; Virtual Private Network (VPN) and AI–big data systems for secure multimedia networks [172]; cybersecurity threat modeling in ICS [173–177]; detection of anomalies in PLC systems [178,179]; secure industrial multi-level architectures [180]; VPN for secure industrial networks with robots [181].	EDT automating the scanning port test by means of the AI classification of threats and attacks and virtually segmenting the ICs network based on the detected risk. Furthermore, the AI-EDT could provide different solution for a reconfiguration of the Information Technology (IT) and the Operational Technology (OT) network, ensuring continuous network security (configuration of each component of the Local Area Network (LAN), including firewall, router, bridge, servers, and switch). The AI-EDTs are also useful to design hybrid virtual networks combing Virtual LAN (VLAN) and VPN channels ensuring data encryption. Furthermore, AI-EDTs are suitable for a continuous update of the classification of threats.
Blockchain	Blockchain integrated in industrial information systems	Blockchain-based Non-Fungible Tokens (NFTs) preventing attacks by ensuring traceability and accountability among distributed nodes [182]; Ethereum blockchain DT modeling [183]; improvement of security by blockchain [184–188] and Web3 technologies [184].	EDT encrypting sensor data to be included in blockchain information structures, certifying the production data and the firm knowledge.
Hardware attacks	Hardware Trojan (HT) attacks in ICS	HT modeling [189–192] and AI predicting hardware attacks [193,194].	AI-EDT predicting and applying real-time corrective actions due to attacks (denoising, elimination of interferences or disturbs, parameter regulation of PID control system, etc.) for the circuits of a production line.

8. Discussion

8.1. Circuit Modeling and AI Data Processing Integration

As a complete example of circuit application and data analysis with AI, a basic example 'transversal' for all the pillars mentioned is proposed. This example concerns the modeling of the most common noise, namely white noise. The example discussed can therefore be integrated with more complex models related to the topics discussed in this paper.

The circuit modeling and simulations represent the first data processing stage of the EDT matching the real industrial environment with machine technologies. The second stage of the EDT data processing considers AI data processing, where the digital output of the circuit (electrical variables such as currents, voltages, and powers estimated by Kirchhoff's circuit laws) is the input of the AI engine. A typical basic EDT example in industrial electronics is the integration of source generators modeling noises [47,49,91] and, consecutively, the simulation of the real behavior of electronic components characterized by the signal over the noise (S/N) signals. The example illustrated in this section considers a complete application the AI-EDT, taking into account a basic element of the circuit model and the white noise typically characterizing the interferences in machine environments: Figure 4 represents the modeling of a white noise generator (B1) connected to a resistor R1 with the function of modulating the signal noise amplitude at the output of the generator. The circuit of Figure 4 can be implemented in different LTSpice EDT circuits influenced by white noise, by coupling the generators in the precise sections where the disturbances are located. The parametric analysis performs different simulations of the voltage output signal, varying the ripple frequency of the noise. Figure 5 illustrates the parametric analysis of the circuit of Figure 4, varying the frequency parameter k (k = 1, 5, and 10): as expected a higher frequency is observed for k=10, preserving the random behavior of the Vout signal (more ripples are observed).

The parametric analysis can be analyzed in the time domain and in the frequency domain by means of the Fast Fourier Transform (FFT). Figure 6 shows the frequency parametric analysis applying the FFT to the signals of Figure 5. Also, for the frequency behavior more ripples are observed for the case of k = 10.

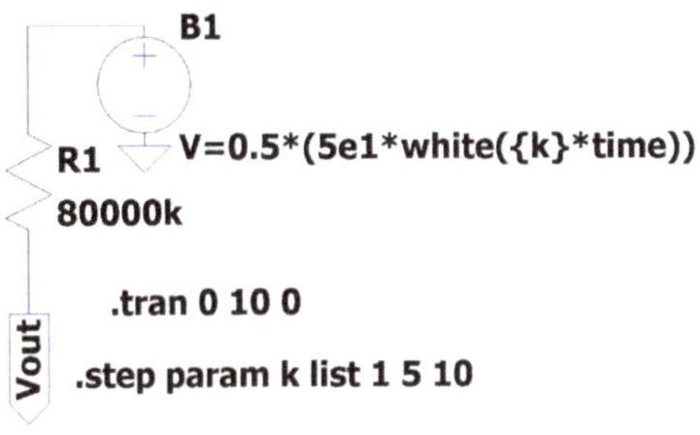

Figure 4. Example of circuit model of white noise source generator. The parameter k varies the ripple frequency of the noise signal generated by voltage generator source B1. The R1 resistance of 80 MΩ is properly selected to visualize the output signal in a range between −12 Volts and 12 Volts.

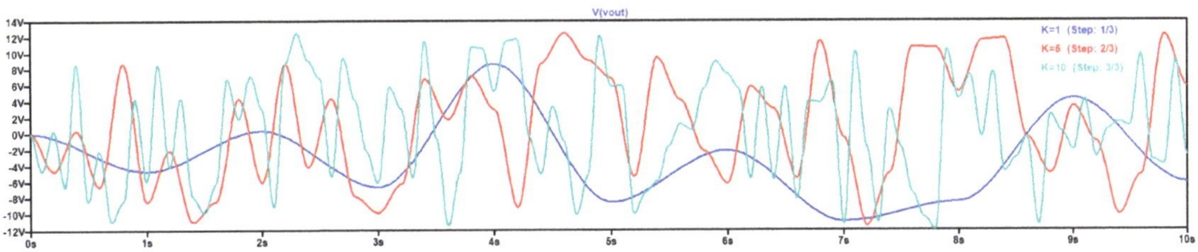

Figure 5. LTSpice parametric analysis: time domain results of the voltage output signal Vout, varying the k parameter of the circuit of Figure 7. More ripples are observed for k = 10.

In order to show how the AI algorithm can be applied for executing the digital data of the circuit (Vout voltage signals), a further signal is added to the signals of Figure 5 to predict the target voltage. The target signal indicates the 'labelled' signal to predict. For the proposed case the target signal is obtained, fixing the parameter value of k = 3 and providing the time domain trend of Figure 7. The AI prediction can be performed by implementing a workflow capable of processing the output data of the circuit of Figure 4. As an example of the AI implementation, the KNIME workflow of Figure 8 implements the supervised Random Forest (RF) algorithm applied to the output results of the circuit of Figure 4. Figure 9 estimates the predicted voltage (prediction of the target signal) as having a trend similar to the target signal, thus proving the good performance of the RF algorithm. This similarity is due to the self-learning of the algorithm. The RF performance of the proposed example is estimated by the following metrics, referring to a normalized dataset: Mean Absolute Error (MAE) = 0.135, Mean Squared Error (MSE) = 0.03, Root Mean Squared Error (RMSE) = 0.174, and Mean Signed Difference = 0.013.

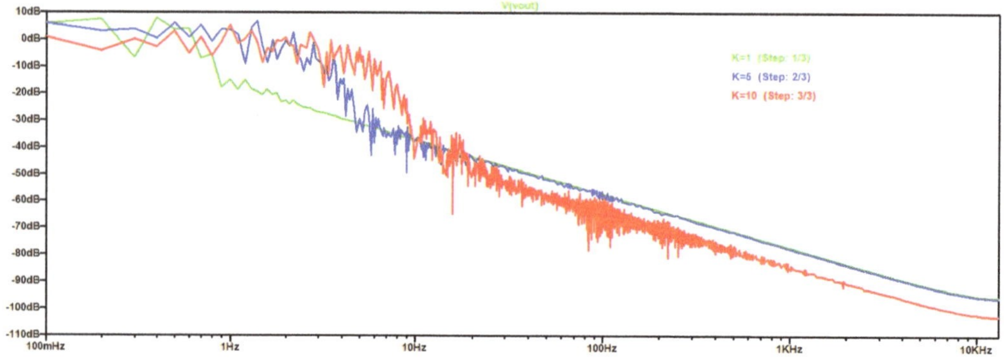

Figure 6. LTSpice parametric simulation FFT: frequency behavior of the circuit of Figure 4. More ripples are also observed in the frequency band for the case of k = 10.

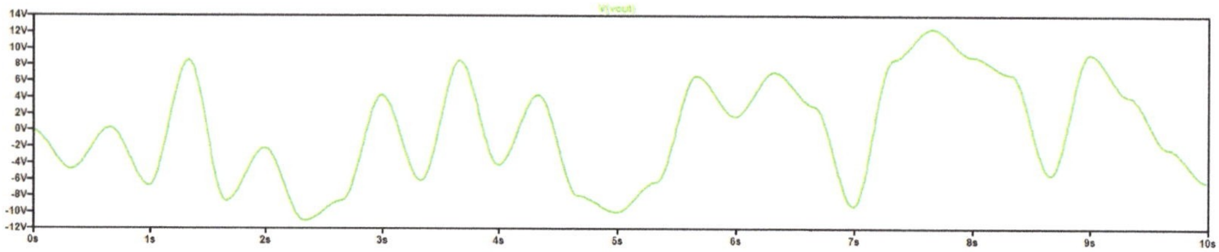

Figure 7. LTSpice output of Figure 4: target signal obtained with k = 3.

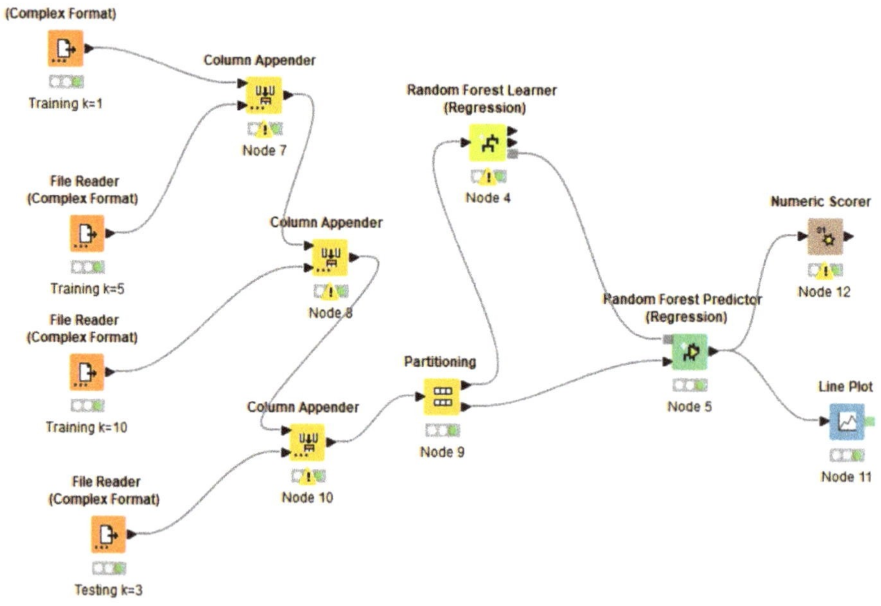

Figure 8. KNIME workflow providing the results of Figure 9.

Other examples of circuit simulations are discussed in Appendix A.

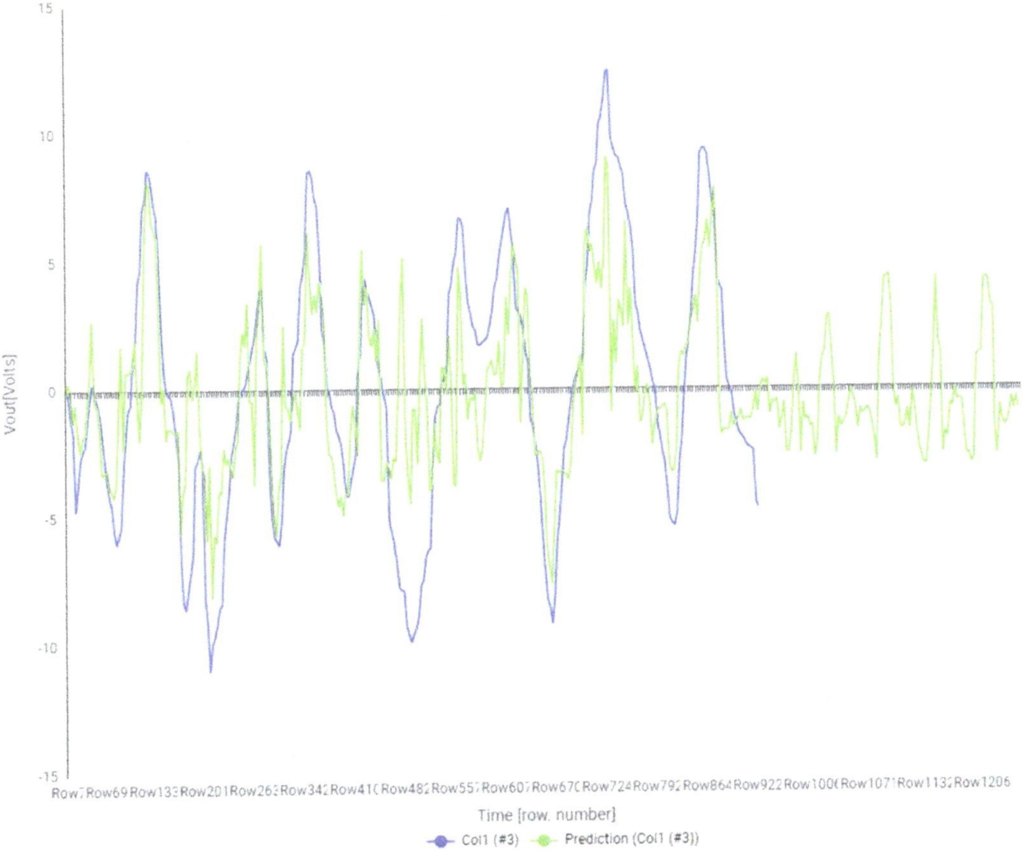

Figure 9. RF output results: comparison between the target signal (Col1(#3)) and the predicted one (Prediction (Col1(#3))). The analyzed dataset is composed of 15,520 samples; the training and the testing dataset are 70% and 30% of the analyzed dataset, respectively; and the hyper-parameters are tree depth = 10, minimum node size = 5, and number of models = 100.

8.2. EDT Framework: Advantages and Disadvantages

The EDT is a sub-class of DTs: specifically, the DT includes the whole digital environment, including the software, while the EDT is the reproduction of circuits and hardware components. Figure 10 illustrates an example of a DT/EDT framework for manufacturing industries, deducted from the literature: AI is executed locally or by cloud computing, for both the DT and EDT, by processing the local data (machine data, ERP data, sensor data, etc.).

Table 8 discusses the main advantages and related disadvantages of the main functionalities of the EDT tools.

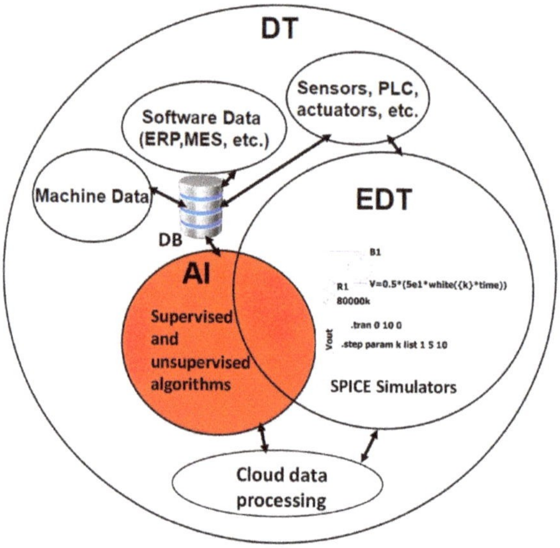

Figure 10. DT and EDT framework for manufacturing industries. AI data processing is performed locally or by cloud computing.

Table 8. Main advantages and disadvantages of the EDT tools.

EDT Tool	Advantages	Disadvantages
Platform integrating different hardware and software models simulating the whole industrial environment	The possibility to integrate many tools into a unique platform allows us to simulate the whole production line [202,203] and extended the simulation case on the whole supply chain (simulations of the behavior of the whole supply chain modeling all the production lines). The platform could operate in the whole process management of industrial production (data layer, model layer, application layer) [204].	High cost for a full integrated EDT: the costs include the human resource costs (working time), computational cost, and economic cost (investment in new technologies). A high computational cost requires new calculus solutions, such as edge computing or quantum computing. Furthermore, the EDT solutions could be characterized by a high impact, integrating the EDT in the operational procedures. This could require a transient to test the adoption of the EDT model.
EDT hierarchical modeling	Possibility to structure a hierarchical [205–208] EDT starting with the material and finishing with logistics, thus defining different operation levels and production priorities based on a risk assessment.	Possible reduced flexibility and ability to adapt to production change. Limited communication between lower and higher levels of the organization.
Big data analytics	Possibility to integrate the EDT big data analytics approaches (for health wearable sensors [209], logistics processes [210], and for the production of electrical vehicles [211]), optimizing quality processes and the machine predictive maintenance.	Big data analytics requires the availability of massive amounts of data of the company's digital knowledge base.
AI training	Possibility to use the output of the electrical variables of the circuit simulation for the construction of the training models of the AI-supervised algorithms [85,91].	The circuital data could be insufficient. This could require the construction of augmented data [212,213] to strengthen the AI training process.

8.3. Limitations, Practical Use Cases, and Workflow Protocol for EDTs Integrating AI

The EDTs discussed in the research pillars are characterized by some important limitations. Table 9 lists the main limitations and the corresponding mitigation procedures.

Table 9. Limitations and related mitigations of EDT with Industry 5.0 facilities.

EDT Limit	Limit Description	Limit Mitigation
EDT real-time analysis	The common technologies are characterized by a time delay of the simulation (computational cost of data processing).	Use of edge computing, GPU boards, or quantum computing to ensure real-time data processing also considering very large datasets to process.
Knowledge base (KB)	The company should have a knowledge base of certified digital data. This requires a careful selection of the data that will be considered useful for data processing purposes.	Implementation of well-defined procedures regarding the extraction of data from company systems (hardware and software) and regarding the fusion of the same to make them usable by AI engines.
Organizational model	The use of advanced EDT requires an upskill or reskill of the human resources involved in the EDT production process.	The company should actuate a continuous training plan for human resources, defining new roles in the organizational model and efficiently allocating the working teams.
Real industrial environment modeling	The model cannot predict all real-world situations, especially for unforeseen events that occur with low probability but can be extremely risky.	For accurate modeling, continuous optimization of the EDT becomes necessary. The AI algorithms could support this optimization.
Ethical aspects	Ethical aspects related to confidentiality and privacy of data.	Composition of an internal scientific ethics committee within the company.
Sustainability	EDT typically does not include all the sustainability aspects [214–218].	Design of EDT based on sustainability aspects.

EDTs integrating AI is the future challenge of Industry 5.0. This integration is to be considered in processes embedded in regular production processes.

Table 10 summarizes all the steps involved in the integration of the EDT in industrial processes according to the Business Process Modeling and Notation (BPMN) scheme of Figure 11. The BPMN is a standard notation useful to verticalize processes associated with the logic of circuits [219].

Table 10. Use criteria of EDT deduced from the analysis.

Function of the Model (Sequential Steps)	Use Criteria	Possible Setting of the EDT
(1) Circuit model	Design of the circuit layout and modeling of the technological elements to simulate the circuit, network, signal, etc., including a.	Primary testing of different parts of the circuit to check each part of the model.
(2a) Circuit simulation	Simulation execution providing digital data (voltage, current, power).	Setting the numerical parameters of the equations, solving the circuital layout (calculus error, precision, etc.).
(2b) Real data monitoring	Sensor data reading the machine and the process parameters.	Adjusting sampling time or other calibration parameters to achieve accurate data.
(3a) AI training	Creation of the ANN training model by considering different voltage outputs and fusing sensor data (2b) with circuital data (2a).	Cleaning of the dataset (filtering of wrong data or missing values) and possible introduction of augmented data (artificial data) to increase the efficiency of the training model.
(3b) AI testing	Selection of the testing dataset of the real physical model.	Change in the testing dataset dimension to increase the AI performance.
(4) Optimization process	Optimization process for the design or the re-design, the machine setting and maintenance, security, and quality.	Processes are re-mapped according to the results of the EDT and setting of the new EDT specifications.
(5) Iteration of circuit modeling and simulation	According to the results of task 4, decide if we need to re-engineer the model or simply re-execute the circuit simulation varying some parameters.	Definition of the time to update the circuital data optimizing the total computational time.
(6) Ending of the EDT use	The process is stopped when the EDT no longer represents the real scenario (machine change, production line, processes, etc.).	The resource and the tools are assigned to other EDT models or facilities.

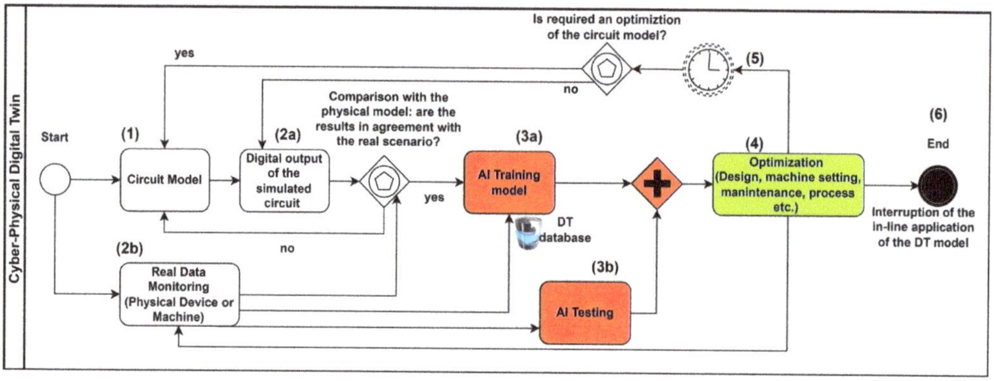

Figure 11. AI-based DT: BPMN explaining the integration of the AI in the circuital model optimizing the manufacturing process.

8.4. Workflow Methodology, Correlation, and Technology Readiness of Research Pillars

The flux diagram in Figure 12 represents the workflow, summarizing the review objectives and describing the adopted methodology. Specifically, the flux indicates the following steps:

- Keyword matching and the use of academic search engines to find papers;
- The verification of the presence of important research features, such as a DT framework architecture, project specifications, AI applications, a circuital approach, and AI-based process integration;
- The definition of perspectives and possible electronic upgrades of the research pillars (last column of Tables 2–7);
- The analysis of relationships between research pillars and the estimation of the related association strengths.

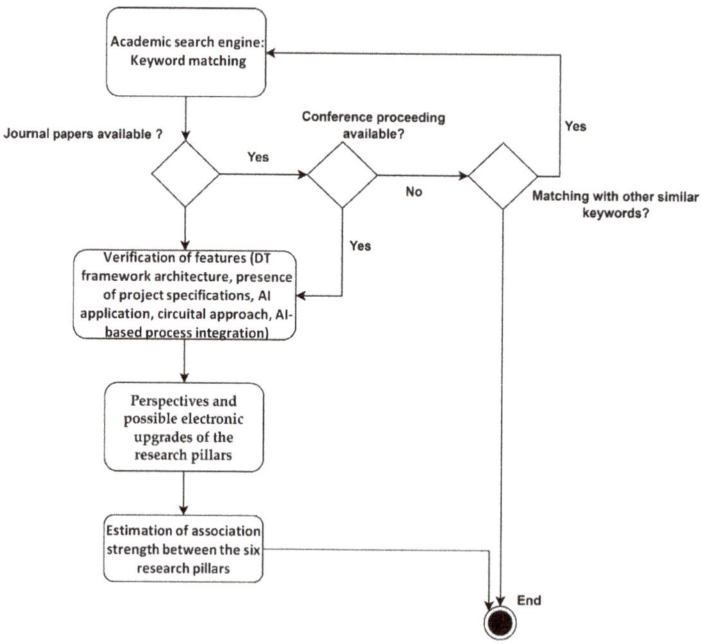

Figure 12. Theoretical workflow representing the review objectives.

An approach to visually map the relationships between research pillars is the Visualization of Similarities (VOS) [220,221]. An open-source tool suitable for this visualization is the VOSviewer. By applying this tool (Version 1.6.20), Figure 13 illustrates the clustering analysis (three clusters grouping pillars 1 and 4 and grouping pillars 3, 5, and 6) and the association strength output, normalizing the strength of the links between pillars (thicker arcs represent higher dependency weights).

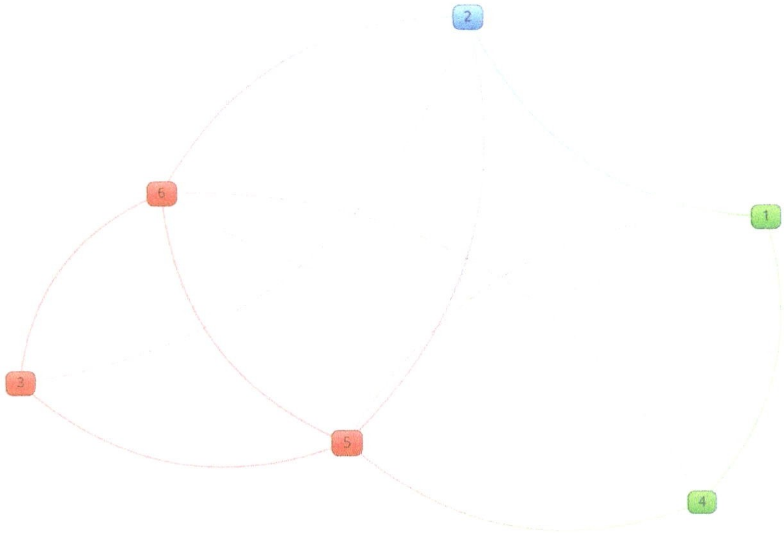

Figure 13. VOSviewer: colored clustering and application of the association strength between the six research pillars (numbered as 1, 2, 3, 4, 5, and 6). The graph provides a meta-level view of how the six pillars interrelate, strengthening this survey's utility for future research planning.

Other information about research maturity of pillars is indicated by the heat map in Figure 14, where a score is assigned for each pillar regarding the scientific impact (deduced from the articles found) and the reachable Technology Readiness Level (TRL).

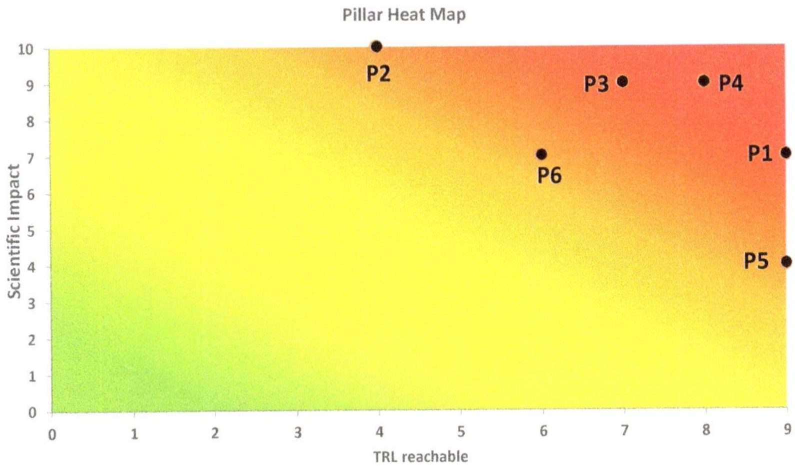

Figure 14. Pillar heat map matching scientific impact and reachable TRL of pillars (P1, P2, P3, P4, P5, and P6).

8.5. AI Importance and Research Gap

Much of the work on AI is focused on creating training dataset models with a good accuracy. Inference acceleration enables AI to solve ever-larger problems on ever-smaller devices, such as for high computing for DNN-based inference [222] and for edge computing approaches [223]. In order to improve the AI computational cost, GPU boards [224] and edge [225] and quantum computing technologies can be adopted [226,227].

Another important aspect about the AI-EDT is the link between the manufacturing processes and the developed models supporting the different production phases, including monitoring, modeling, optimization, design and preparation, control, and operation [228–231]. Important advances for AI-based EDTs are the application of Convolutional Neural Network (CNN) matching the requirements for ultra-high voltage direct current (UHVDC) projects [232], Long Short-Term Memory (LSTM) predicting the health condition of parts of Computer Numerical Control (CNC) machine tools [233], or hybrid CNN-LSTM networks improving AI algorithm performances [234].

Circuit simulators and the related output data in EDT models represent the AI datasets, predicting the real behavior of the production machine with a good accuracy. In particular, for the optocoupler switch of a PLC, circuit simulations process different white noises and interferences (pulsed signals) of the 3D environment modeled by input ports: the AI-RF algorithm provides good matching between the simulated and the predicted trend of the optocoupler output voltage with an MAE of 0.8% [47]. Concerning amplification circuits, the real machine environment is modeled by the interference of sinusoidal pulse signals with different carriers (each carrier frequency represents an interference, and two interferences are simultaneously coupled into the amplification system): the AI-ANN-MLP algorithm provides the prediction of the amplified signal with an MAE of 2% [84]. Passive RC filters are modeled to integrate electronic noises (Johnson noise or Schottky noise) and switching circuits, selecting the best filtering network according to the ANN classification with an MAE of 6% [83]. The simulations of PID control circuits are used to train ANNs for predicting anomalies by observing when the predicted values exceed the envelope of the input signal and constructing precise risk maps (k-means algorithm) [85].

In order to highlight the completeness of the methodology followed in this review, Table 11 lists the research gaps in the prior selected research reviews according to the main features, such as the framework, project matching, AI prediction or classification, circuital approach, and DT-AI process integration.

Table 11. Research gap in the prior selected research reviews.

Reference	DT Framework Based on an Architecture	Topics Matching with Projects	AI Prediction/Classification	Circuital Approach	Procedure Integrating DT and AI in Production Processes
[13]	√	√			
[15]	√				
[18]	√				
[77]	√		√		√
[113]	√	√		√	
[123]	√		√		
[124]	√	√	√		
[185]	√		√		
[203]	√	√			
Proposed work	√	√	√	√	√

9. Conclusions

This review discusses the recent topics on DTs and EDTs, with a particular focus on the development of the use of models in industrial scenarios, by selecting six main research pillars suggested by projects. Specifically, the pillars are related to Industry 4.0

manufacturing, Industry 5.0 facilities, advanced healthcare, smart energy, inspection and monitoring systems, and cybersecurity. Each research pillar is classified by grouping different research sub-topics according to the output provided by the research engine portals of scientific publications. This review proposes some examples of EDT models to facilitate the comprehension of the modeling approach and possible implementations of AI related to the discussed pillars. This review focuses on the following consequential items:

- Finding research topics on DT matching in recent industrial research projects using a defined methodological approach (see Figures 1 and 12) and defining the relationships and impacts of the identified research pillars (see Figures 13 and 14);
- Finding useful elements to define possible upgrades of DTs and EDTs, including AI (see last columns of Tables 2–7);
- Suggesting a methodological approach for implementing EDTs based on the use of circuit models (see Section 8.1 and the Appendix A);
- Proposing the integration of EDT models into manufacturing processes (see Figure 11).

Finally, this review highlights perspectives, advantages, disadvantages, and limitations for the future development of the EDT models in industrial processes. This study is also suitable for researchers to define possible new projects, including parts of the developed discussion, and finding possible correlations and relationships between the discussed topics.

Review's Limitations and Future Research Directions

This review is focused on identifying some topics that can be associated with DTs, using a well-defined research methodology. Specifically, the European research context has been considered as a 'starting point'. This certainly limits the selection of other sub-topics that could be particularly important in the international design scenario. In this direction, future works will be oriented to find possible correlations between different project portals of multiple continental areas.

Furthermore, many weaknesses of the cited publications are to be identified in the discussion of the full integration of circuit models with AI. For this reason, in addition to an effort to understand how AI can be implemented in the perspective of use, a further step has been taken in defining, in Figure 11, the integration procedure between AI and EDTs specifically for manufacturing processes.

This review suggests different advanced industrial frameworks, including AI, for the improvement of product quality, simulating the behavior in real scenarios by means of the EDT. The EDT models are useful for simulating the whole product life cycle and processes of production monitoring, product implementation, and deployment. A new research direction is the AI data-driven processes supported by EDTs capable of dynamically adjusting the production or the implementation process, thus increasing the quality and the efficiency.

Funding: This research received no external funding.

Data Availability Statement: Data are included in the paper.

Conflicts of Interest: The authors declare no conflicts of interest.

Appendix A

Design and production in Industry 4.0 environments are supported by circuit modeling and simulation. This Appendix demonstrates some EDT examples. Since three-phase motors are commonly used in industrial environments, it is useful to model the related circuit systems. As a circuit example, the model in Figure A1a shows a typical layout of the power supply circuit of a star-connected network, where the input voltage signals modeled by voltage generators V1, V2, and V3 are mismatched by a phase of 120° (see in Figure A1b the time domain voltage signals corresponding to the labels V1out, V2out, and V3out). The circuit model is useful for understanding how the electrical signals are applied to specific load networks to ensure a balanced power supply and to know the distribution of the electrical power on the different loads.

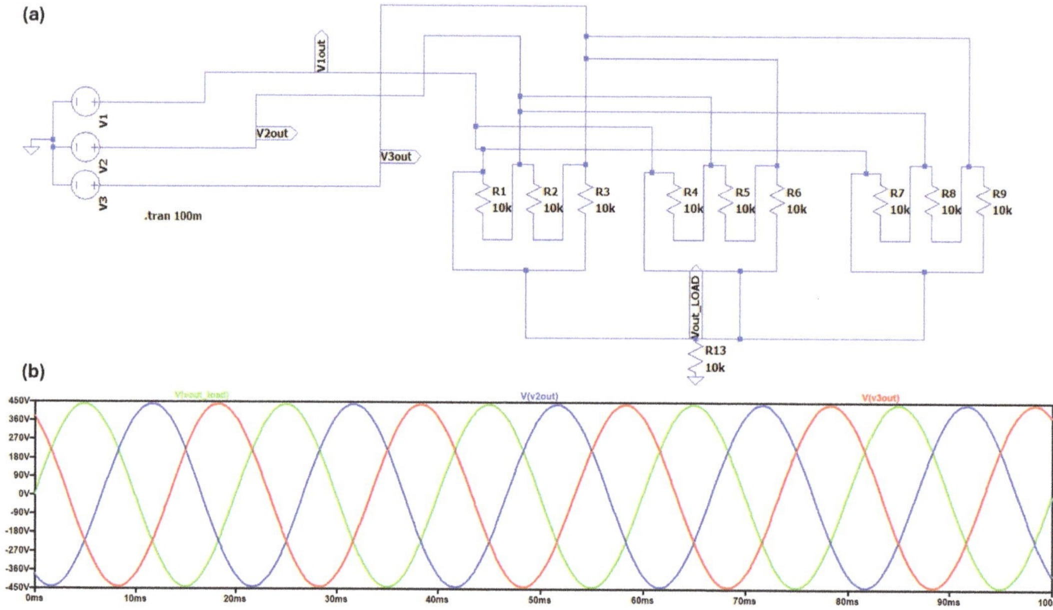

Figure A1. LTSpice simulation: (**a**) circuital EDT modeling a three-phase network powering a star-connected balanced three-phase load. (**b**) Time domain transient analysis calculating the input voltage signals of the electrical three-phase network (phase mismatch of 120° between the voltage signals). The resistances of 10 kΩ are able to provide an output voltage of 450 Volts.

The second example in Figure A2 simulates the transmission of a pulsed signal traveling on a transmission line, which can be a coaxial cable typically implemented in standard Ethernet industrial networks. The simulation allows for the comprehension of the effect of the time delay during the data protocol transmission, thus simulating the real network behavior and taking into account the delay occurring at the physical level (first level of the ISO/OSI scale related the transmission medium). The example proves that it is possible to simulate both the electrical signal and the information signal that travels in an industrial network.

The third example considers an EDT model detecting Electroencephalogram (EEG) signals in medical devices. The design of new healthcare hardware sensors is typically performed by the circuit modeling and simulations. The difficulty of the EDT implementation is mainly the circuit modeling of physical interactions between sensors and human body parts. In this direction, in order to test the efficiency of a real medical device, a good approach is to simulate the circuit by considering a real physiological signal as the input signal. Figure A3 illustrates the example of an EEG electrode, where the brain signal is 'imported' in the circuit by the V2 generator reading a text file (brain signal exported by a real electrode). The model simulates the signal coupling in the skin region and at the skin/electrode interface (gel materials), using different resistances and capacitances. A possible noisy output can be analyzed by AI algorithms extracting the correct trend of the brain signal and correctly processing the output of the EEG amplification system [91]. Other medical devices detecting physiological parameters can be modeled by similar electrical components, reproducing the interaction of the electronic components with the human body tissues characterized by other specific resistances, capacitances, and inductance parameters modeling the physiological domain.

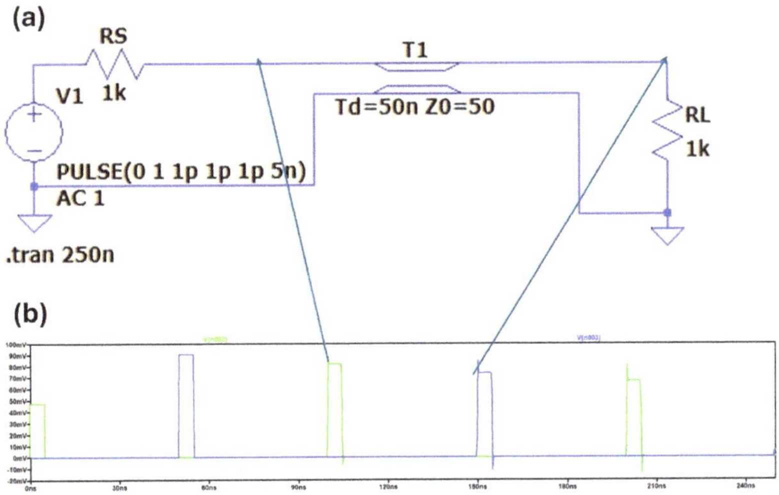

Figure A2. LTSpice simulation: (**a**) circuital EDT modeling a transmission line of an industrial network with a characteristic impedance of 50 Ω. (**b**) Simulation illustrating the delay between the input pulse signal and the output one due to the crossing of the transmission line. The input resistance of RS = 1 kΩ and load resistance of RL = 1 kΩ are able to provide a voltage signal between 50 mV and 80 mV in the observed time window.

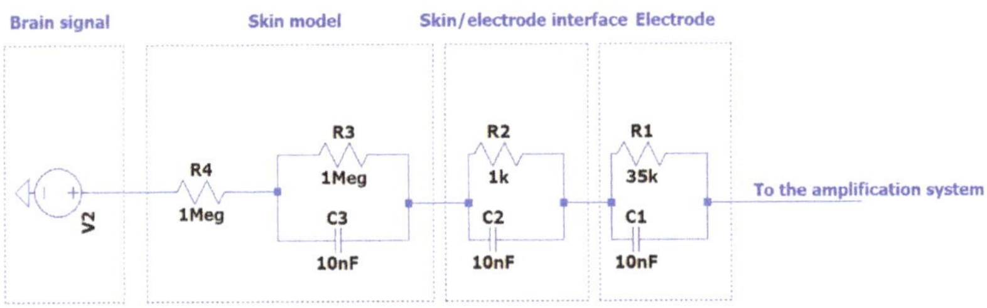

Figure A3. Example of circuit LTSpice model of an EEG electrode connecting all the regions and the human body interfaces characterizing the EEG signal transmission. The resistances R1, R2, R3, and R4 and the capacitances C1, C2, and C3 are related to a physical model simulating the skin/electrode interface [91].

References

1. Kohesio: Discover EU Projects in Your Region. Available online: https://kohesio.ec.europa.eu/en/projects (accessed on 17 May 2025).
2. Wang, M.; Wang, C.; Hnydiuk-Stefan, A.; Feng, S.; Atilla, I.; Li, Z. Recent Progress on Reliability Analysis of Offshore Wind Turbine Support Structures Considering Digital Twin Solutions. *Ocean Eng.* **2021**, *232*, 109168. [CrossRef]
3. Olatunji, O.O.; Adedeji, P.A.; Madushele, N.; Jen, T.-C. Overview of Digital Twin Technology in Wind Turbine Fault Diagnosis and Condition Monitoring. In Proceedings of the 2021 IEEE 12th International Conference on Mechanical and Intelligent Manufacturing Technologies (ICMIMT), Cape Town, South Africa, 13–15 May 2021; pp. 201–207.

4. Iosifidis, N.; Zhong, Y.; Hu, B.; Chen, B.; Ran, L.; Lakshminarayana, S.; Jia, C.; McKeever, P.; Ng, C. Reliability of Wind Turbine Power Modules Using High-Resolution Wind Data Reconstruction: A Digital Twin Concept. In Proceedings of the 2021 IEEE Energy Conversion Congress and Exposition (ECCE), Virtual, 10–14 October 2021; pp. 3630–3637.
5. Malik, A.A. Digital Twin Based Development of Mobile Robot Assistant in Wind Turbines Manufacturing. In *Volume 2: 43rd Computers and Information in Engineering Conference (CIE)*; American Society of Mechanical Engineers: New York, NY, USA, 2023.
6. Di Nezio, G.; Di Benedetto, M.; Lidozzi, A.; Solero, L. Digital-Twin Based Health Monitoring for Multi-Phase Boost Rectifier in Wind Offshore Applications. *IEEE J. Emerg. Sel. Top. Power Electron.* **2024**, *13*, 2738–2748. [CrossRef]
7. Mahmoud, M.; Semeraro, C.; Abdelkareem, M.A.; Olabi, A.G. Designing and Prototyping the Architecture of a Digital Twin for Wind Turbine. *Int. J. Thermofluids* **2024**, *22*, 100622. [CrossRef]
8. Chetan, M.; Yao, S.; Griffith, D.T. Multi-fidelity Digital Twin Structural Model for a Sub-scale Downwind Wind Turbine Rotor Blade. *Wind Energy* **2021**, *24*, 1368–1387. [CrossRef]
9. Fu, Y.; Zhu, G.; Zhu, M.; Xuan, F. Digital Twin for Integration of Design-Manufacturing-Maintenance: An Overview. *Chin. J. Mech. Eng.* **2022**, *35*, 80. [CrossRef]
10. Ebni, M.; Hosseini Bamakan, S.M.; Qu, Q. Digital Twin Based Smart Manufacturing; From Design to Simulation and Optimization Schema. *Procedia Comput. Sci.* **2023**, *221*, 1216–1225. [CrossRef]
11. O'Connell, E.; O'Brien, W.; Bhattacharya, M.; Moore, D.; Penica, M. Digital Twins: Enabling Interoperability in Smart Manufacturing Networks. *Telecom* **2023**, *4*, 265–278. [CrossRef]
12. Xu, J.; Guo, T. Application and Research on Digital Twin in Electronic Cam Servo Motion Control System. *Int. J. Adv. Manuf. Technol.* **2021**, *112*, 1145–1158. [CrossRef]
13. Onaji, I.; Tiwari, D.; Soulatiantork, P.; Song, B.; Tiwari, A. Digital Twin in Manufacturing: Conceptual Framework and Case Studies. *Int. J. Comput. Integr. Manuf.* **2022**, *35*, 831–858. [CrossRef]
14. Li, L.; Lei, B.; Mao, C. Digital Twin in Smart Manufacturing. *J. Ind. Inf. Integr.* **2022**, *26*, 100289. [CrossRef]
15. He, B.; Bai, K.-J. Digital Twin-Based Sustainable Intelligent Manufacturing: A Review. *Adv. Manuf.* **2021**, *9*, 1–21. [CrossRef]
16. Gunes, V. A Digital Twin Design Methodology for Control, Simulation, and Monitoring of Fluidic Circuits. *Int. J. Adv. Manuf. Technol.* **2024**, *134*, 3863–3875. [CrossRef]
17. Chen, R.; Chen, X.; Wang, L.; Li, J. The Core Industry Manufacturing Process of Electronics Assembly Based on Smart Manufacturing. *ACM Trans. Manag. Inf. Syst.* **2022**, *13*, 1–19. [CrossRef]
18. Lo, C.K.; Chen, C.H.; Zhong, R.Y. A Review of Digital Twin in Product Design and Development. *Adv. Eng. Inform.* **2021**, *48*, 101297. [CrossRef]
19. Gopal, L.; Singh, H.; Mounica, P.; Mohankumar, N.; Challa, N.P.; Jayaraman, P. Digital Twin and IOT Technology for Secure Manufacturing Systems. *Measur. Sens.* **2023**, *25*, 100661. [CrossRef]
20. Su, J.; Mao, Q.; Liao, Z.; Sheng, Z.; Huang, C.; Zhang, X. A Real-Time Cross-Domain WI-Fi-Based Gesture Recognition System for Digital Twins. *IEEE J. Sel. Areas Commun.* **2023**, *41*, 3690–3701. [CrossRef]
21. Wang, H.; Yang, Z.; Zhang, Q.; Sun, Q.; Lim, E. A Digital Twin Platform Integrating Process Parameter Simulation Solution for Intelligent Manufacturing. *Electronics* **2024**, *13*, 802. [CrossRef]
22. Azangoo, M.; Taherkordi, A.; Blech, J.O.; Vyatkin, V. Digital Twin-Assisted Controlling of AGVs in Flexible Manufacturing Environments. In Proceedings of the 2021 IEEE 30th International Symposium on Industrial Electronics (ISIE), Kyoto, Japan, 20–23 June 2021; pp. 1–7.
23. Liang, X.; Xiao, R.; Zhang, J. A Review on Digital Twin for Robotics in Smart Manufacturing. In Proceedings of the 2022 IEEE 17th Conference on Industrial Electronics and Applications (ICIEA), Chengdu, China, 16–19 December 2022; pp. 1510–1515.
24. Misra, A.; Agrawal, A.; Misra, V. Robotics in Industry 4.0. In *Handbook of Smart Materials, Technologies, and Devices*; Springer International Publishing: Cham, Switzerland, 2021; pp. 1–35.
25. Yan, D.; Sha, W.; Wang, D.; Yang, J.; Zhang, S. Digital Twin-Driven Variant Design of a 3C Electronic Product Assembly Line. *Sci. Rep.* **2022**, *12*, 3846. [CrossRef]
26. Li, X.; Niu, W.; Tian, H. Application of Digital Twin in Electric Vehicle Powertrain: A Review. *World Electr. Veh. J.* **2024**, *15*, 208. [CrossRef]
27. Wu, Q.; Mao, Y.; Chen, J.; Wang, C. Application Research of Digital Twin-Driven Ship Intelligent Manufacturing System: Pipe Machining Production Line. *J. Mar. Sci. Eng.* **2021**, *9*, 338. [CrossRef]
28. Mendi, A.F. A Digital Twin Case Study on Automotive Production Line. *Sensors* **2022**, *22*, 6963. [CrossRef]
29. Yang, J.; Son, Y.H.; Lee, D.; Noh, S.D. Digital Twin-Based Integrated Assessment of Flexible and Reconfigurable Automotive Part Production Lines. *Machines* **2022**, *10*, 75. [CrossRef]
30. Krenczyk, D.; Paprocka, I. Integration of Discrete Simulation, Prediction, and Optimization Methods for a Production Line Digital Twin Design. *Materials* **2023**, *16*, 2339. [CrossRef]
31. Yuan, G.; Liu, X.; Zhu, C.; Wang, C.; Zhu, M.; Sun, Y. Multi-Objective Coupling Optimization of Electrical Cable Intelligent Production Line Driven by Digital Twin. Robot. *Comput. Integr. Manuf.* **2024**, *86*, 102682. [CrossRef]

32. Guo, K.; Wan, X.; Liu, L.; Gao, Z.; Yang, M. Fault Diagnosis of Intelligent Production Line Based on Digital Twin and Improved Random Forest. *Appl. Sci.* **2021**, *11*, 7733. [CrossRef]
33. Lee, D.; Kim, C.-K.; Yang, J.; Cho, K.-Y.; Choi, J.; Noh, S.-D.; Nam, S. Digital Twin-Based Analysis and Optimization for Design and Planning of Production Lines. *Machines* **2022**, *10*, 1147. [CrossRef]
34. Kousi, N.; Gkournelos, C.; Aivaliotis, S.; Lotsaris, K.; Bavelos, A.C.; Baris, P.; Michalos, G.; Makris, S. Digital Twin for Designing and Reconfiguring Human–Robot Collaborative Assembly Lines. *Appl. Sci.* **2021**, *11*, 4620. [CrossRef]
35. Fan, Y.; Yang, J.; Chen, J.; Hu, P.; Wang, X.; Xu, J.; Zhou, B. A Digital-Twin Visualized Architecture for Flexible Manufacturing System. *J. Manuf. Syst.* **2021**, *60*, 176–201. [CrossRef]
36. Ricondo, I.; Porto, A.; Ugarte, M. A Digital Twin Framework for the Simulation and Optimization of Production Systems. *Procedia CIRP* **2021**, *104*, 762–767. [CrossRef]
37. Alam, M.D.; Kabir, G.; Mirmohammadsadeghi, S. A Digital Twin Framework Development for Apparel Manufacturing Industry. *Decis. Anal. J.* **2023**, *7*, 100252. [CrossRef]
38. Seidel, R.; Rachinger, B.; Thielen, N.; Schmidt, K.; Meier, S.; Franke, J. Development and Validation of a Digital Twin Framework for SMT Manufacturing. *Comput. Ind.* **2023**, *145*, 103831. [CrossRef]
39. Moeller, H.; Inamdar, A.; van Driel, W.D.; Bredberg, J.; Hille, P.; Knoll, H.; Vandevelde, B. Digital Twin Technology in Electronics. In *Recent Advances in Microelectronics Reliability*; Springer International Publishing: Cham, Switzerland, 2024; pp. 283–321.
40. Chen, H.; Zhang, Z.; Karamanakos, P.; Rodriguez, J. Digital Twin Techniques for Power Electronics-Based Energy Conversion Systems: A Survey of Concepts, Application Scenarios, Future Challenges, and Trends. *IEEE Ind. Electron. Mag.* **2023**, *17*, 20–36. [CrossRef]
41. Mansour, D.-E.A.; Numair, M.; Zalhaf, A.S.; Ramadan, R.; Darwish, M.M.F.; Huang, Q.; Hussien, M.G.; Abdel-Rahim, O. Applications of IoT and Digital Twin in Electrical Power Systems: A Comprehensive Survey. *IET Gener. Transm. Distrib.* **2023**, *17*, 4457–4479. [CrossRef]
42. Agarwal, R.; Bhatti, G.; Singh, R.R.; Indragandhi, V.; Suresh, V.; Jasinska, L.; Leonowicz, Z. Intelligent Fault Detection in Hall-Effect Rotary Encoders for Industry 4.0 Applications. *Electronics* **2022**, *11*, 3633. [CrossRef]
43. Lee, H.; Kim, S.D.; Amin, M.A.U.A. Control Framework for Collaborative Robot Using Imitation Learning-Based Teleoperation from Human Digital Twin to Robot Digital Twin. *Mechatronics* **2022**, *85*, 102833. [CrossRef]
44. Sun, Y.; Shi, Y.; Hu, Q.; Xie, C.; Su, T. DTformer: An Efficient Digital Twin Model for Loss Measurement in UHVDC Transmission Systems. *IEEE Trans. Power Syst.* **2024**, *39*, 3548–3559. [CrossRef]
45. Cai, W.; Zhang, Q.; Cui, J. A Novel Fault Diagnosis Method for Denoising Autoencoder Assisted by Digital Twin. *Comput. Intell. Neurosci.* **2022**, *2022*, 1–8. [CrossRef] [PubMed]
46. Ciganek, J.; Zemla, F. Design of Digital Twin for PLC System. In Proceedings of the 2022 Cybernetics & Informatics (K&I), Visegrad, Hungary, 11–14 September 2022; pp. 1–6.
47. Massaro, A. Artificial Intelligence Signal Control in Electronic Optocoupler Circuits Addressed on Industry 5.0 Digital Twin. *Electronics* **2024**, *13*, 4543. [CrossRef]
48. Zhang, Q.; Shen, S.; Li, H.; Cao, W.; Tang, W.; Jiang, J.; Deng, M.; Zhang, Y.; Gu, B.; Wu, K.; et al. Digital Twin-Driven Intelligent Production Line for Automotive MEMS Pressure Sensors. *Adv. Eng. Inform.* **2022**, *54*, 101779. [CrossRef]
49. Massaro, A. Advanced Electronic and Optoelectronic Sensors, Applications, Modelling and Industry 5.0 Perspectives. *Appl. Sci.* **2023**, *13*, 4582. [CrossRef]
50. Ali, A.R.; Ramadan, M.W.A. Artificial Neural Networks as Digital Twins for Whispering Gallery Mode Optical Sensors in Robotics Applications. *Photonic Sens.* **2025**, *15*, 250206. [CrossRef]
51. Massaro, A. *Electronics in Advanced Research Industries: Industry 4.0 to Industry 5.0 Advances*; Wiley: Hoboken, NJ, USA, 2021.
52. Oguz, I.; Suter, L.J.E.; Hsieh, J.-L.; Yildirim, M.; Dinc, N.U.; Moser, C.; Psaltis, D. Training Hybrid Neural Networks with Multimode Optical Nonlinearities Using Digital Twins. *Nanophotonics* **2025**, *14*, 2787–2797. [CrossRef]
53. Zhuge, Q. AI-Driven Digital Twin for Optical Networks. In Proceedings of the European Conference on Optical Communication (ECOC), Basel, Switzerland, 18–22 September 2022; pp. 1–62.
54. Lay-Ekuakille, A.; Massaro, A.; Singh, S.P.; Jabłoński, I.; Rahman, M.Z.U.; Spano, F. Optoelectronic and Nanosensors Detection Systems: A Review. *IEEE Sens. J.* **2021**, *21*, 12645–12653. [CrossRef]
55. Faruk, M.S.; Savory, S.J. Measurement Informed Models and Digital Twins for Optical Fiber Communication Systems. *J. Light. Technol.* **2024**, *42*, 1016–1030. [CrossRef]
56. Zhuge, Q.; Liu, X.; Zhang, Y.; Cai, M.; Liu, Y.; Qiu, Q.; Zhong, X.; Wu, J.; Gao, R.; Yi, L.; et al. Building a Digital Twin for Intelligent Optical Networks [Invited Tutorial]. *J. Opt. Commun. Netw.* **2023**, *15*, C242. [CrossRef]
57. Wang, D.; Zhang, Z.; Zhang, M.; Fu, M.; Li, J.; Cai, S.; Zhang, C.; Chen, X. The Role of Digital Twin in Optical Communication: Fault Management, Hardware Configuration, and Transmission Simulation. *IEEE Commun. Mag.* **2021**, *59*, 133–139. [CrossRef]

58. Borraccini, G.; Straullu, S.; Giorgetti, A.; Ambrosone, R.; Virgillito, E.; D'Amico, A.; D'Ingillo, R.; Aquilino, F.; Nespola, A.; Sambo, N.; et al. Experimental Demonstration of Partially Disaggregated Optical Network Control Using the Physical Layer Digital Twin. *IEEE Trans. Netw. Serv. Manag.* **2023**, *20*, 2343–2355. [CrossRef]
59. Koulountzios, P.; Rymarczyk, T.; Soleimani, M. A 4-D Ultrasound Tomography for Industrial Process Reactors Investigation. *IEEE Trans. Instrum. Meas.* **2022**, *71*, 1–14. [CrossRef]
60. Boparai, K.S.; Kumar, A.; Singh, R. Integration of 4D Imaging with 4D Printing. In *4D Imaging to 4D Printing*; CRC Press: Boca Raton, FL, USA, 2022; pp. 1–26, ISBN 9781003205531.
61. Wang, F.; Yue, Z.; Liu, J.; Qi, H.; Sun, W.; Chen, M.; Wang, Y.; Yue, H. Quantitative Imaging of Printed Circuit Board (PCB) Delamination Defects Using Laser-Induced Ultrasound Scanning Imaging. *J. Appl. Phys.* **2022**, *131*, 53101. [CrossRef]
62. Dai, L.-N.; Ni, C.-Y.; Ying, K.-N.; Yuan, L.; Kan, W.-W.; Shen, Z.-H. Defect Imaging Based on Laser Ultrasonic Frequency Domain Synthetic Aperture Focusing Technology with Separated Generation–Detection and 2-D Equivalent Velocity Mapping. *Opt. Laser Technol.* **2022**, *156*, 108485. [CrossRef]
63. Chen, L.; Liu, Z.; Tang, Z.; Duan, J.; Zhu, Y.; Zhang, Z.; Liu, X.; He, C. Optimization of Selection Matrix Capture for Micro Defects Laser Ultrasound Imaging Using Multi-Parameter Genetic Algorithm. *NDT E Int.* **2025**, *152*, 103325. [CrossRef]
64. Cieślak, S.; Zawadzki, P.; Gapsa, J. Digital Twin Application for Vision Control in the Production of Mechatronic Ladders. In *Lecture Notes in Mechanical Engineering*; Springer Nature Switzerland: Cham, Switzerland, 2024; pp. 216–227, ISBN 9783031564697.
65. Li, Q.; Ye, Q.J.; Zhang, N.; Zhang, W.; Hu, F. Digital-Twin-Enabled Industrial IoT: Vision, Framework, and Future Directions. *IEEE Wirel. Commun.* **2025**, 1–9. [CrossRef]
66. Salini, S.; Ivy, B.P.U. Digital Twin and Artificial Intelligence in Industries. In *Digital Twin for Smart Manufacturing*; Elsevier: Amsterdam, The Netherlands, 2023; pp. 35–58, ISBN 9780323992053.
67. Wang, S.; Zhang, J.; Wang, P.; Law, J.; Calinescu, R.; Mihaylova, L. A Deep Learning-Enhanced Digital Twin Framework for Improving Safety and Reliability in Human–Robot Collaborative Manufacturing. *Robot. Comput. Integr. Manuf.* **2024**, *85*, 102608. [CrossRef]
68. Dosoftei, C.-C. Simulation Power vs. Immersive Capabilities: Enhanced Understanding and Interaction with Digital Twin of a Mechatronic System. *Appl. Sci.* **2023**, *13*, 6463. [CrossRef]
69. Wei, Y.; Hu, T.; Yue, P.; Luo, W.; Ma, S. Study on the Construction Theory of Digital Twin Mechanism Model for Mechatronics Equipment. *Int. J. Adv. Manuf. Technol.* **2022**, *131*, 5383–5401. [CrossRef]
70. Classens, K.; Heemels, W.P.M.H.M.; Oomen, T. Digital Twins in Mechatronics: From Model-Based Control to Predictive Maintenance. In Proceedings of the 2021 IEEE 1st International Conference on Digital Twins and Parallel Intelligence (DTPI), Beijing, China, 15 July–15 August 2021; pp. 336–339.
71. Sturm, C.; Steck, M.; Bremer, F.; Revfi, S.; Nelius, T.; Gwosch, T.; Albers, A.; Matthiesen, S. Creation of Digital Twins—Key Characteristics of Physical to Virtual Twinning in Mechatronic Product Development. *Proc. Des. Soc.* **2021**, *1*, 781–790. [CrossRef]
72. Nezzi, C.; Gufler, V.; Vidoni, R.; Rauch, E. Kinematic and Dynamic Modeling of Mechanical Systems towards Digital Twins. *Results Eng.* **2025**, *26*, 104874. [CrossRef]
73. Chaiprabha, K.; Chancharoen, R. A Deep Trajectory Controller for a Mechanical Linear Stage Using Digital Twin Concept. *Actuators* **2023**, *12*, 91. [CrossRef]
74. Braun, D.; Riedhammer, M.; Jazdi, N.; Schloegl, W.; Weyrich, M. A Methodology for the Detection of Functional Relations of Mechatronic Components and Assemblies in Brownfield Systems. *Procedia CIRP* **2022**, *107*, 119–124. [CrossRef]
75. Jhunjhunwala, P.; Atmojo, U.D.; Vyatkin, V. Applying Skill-Based Engineering Using OPC-UA in Production System with a Digital Twin. In Proceedings of the 2021 IEEE 30th International Symposium on Industrial Electronics (ISIE), Kyoto, Japan, 20–23 June 2021; pp. 1–6.
76. Wang, J.; Niu, X.; Gao, R.X.; Huang, Z.; Xue, R. Digital Twin-Driven Virtual Commissioning of Machine Tool. *Robot. Comput. Integr. Manuf.* **2023**, *81*, 102499. [CrossRef]
77. Wu, L.; Leng, J.; Ju, B. Digital Twins-Based Smart Design and Control of Ultra-Precision Machining: A Review. *Symmetry* **2021**, *13*, 1717. [CrossRef]
78. Zaidi, A.; Keshta, I.; Gupta, Z.; Pundhir, P.; Pandey, T.; Rai, P.K.; Shabaz, M.; Soni, M. Smart Implementation of Industrial Internet of Things Using Embedded Mechatronic System. *IEEE Embed. Syst. Lett.* **2024**, *16*, 190–193. [CrossRef]
79. Ebadpour, M.; Jamshidi, M.; Talla, J.; Hashemi-Dezaki, H.; Peroutka, Z. Digital Twin Model of Electric Drives Empowered by EKF. *Sensors* **2023**, *23*, 2006. [CrossRef]
80. Liu, J.; Zhang, K. Design and Simulation Debugging of Automobile Connecting Rod Production Line Based on the Digital Twin. *Appl. Sci.* **2023**, *13*, 4919. [CrossRef]
81. Mazumder, A.; Sahed, M.F.; Tasneem, Z.; Das, P.; Badal, F.R.; Ali, M.F.; Ahamed, M.H.; Abhi, S.H.; Sarker, S.K.; Das, S.K.; et al. Towards next Generation Digital Twin in Robotics: Trends, Scopes, Challenges, and Future. *Heliyon* **2023**, *9*, e13359. [CrossRef] [PubMed]

82. Viola, J.; Chen, Y. Parallel Self Optimizing Control Framework for Digital Twin Enabled Smart Control Engineering. In Proceedings of the 2021 IEEE 1st International Conference on Digital Twins and Parallel Intelligence (DTPI), Beijing, China, 15 July–15 August 2021; pp. 358–361.
83. Massaro, A. Artificial Intelligence Enabling Denoising in Passive Electronic Filtering Circuits for Industry 5.0 Machines. *Machines* **2024**, *12*, 551. [CrossRef]
84. Massaro, A. ANNs Predicting Noisy Signals in Electronic Circuits: A Model Predicting the Signal Trend in Amplification Systems. *AI* **2024**, *5*, 533–549. [CrossRef]
85. Massaro, A.; Santarsiero, F.; Schiuma, G. Advanced Electronic Controller Circuits Enabling Production Processes and AI-Driven KM in Industry 5.0. *J. Ind. Inf. Integr.* **2025**, *45*, 100841. [CrossRef]
86. Massaro, A. Advanced Control Systems in Industry 5.0 Enabling Process Mining. *Sensors* **2022**, *22*, 8677. [CrossRef]
87. Keshmiri Neghab, H.; Jamshidi, M.; Keshmiri Neghab, H. Digital Twin of a Magnetic Medical Microrobot with Stochastic Model Predictive Controller Boosted by Machine Learning in Cyber-Physical Healthcare Systems. *Information* **2022**, *13*, 321. [CrossRef]
88. Alazab, M.; Khan, L.U.; Koppu, S.; Ramu, S.P.; Iyapparaja; Boobalan, P.; Baker, T.; Maddikunta, P.K.R.; Gadekallu, T.R.; Aljuhani, A. Digital Twins for Healthcare 4.0—Recent Advances, Architecture, and Open Challenges. *IEEE Consum. Electron. Mag.* **2023**, *12*, 29–37. [CrossRef]
89. Katsoulakis, E.; Wang, Q.; Wu, H.; Shahriyari, L.; Fletcher, R.; Liu, J.; Achenie, L.; Liu, H.; Jackson, P.; Xiao, Y.; et al. Digital Twins for Health: A Scoping Review. *NPJ Digit. Med.* **2024**, *7*, 77. [CrossRef] [PubMed]
90. Khan, S.; Saied, I.M.; Ratnarajah, T.; Arslan, T. Evaluation of Unobtrusive Microwave Sensors in Healthcare 4.0—Toward the Creation of Digital-Twin Model. *Sensors* **2022**, *22*, 8519. [CrossRef]
91. Massaro, A. Electronic Artificial Intelligence–Digital Twin Model for Optimizing Electroencephalogram Signal Detection. *Electronics* **2025**, *14*, 1122. [CrossRef]
92. Johnson, Z.; Saikia, M.J. Digital Twins for Healthcare Using Wearables. *Bioengineering* **2024**, *11*, 606. [CrossRef] [PubMed]
93. Shafik, W. The Next Generation of Health Monitoring: Digital Twins and Medical Wearables. In *AI-Powered Digital Twins for Predictive Healthcare: Creating Virtual Replicas of Humans*; Balasubramaniam, S., Kadry, S., Eds.; IGI Global Scientific Publishing: Hershey, PA, USA, 2025; pp. 69–106. [CrossRef]
94. Uhlenberg, L.; Derungs, A.; Amft, O. Co-Simulation of Human Digital Twins and Wearable Inertial Sensors to Analyse Gait Event Estimation. *Front. Bioeng. Biotechnol.* **2023**, *11*, 1104000. [CrossRef] [PubMed]
95. Yu, F.; Chen, Z.; Jiang, M.; Tian, Z.; Peng, T.; Hu, X. Smart Clothing System with Multiple Sensors Based on Digital Twin Technology. *IEEE Internet Things J.* **2023**, *10*, 6377–6387. [CrossRef]
96. Lyu, Z. Multisensor Data Fusion in Digital Twins for Smart Healthcare. In *Data Fusion Techniques and Applications for Smart Healthcare*; Elsevier: Amsterdam, The Netherlands, 2024; pp. 21–44, ISBN 9780443132339.
97. Sharma, V.; Sharma, K.; Kumar, A. AI and Digital Twins Transforming Healthcare IoT. In Proceedings of the 2024 14th International Conference on Cloud Computing, Data Science & Engineering (Confluence), Noida, India, 18–19 January 2024; pp. 6–11.
98. Elayan, H.; Aloqaily, M.; Guizani, M. Digital Twin for Intelligent Context-Aware IoT Healthcare Systems. *IEEE Internet Things J.* **2021**, *8*, 16749–16757. [CrossRef]
99. Rashid, S.M.; Ghiasi, A.R. Adaptive Digital Twin Integration with Multilevel Inverter Control for Energy Efficient Smart Rehabilitation Systems. *Sci. Rep.* **2025**, *15*, 8511. [CrossRef]
100. Lauer-Schmaltz, M.W.; Cash, P.; Paulin Hansen, J.; Das, N. Human Digital Twins in Rehabilitation: A Case Study on Exoskeleton and Serious-Game-Based Stroke Rehabilitation Using the ETHICA Methodology. *IEEE Access* **2024**, *12*, 180968–180991. [CrossRef]
101. Khan, M.M.R.; Sunny, M.S.H.; Ahmed, T.; Shahria, M.T.; Modi, P.P.; Zarif, M.I.I.; De Caro, J.D.S.; Ahamed, S.I.; Ahmed, H.U.; Ketchum, E.M.M.; et al. Development of a Robot-Assisted Telerehabilitation System with Integrated IIoT and Digital Twin. *IEEE Access* **2023**, *11*, 70174–70189. [CrossRef]
102. Harlan, B.; Tjahyadi, H. Minimally Invasive Motor Function Rehabilitation through Digital Twin Technology: A Review. In Proceedings of the 2024 2nd International Conference on Technology Innovation and Its Applications (ICTIIA), Medan, Indonesia, 12–13 September 2024; pp. 1–5.
103. Xie, S.; Zhan, M.; Li, Y.; Xi, F. The Virtual-Real Interaction System Design and Interaction Characteristics Research of an Ankle Rehabilitation Robot Based on Digital Twin. *Technol. Health Care* **2025**, 9287329251337237. [CrossRef]
104. De Fazio, R.; Mastronardi, V.M.; De Vittorio, M.; Visconti, P. Wearable Sensors and Smart Devices to Monitor Rehabilitation Parameters and Sports Performance: An Overview. *Sensors* **2023**, *23*, 1856. [CrossRef]
105. Dziurdzia, P.; Bratek, P.; Markiewicz, M. An Efficient Electrothermal Model of a Thermoelectric Converter for a Thermal Energy Harvesting Process Simulation and Electronic Circuits Powering. *Energies* **2023**, *17*, 204. [CrossRef]
106. Tabaie, Z.; Omidvar, A. Human Body Heat-Driven Thermoelectric Generators as a Sustainable Power Supply for Wearable Electronic Devices: Recent Advances, Challenges, and Future Perspectives. *Heliyon* **2023**, *9*, e14707. [CrossRef]
107. Zou, Y.; Bo, L.; Li, Z. Recent Progress in Human Body Energy Harvesting for Smart Bioelectronic System. *Fundam. Res.* **2021**, *1*, 364–382. [CrossRef]

108. Yang, Y.; Chen, Z.; Kuai, Q.; Liang, J.; Liu, J.; Zeng, X. Circuit Techniques for High Efficiency Piezoelectric Energy Harvesting. *Micromachines* **2022**, *13*, 1044. [CrossRef] [PubMed]
109. Haroun, A.; Le, X.; Gao, S.; Dong, B.; He, T.; Zhang, Z.; Wen, F.; Xu, S.; Lee, C. Progress in Micro/Nano Sensors and Nanoenergy for Future AIoT-Based Smart Home Applications. *Nano Ex.* **2021**, *2*, 22005. [CrossRef]
110. Jettanasen, C.; Songsukthawan, P.; Ngaopitakkul, A. Conversion of Mechanical Energy to Electrical Energy Using Piezoelectric Materials for Bicycle Lane Lighting Systems. *Appl. Sci.* **2022**, *12*, 7237. [CrossRef]
111. Carneiro, P.; Soares dos Santos, M.P.; Rodrigues, A.; Ferreira, J.A.F.; Simões, J.A.O.; Marques, A.T.; Kholkin, A.L. Electromagnetic Energy Harvesting Using Magnetic Levitation Architectures: A Review. *Appl. Energy* **2020**, *260*, 114191. [CrossRef]
112. Hao, D.; Fan, C.; Xia, X.; Zhang, Z.; Yang, Y. Hybrid Electromagnetic-Triboelectric Hip Energy Harvester for Wearables and AI-Assisted Motion Monitoring. *Small* **2025**, *21*, e2500643. [CrossRef]
113. Pang, Y.; He, T.; Liu, S.; Zhu, X.; Lee, C. Triboelectric Nanogenerator-Enabled Digital Twins in Civil Engineering Infrastructure 4.0: A Comprehensive Review. *Adv. Sci.* **2024**, *11*, e2306574. [CrossRef]
114. Jiang, J.; Fan, C.; Chen, H.; Wu, F.; Feng, X.; Xiao, C.; Pan, H.; Wu, X.; Zhang, Z. A Self-Powered Triboelectric Nano-Sensor Enabled Digital Twin for Self-Sustained Machine Monitoring in Smart Mine. *Nano Res.* **2025**, *18*, 94907287. [CrossRef]
115. Artetxe, E.; Uralde, J.; Barambones, O.; Calvo, I.; Martin, I. Maximum Power Point Tracker Controller for Solar Photovoltaic Based on Reinforcement Learning Agent with a Digital Twin. *Mathematics* **2023**, *11*, 2166. [CrossRef]
116. Chalal, L.; Saadane, A.; Rachid, A. Unified Environment for Real Time Control of Hybrid Energy System Using Digital Twin and IoT Approach. *Sensors* **2023**, *23*, 5646. [CrossRef] [PubMed]
117. Kabir, M.R.; Halder, D.; Ray, S. Digital Twins for IoT-Driven Energy Systems: A Survey. *IEEE Access* **2024**, *12*, 177123–177143. [CrossRef]
118. Hossen, T.; Gursoy, M.; Mirafzal, B. Digital Twin for Self-Security of Smart Inverters. In Proceedings of the 2021 IEEE Energy Conversion Congress and Exposition (ECCE), Virtual, 10–14 October 2021; pp. 713–718.
119. Numair, M.; Aboushady, A.A.; Arraño-Vargas, F.; Farrag, M.E.; Elyan, E. Fault Detection and Localisation in LV Distribution Networks Using a Smart Meter Data-Driven Digital Twin. *Energies* **2023**, *16*, 7850. [CrossRef]
120. Yue, G.; Du, Z.; Xiu, L.; Meng, S.; Ma, Z.; Xie, Y. Electric Field Equivalent Method for AC/DC Hybrid Electric Equipment Adapted to Digital Twin with Low Latency. *Electr. Power Syst. Res.* **2022**, *213*, 108799. [CrossRef]
121. Yuan, Z.; Pan, Y.; Wang, H.; Wang, S.; Peng, Y.; Jin, C.; Xu, C.; Feng, X.; Shen, K.; Zheng, Y.; et al. Fault Data Generation of Lithium Ion Batteries Based on Digital Twin: A Case for Internal Short Circuit. *J. Energy Storage* **2023**, *64*, 107113. [CrossRef]
122. Li, Q.; He, Y. An Overview of Digital Twin Concept for Key Components of Renewable Energy Systems. *Int. J. Robot. Autom. Technol.* **2021**, *8*, 29–47. [CrossRef]
123. Das, O.; Zafar, M.H.; Sanfilippo, F.; Rudra, S.; Kolhe, M.L. Advancements in Digital Twin Technology and Machine Learning for Energy Systems: A Comprehensive Review of Applications in Smart Grids, Renewable Energy, and Electric Vehicle Optimisation. *Energy Convers. Manag. X* **2024**, *24*, 100715. [CrossRef]
124. Massaro, A.; Starace, G. Advanced and Complex Energy Systems Monitoring and Control: A Review on Available Technologies and Their Application Criteria. *Sensors* **2022**, *22*, 4929. [CrossRef]
125. Bayer, D.; Pruckner, M. A Digital Twin of a Local Energy System Based on Real Smart Meter Data. *Energy Inform.* **2023**, *6*, 8. [CrossRef]
126. Meyer, L.; Jorstad, K.L.; Mortensen, L.K.; Santos, A.Q.; Handrup, K.; Shaker, H.R. Digital Twin of Radio-Mesh Smart Meter Communication for Outage Management Using a Multi-Agent System. In Proceedings of the 2022 IEEE 10th International Conference on Smart Energy Grid Engineering (SEGE), Oshawa, ON, Canada, 10–12 August 2022; pp. 119–124.
127. Lopez-Lorente, J.; Xydas, C.M.; Makrides, G.; Georghiou, G.E. Evaluating Voltage Estimation in a Nanogrid Using Digital Twin Models and Real-Time Smart Meter Data. In Proceedings of the 2022 International Conference on Smart Energy Systems and Technologies (SEST), Eindhoven, The Netherlands, 5–7 September 2022; pp. 1–6.
128. Piras, G.; Agostinelli, S.; Muzi, F. Smart Buildings and Digital Twin to Monitoring the Efficiency and Wellness of Working Environments: A Case Study on IoT Integration and Data-Driven Management. *Appl. Sci.* **2025**, *15*, 4939. [CrossRef]
129. Starace, G.; Tiwari, A.; Colangelo, G.; Massaro, A. Advanced Data Systems for Energy Consumption Optimization and Air Quality Control in Smart Public Buildings Using a Versatile Open Source Approach. *Electronics* **2022**, *11*, 3904. [CrossRef]
130. Eneyew, D.D.; Capretz, M.A.M.; Bitsuamlak, G.T. Toward Smart-Building Digital Twins: BIM and IoT Data Integration. *IEEE Access* **2022**, *10*, 130487–130506. [CrossRef]
131. Abdullahi, I.; Longo, S.; Samie, M. Towards a Distributed Digital Twin Framework for Predictive Maintenance in Industrial Internet of Things (IIoT). *Sensors* **2024**, *24*, 2663. [CrossRef]
132. Kherbache, M.; Maimour, M.; Rondeau, E. Network Digital Twin for the Industrial Internet of Things. In Proceedings of the 2022 IEEE 23rd International Symposium on a World of Wireless, Mobile and Multimedia Networks (WoWMoM), Belfast, UK, 14–17 June 2022; pp. 573–578.

133. Santos, J.F.D.; Tshoombe, B.K.; Santos, L.H.B.; Araujo, R.C.F.; Manito, A.R.A.; Fonseca, W.S.; Silva, M.O. Digital Twin-Based Monitoring System of Induction Motors Using IoT Sensors and Thermo-Magnetic Finite Element Analysis. *IEEE Access* **2023**, *11*, 1682–1693. [CrossRef]
134. Caiza, G.; Sanz, R. Digital Twin to Control and Monitor an Industrial Cyber-Physical Environment Supported by Augmented Reality. *Appl. Sci.* **2023**, *13*, 7503. [CrossRef]
135. Wang, K.-J.; Lee, Y.-H.; Angelica, S. Digital Twin Design for Real-Time Monitoring—A Case Study of Die Cutting Machine. *Int. J. Prod. Res.* **2021**, *59*, 6471–6485. [CrossRef]
136. Ranpariya, A.; Sharma, S. A Digital Twin Solution for Fault Detection in Time-Critical IIoT Applications. *J. Simul.* **2025**, *19*, 441–454. [CrossRef]
137. Wu, W.; Zhao, Z.; Shen, L.; Kong, X.T.R.; Guo, D.; Zhong, R.Y.; Huang, G.Q. Just Trolley: Implementation of Industrial IoT and Digital Twin-Enabled Spatial-Temporal Traceability and Visibility for Finished Goods Logistics. *Adv. Eng. Inform.* **2022**, *52*, 101571. [CrossRef]
138. Guo, D.; Zhong, R.Y.; Rong, Y.; Huang, G.G.Q. Synchronization of Shop-Floor Logistics and Manufacturing under IIoT and Digital Twin-Enabled Graduation Intelligent Manufacturing System. *IEEE Trans. Cybern.* **2023**, *53*, 2005–2016. [CrossRef]
139. Hu, P.; He, C.; Zhu, Y.; Li, T. The Product Quality Inspection Scheme Based on Software-Defined Edge Intelligent Controller in Industrial Internet of Things. *J. Cloud Comput. Adv. Syst. Appl.* **2023**, *12*, 113. [CrossRef]
140. Schrader, B.; Guerra-Zubiaga, D.; McMichael, G. Soft Sensor Digital Twin Implementation of a Pick-and-Place Operation. In Proceedings of the Volume 3: Advanced Manufacturing, New Orleans, LA, USA, 29 October–2 November 2023; American Society of Mechanical Engineers: New York, NY, USA, 2023.
141. Bolbotinović, Ž.; Milić, S.D.; Janda, Ž.; Vukmirović, D. Ai-Powered Digital Twin in the Industrial IoT. *Int. J. Electr. Power Energy Syst.* **2025**, *167*, 110656. [CrossRef]
142. Wang, K.; Wang, X.; Tan, C.; Dong, S.; Zhao, F.; Lian, S. Digital Twin-Enhanced Robotic System for Remote Diesel Engine Assembly Defect Inspection. *Ind. Rob.* **2025**, *52*, 266–276. [CrossRef]
143. Rodriguez-Aguilar, R.; Marmolejo-Saucedo, J.-A.; Köse, U. Development of a Digital Twin Driven by a Deep Learning Model for Fault Diagnosis of Electro-Hydrostatic Actuators. *Mathematics* **2024**, *12*, 3124. [CrossRef]
144. Sheraz, M.; Chuah, T.C.; Lee, Y.L.; Alam, M.M.; Al-Habashna, A.; Han, Z. A Comprehensive Survey on Revolutionizing Connectivity through Artificial Intelligence-Enabled Digital Twin Network in 6G. *IEEE Access* **2024**, *12*, 49184–49215. [CrossRef]
145. Alcaraz, C.; Lopez, J. Protecting Digital Twin Networks for 6G-Enabled Industry 5.0 Ecosystems. *IEEE Netw.* **2023**, *37*, 302–308. [CrossRef]
146. Dos Santos, J.F.; Lisboa, Y.M.; Barbosa, E.F.; Santos, L.H.; Manito, A.R.; Silva, M.O.; Fonseca, W.S. Motor Monitoring and Analysis System (SMAM) as a Resource for Predictive Maintenance in Industrial Environments through Digital Twin and IIoT. In *Induction Motors—Latest Research and Applications [Working Title]*; El-Shahat, A., Ed.; IntechOpen: London, UK, 2024.
147. Tang, Y.M.; Kuo, W.T.; Lee, C.K.M. Real-Time Mixed Reality (MR) and Artificial Intelligence (AI) Object Recognition Integration for Digital Twin in Industry 4.0. *Internet Things* **2023**, *23*, 100753. [CrossRef]
148. Min, Z.; Zhou, S.; Kang, Z.; Shekhar, S.; Mahmoudi, C.; Gokhale, S.; Gokhale, A. Managing and Optimizing 5G & beyond Network Resources for Multi-Task Digital Twin Applications in Industry 4.0. In Proceedings of the 2023 IEEE 26th International Symposium on Real-Time Distributed Computing (ISORC), Nashville, TN, USA, 23–25 May 2023; pp. 220–223.
149. Arbabian, A.; Pezeshk, A.; Yang, K. Productivity Enhancement in Semiconductor Manufacturing with Ai-Enabled Operations Digital Twin Platform. In Proceedings of the 2024 19th International Microsystems, Packaging, Assembly and Circuits Technology Conference (IMPACT), Taipei, Taiwan, 22–25 October 2024; pp. 280–283.
150. Massaro, A. Intelligent Materials and Nanomaterials Improving Physical Properties and Control Oriented on Electronic Implementations. *Electronics* **2023**, *12*, 3772. [CrossRef]
151. Corradini, F.; Silvestri, M. Design and Testing of a Digital Twin for Monitoring and Quality Assessment of Material Extrusion Process. *Addit. Manuf.* **2022**, *51*, 102633. [CrossRef]
152. Lugaresi, G.; Matta, A. Automated Digital Twin Generation of Manufacturing Systems with Complex Material Flows: Graph Model Completion. *Comput. Ind.* **2023**, *151*, 103977. [CrossRef]
153. Turan, E.; Konuşkan, Y.; Yıldırım, N.; Tunçalp, D.; İnan, M.; Yasin, O.; Turan, B.; Kerimoğlu, V. Digital Twin Modelling for Optimizing the Material Consumption: A Case Study on Sustainability Improvement of Thermoforming Process. *Sustain. Comput. Inform. Syst.* **2022**, *35*, 100655. [CrossRef]
154. Modoni, G.E.; Stampone, B.; Trotta, G. Application of the Digital Twin for in Process Monitoring of the Micro Injection Moulding Process Quality. *Comput. Ind.* **2022**, *135*, 103568. [CrossRef]
155. Psarommatis, F. A Generic Methodology and a Digital Twin for Zero Defect Manufacturing (ZDM) Performance Mapping towards Design for ZDM. *J. Manuf. Syst.* **2021**, *59*, 507–521. [CrossRef]
156. Henson, C.M.; Decker, N.I.; Huang, Q. A Digital Twin Strategy for Major Failure Detection in Fused Deposition Modeling Processes. *Procedia Manuf.* **2021**, *53*, 359–367. [CrossRef]

157. Rachmawati, S.M.; Putra, M.A.P.; Lee, J.M.; Kim, D.S. Digital Twin-Enabled 3D Printer Fault Detection for Smart Additive Manufacturing. *Eng. Appl. Artif. Intell.* **2023**, *124*, 106430. [CrossRef]
158. Chen, L.; Bi, G.; Yao, X.; Tan, C.; Su, J.; Ng, N.P.H.; Chew, Y.; Liu, K.; Moon, S.K. Multisensor Fusion-Based Digital Twin for Localized Quality Prediction in Robotic Laser-Directed Energy Deposition. *Robot. Comput. Integr. Manuf.* **2023**, *84*, 102581. [CrossRef]
159. Kim, W.; Lee, G.; Son, H.; Choi, H.; Youn, B.D. Estimation of Fatigue Crack Initiation and Growth in Engineering Product Development Using a Digital Twin Approach. *Reliab. Eng. Syst. Saf.* **2022**, *226*, 108721. [CrossRef]
160. Li, X.; Li, G.; Bi, Z. Digital Twins to Predict Crack Propagation of Sustainable Engineering Materials under Different Loads. *Machines* **2024**, *12*, 125. [CrossRef]
161. Fang, X.; Liu, G.; Wang, H.; Tian, X. A Digital Twin Modeling Method Based on Multi-Source Crack Growth Prediction Data Fusion. *Eng. Fail. Anal.* **2023**, *154*, 107645. [CrossRef]
162. Piltan, F.; Kim, J.-M. Crack Size Identification for Bearings Using an Adaptive Digital Twin. *Sensors* **2021**, *21*, 5009. [CrossRef] [PubMed]
163. Fang, X.; Wang, H.; Li, W.; Liu, G.; Cai, B. Fatigue Crack Growth Prediction Method for Offshore Platform Based on Digital Twin. *Ocean Eng.* **2022**, *244*, 110320. [CrossRef]
164. Qiu, A.; Aakyiir, M.; Wang, R.; Yang, Z.; Umer, A.; Lee, I.; Hsu, H.-Y.; Ma, J. Stretchable and Calibratable Graphene Sensors for Accurate Strain Measurement. *Mater. Adv.* **2020**, *1*, 235–243. [CrossRef]
165. Song, C.; Lee, H.; Park, C.; Lee, B.; Kim, J.; Park, C.; Lai, C.H.; Cho, S.J. Advances in Crack-Based Strain Sensors on Stretchable Polymeric Substrates: Crack Mechanisms, Geometrical Factors, and Functional Structures. *Polymers* **2025**, *17*, 941. [CrossRef] [PubMed]
166. Al-Hawawreh, M.; Hossain, M.S. Digital Twin-Driven Secured Edge-Private Cloud Industrial Internet of Things (IIoT) Framework. *J. Netw. Comput. Appl.* **2024**, *226*, 103888. [CrossRef]
167. Alcaraz, C.; Lopez, J. Digital Twin: A Comprehensive Survey of Security Threats. *IEEE Commun. Surv. Tutor.* **2022**, *24*, 1475–1503. [CrossRef]
168. Melis, A.; Piroddi, A.; Kaya, O.; Girau, R. An IIoT Digital Twin Architecture for Anomaly Traffic Analysis of HWIL Devices Using Neural Network. In Proceedings of the 2024 IEEE 29th International Workshop on Computer Aided Modeling and Design of Communication Links and Networks (CAMAD), Athens, Greece, 21–23 October 2024; pp. 1–6.
169. Soudan, B. Cybersecurity of Digital Twins in Industrial IoT Environments. In Proceedings of the 2024 Advances in Science and Engineering Technology International Conferences (ASET), Abu Dhabi, United Arab Emirates, 3–5 June 2024; pp. 1–6.
170. Ahmed, M.A.; Alnatheer, S. Intrusion Detection in a Digital Twin-Enabled Secure Industrial Internet of Things Environment for Industrial Sustainability. *Eng. Technol. Appl. Sci. Res.* **2025**, *15*, 21263–21269. [CrossRef]
171. Balta, E.C.; Pease, M.; Moyne, J.; Barton, K.; Tilbury, D.M. Digital Twin-Based Cyber-Attack Detection Framework for Cyber-Physical Manufacturing Systems. *IEEE Trans. Autom. Sci. Eng.* **2024**, *21*, 1695–1712. [CrossRef]
172. Massaro, A.; Gargaro, M.; Dipierro, G.; Galiano, A.M.; Buonopane, S. Prototype Cross Platform Oriented on Cybersecurity, Virtual Connectivity, Big Data and Artificial Intelligence Control. *IEEE Access* **2020**, *8*, 197939–197954. [CrossRef]
173. Harrison, L. Cybersecurity Threat Modeling and Mitigation Using the Digital Twin. *ATZ Electron Worldw* **2022**, *17*, 40–43. [CrossRef]
174. Dietz, M.; Schlette, D.; Pernul, G. Harnessing Digital Twin Security Simulations for Systematic Cyber Threat Intelligence. In Proceedings of the 2022 IEEE 46th Annual Computers, Software, and Applications Conference (COMPSAC), Los Alamitos, CA, USA, 27 June–1 July 2022; pp. 789–797.
175. Vasilica, B.-V.; Anton, F.-D.; Pietraru, R.; Anton, S.-O.; Chiriac, B.-N. Enhancing Security in Smart Robot Digital Twins Through Intrusion Detection Systems. *Appl. Sci.* **2025**, *15*, 4596. [CrossRef]
176. Atal, D.K.; Tiwari, V.; Kumar, D. Recent Advances in Cybersecurity in Smart Manufacturing Systems in the Industry. In *Handbook of Smart Manufacturing*; CRC Press: Boca Raton, FL, USA, 2023; pp. 41–62, ISBN 9781003333760.
177. Jeffrey, N.; Tan, Q.; Villar, J.R. A Review of Anomaly Detection Strategies to Detect Threats to Cyber-Physical Systems. *Electronics* **2023**, *12*, 3283. [CrossRef]
178. Xu, Z.; Zhang, Z.; Zheqiu, H. PLC-MDT: A Framework for Detecting Anomalies with Digital Twins of Industrial Control Systems. *IEEE Sens. J.* **2025**, *25*, 17739–17749. [CrossRef]
179. Francia, G., III; El-Sheikh, E. Open Platform Infrastructure for Industrial Control Systems Security. In *Lecture Notes in Networks and Systems*; Springer Nature Switzerland: Cham, Switzerland, 2024; pp. 233–243, ISBN 9783031569494.
180. Filipescu, A.; Simion, G.; Ionescu, D.; Filipescu, A. IoT-Cloud, VPN, and Digital Twin-Based Remote Monitoring and Control of a Multifunctional Robotic Cell in the Context of AI, Industry, and Education 4.0 and 5.0. *Sensors* **2024**, *24*, 7451. [CrossRef]
181. Simion, G.; Filipescu, A.; Ionescu, D.; Filipescu, A. Cloud/VPN-Based Remote Control of a Modular Production System Assisted by a Mobile Cyber–Physical Robotic System—Digital Twin Approach. *Sensors* **2025**, *25*, 591. [CrossRef]

182. Prathiba, S.B.; Govindarajan, Y.; Pranav Amirtha Ganesan, V.; Ramachandran, A.; Selvaraj, A.K.; Kashif Bashir, A.; Reddy Gadekallu, T. Fortifying Federated Learning in IIoT: Leveraging Blockchain and Digital Twin Innovations for Enhanced Security and Resilience. *IEEE Access* **2024**, *12*, 68968–68980. [CrossRef]
183. Onwubiko, A.; Singh, R.; Awan, S.; Pervez, Z.; Ramzan, N. Enabling Trust and Security in Digital Twin Management: A Blockchain-Based Approach with Ethereum and IPFS. *Sensors* **2023**, *23*, 6641. [CrossRef] [PubMed]
184. Hosseini Bamakan, S.M.; Banaeian Far, S. Distributed and Trustworthy Digital Twin Platform Based on Blockchain and Web3 Technologies. *Cyber Secur. Appl.* **2025**, *3*, 100064. [CrossRef]
185. Homaei, M.; Mogollón-Gutiérrez, Ó.; Sancho, J.C.; Ávila, M.; Caro, A. A Review of Digital Twins and Their Application in Cybersecurity Based on Artificial Intelligence. *Artif. Intell. Rev.* **2024**, *57*, 201. [CrossRef]
186. Liu, J.; Yeoh, W.; Qu, Y.; Gao, L. Blockchain-Based Digital Twin for Supply Chain Management: State-of-the-Art Review and Future Research Directions. *arXiv* **2022**, arXiv:2202.03966. [CrossRef]
187. Liu, J.; Yeoh, W.; Gao, S.; Qu, Y.; Gao, L. Blockchain-Based Digital Twin for Supply Chain Management: A Survey. In *Digital Twins for Smart Cities and Villages*; Elsevier: Amsterdam, The Netherlands, 2025; pp. 581–600, ISBN 9780443288845.
188. Awotunde, J.B.; Adeniyi, A.E.; Falola, P.B.; Adeniran, T.C. A Blockchain-Enabled Framework to Enhance Cybersecurity Challenges in Smart Agriculture. In *Blockchain and Digital Twin Applications in Smart Agriculture*; Auerbach Publications: New York, NY, USA, 2025; pp. 69–91, ISBN 9781003507390.
189. Hayashi, V.T.; Vicente Ruggiero, W. Hardware Trojan Detection in Open-Source Hardware Designs Using Machine Learning. *IEEE Access* **2025**, *13*, 37771–37788. [CrossRef]
190. Monjur, M.M.R.; Heacock, J.; Calzadillas, J.; Mahmud, M.D.S.; Roth, J.; Mankodiya, K.; Sazonov, E.; Yu, Q. Hardware Security in Sensor and Its Networks. *Front. Sens.* **2022**, *3*, 850056. [CrossRef]
191. Monjur, M.R.; Sunkavilli, S.; Yu, Q. ADobf: Obfuscated Detection Method against Analog Trojans on I^{2}C Master-Slave Interface. In Proceedings of the 2020 IEEE 63rd International Midwest Symposium on Circuits and Systems (MWSCAS), Springfield, IL, USA, 9–12 August 2020; pp. 1064–1067.
192. Monjur, M.; Calzadillas, J.; Yu, Q. Hardware Security Risks and Threat Analyses in Advanced Manufacturing Industry. *ACM Trans. Des. Automat. Electron. Syst.* **2023**, *28*, 1–22. [CrossRef]
193. Massaro, A.; Epicoco, N.; Loseto, G.; Ardito, C.A. AI-Based Detection and Simulation of Hardware Trojan Attacks in Industrial Applications. *IEEE Access* **2025**, *13*, 50967–50979. [CrossRef]
194. Massaro, A.; Epicoco, N.; Loseto, G.; Ardito, C. IIOT Cyberattacks in Industrial Field Control Systems: PID Attack and Hardware Trojan Effects. In Proceedings of the 2024 IEEE 20th International Conference on Automation Science and Engineering (CASE), Bari, Italy, 28 August–1 September 2024; pp. 1702–1707.
195. Zhao, F.; Wu, Y.; Hu, M.; Chang, C.-W.; Liu, R.; Qiu, R.; Yang, X. Current Progress of Digital Twin Construction Using Medical Imaging. *arXiv* **2024**, arXiv:2411.08173. [CrossRef]
196. Langton, C.M.; Grimm, A.; Lloyd, D.G.; Frossard, L.A. A 3D-Printed Phantom Twin and Multi-Transducer Holder for Dynamic Anatomical Ultrasonography of the Lower Limb. *J. 3D Print. Med.* **2023**, *7*, 3DP009. [CrossRef]
197. Werner, H.; Tonni, G.; Lopes, J. *3D Physical and Virtual Models in Fetal Medicine: Applications and Procedures*; Springer International Publishing: Cham, Switzerland, 2023; ISBN 9783031148545.
198. Jost, E.; Kosian, P.; Jimenez Cruz, J.; Albarqouni, S.; Gembruch, U.; Strizek, B.; Recker, F. Evolving the Era of 5D Ultrasound? A Systematic Literature Review on the Applications for Artificial Intelligence Ultrasound Imaging in Obstetrics and Gynecology. *J. Clin. Med.* **2023**, *12*, 6833. [CrossRef] [PubMed]
199. Manohar, S.; Sechopoulos, I.; Anastasio, M.A.; Maier-Hein, L.; Gupta, R. Super Phantoms: Advanced Models for Testing Medical Imaging Technologies. *Commun. Eng.* **2024**, *3*, 73. [CrossRef]
200. Sermesant, M.; Delingette, H.; Cochet, H.; Jaïs, P.; Ayache, N. Applications of Artificial Intelligence in Cardiovascular Imaging. *Nat. Rev. Cardiol.* **2021**, *18*, 600–609. [CrossRef]
201. Allenby, M.C.; Woodruff, M.A. Image Analyses for Engineering Advanced Tissue Biomanufacturing Processes. *Biomaterials* **2022**, *284*, 121514. [CrossRef]
202. Martinez-Ruedas, C.; Flores-Arias, J.-M.; Moreno-Garcia, I.M.; Linan-Reyes, M.; Bellido-Outeiriño, F.J. A Cyber–Physical System Based on Digital Twin and 3D SCADA for Real-Time Monitoring of Olive Oil Mills. *Technologies* **2024**, *12*, 60. [CrossRef]
203. Javaid, M.; Haleem, A.; Suman, R. Digital Twin Applications toward Industry 4.0: A Review. *Cogn. Robot.* **2023**, *3*, 71–92. [CrossRef]
204. Lv, Z. Digital Twins in Industry 5.0. *Research* **2023**, *6*, 71. [CrossRef]
205. Wang, Y.; Wang, S.; Yang, B.; Zhu, L.; Liu, F. Big Data Driven Hierarchical Digital Twin Predictive Remanufacturing Paradigm: Architecture, Control Mechanism, Application Scenario and Benefits. *J. Clean. Prod.* **2020**, *248*, 119299. [CrossRef]
206. Ruhe, S.; Schaefer, K.; Branz, S.; Nicolai, S.; Bretschneider, P.; Westermann, D. Design and Implementation of a Hierarchical Digital Twin for Power Systems Using Real-Time Simulation. *Electronics* **2023**, *12*, 2747. [CrossRef]

207. Martinelli, M.; Zhang, J.; Splettstoßer, A.-K.; Picone, M.; Lippi, M.; Wortmann, A. Hierarchical Digital Twin Ecosystem for Industrial Manufacturing Scenarios. In Proceedings of the 2024 50th Euromicro Conference on Software Engineering and Advanced Applications (SEAA), Paris, France, 28–30 August 2024; pp. 56–63.
208. Finke, C.; Groth, M.; Schumann, M.; Dewitz, P.; Gehrke, J.; Marahrens, T. Design and Implementation of Hierarchical Digital Twins in Industrial Production Environments. In Proceedings of the Hawaii International Conference on System Sciences, Maui, HI, USA, 3–6 January 2023; pp. 1448–1457. [CrossRef]
209. Massaro, A.; Ricci, G.; Selicato, S.; Raminelli, S.; Galiano, A. Decisional Support System with Artificial Intelligence Oriented on Health Prediction Using a Wearable Device and Big Data. In Proceedings of the 2020 IEEE International Workshop on Metrology for Industry 4.0 & IoT, Rome, Italy, 3–5 June 2020; pp. 718–723.
210. Figueiras, P.; Lourenco, L.; Costa, R.; Graca, D.; Garcia, G.; Jardim-Goncalves, R. Big Data Provision for Digital Twins in Industry 4.0 Logistics Processes. In Proceedings of the 2021 IEEE International Workshop on Metrology for Industry 4.0 & IoT (MetroInd4.0&IoT), Virtual, 7–9 June 2021; pp. 516–521.
211. Zhou, Q.; Li, J.; Zhang, C.; Xu, H. Digital Twin. In *Big Data and Electric Mobility*; CRC Press: Boca Raton, FL, USA, 2025; pp. 74–98, ISBN 9781003302827.
212. Aublet, A.; N'Guyen, F.; Proudhon, H.; Ryckelynck, D. Multimodal Data Augmentation for Digital Twining Assisted by Artificial Intelligence in Mechanics of Materials. *Front. Mater.* **2022**, *9*, 971816. [CrossRef]
213. Lyu, P.; Zhang, H.; Yu, W.; Liu, C. A Novel Model-Independent Data Augmentation Method for Fault Diagnosis in Smart Manufacturing. *Procedia CIRP* **2022**, *107*, 949–954. [CrossRef]
214. Kies, A.D.; Krauß, J.; Schmetz, A.; Schmitt, R.H.; Brecher, C. Interaction of Digital Twins in a Sustainable Battery Cell Production. *Procedia CIRP* **2022**, *107*, 1216–1220. [CrossRef]
215. Kamble, S.S.; Gunasekaran, A.; Parekh, H.; Mani, V.; Belhadi, A.; Sharma, R. Digital Twin for Sustainable Manufacturing Supply Chains: Current Trends, Future Perspectives, and an Implementation Framework. *Technol. Forecast. Soc. Change* **2022**, *176*, 121448. [CrossRef]
216. Davila, R.M.F.; Schwark, F.; Dawel, L.; Pehlken, A. Sustainability Digital Twin: A Tool for the Manufacturing Industry. *Procedia CIRP* **2023**, *116*, 143–148. [CrossRef]
217. Ma, S.; Ding, W.; Liu, Y.; Ren, S.; Yang, H. Digital Twin and Big Data-Driven Sustainable Smart Manufacturing Based on Information Management Systems for Energy-Intensive Industries. *Appl. Energy* **2022**, *326*, 119986. [CrossRef]
218. Franciosi, C.; Miranda, S.; Veneroso, C.R.; Riemma, S. Improving Industrial Sustainability by the Use of Digital Twin Models in Maintenance and Production Activities. *IFAC-Pap.* **2022**, *55*, 37–42. [CrossRef]
219. Massaro, A. Process Mining in Production Management, Intelligent Control, and Advanced KPI for Dynamic Process Optimization: Industry 5.0 Production Processes. In *IGI Global, Perspectives and Considerations on the Evolution of Smart Systems*; IGI: Hershey, PA, USA, 2023; Chapter 1, pp. 1–17.
220. Fakhar Manesh, M.; Pellegrini, M.M.; Marzi, G.; Dabic, M. Knowledge Management in the Fourth Industrial Revolution: Mapping the Literature and Scoping Future Avenues. *IEEE Trans. Eng. Manage.* **2021**, *68*, 289–300. [CrossRef]
221. van Eck, N.; Waltman, L.; van den Berg, J.; Kaymak, U. Visualizing the Computational Intelligence Field [Application Notes]. *IEEE Comput. Intell. Mag.* **2006**, *1*, 6–10. [CrossRef]
222. Dong, C.; Shafiq, M.; Dabel, M.M.A.; Sun, Y.; Tian, Z. DNN Inference Acceleration for Smart Devices in Industry 5.0 by Decentralized Deep Reinforcement Learning. *IEEE Trans. Consum. Electron.* **2024**, *70*, 1519–1530. [CrossRef]
223. Sharma, M.; Tomar, A.; Hazra, A. Edge Computing for Industry 5.0: Fundamental, Applications, and Research Challenges. *IEEE Internet Things J.* **2024**, *11*, 19070–19093. [CrossRef]
224. Zhang, Y.; He, G.; Ma, L.; Liu, X.; Hjorth, J.J.J.; Kozlov, A.; He, Y.; Zhang, S.; Kotaleski, J.H.; Tian, Y.; et al. A GPU-Based Computational Framework That Bridges Neuron Simulation and Artificial Intelligence. *Nat. Commun.* **2023**, *14*, 5798. [CrossRef] [PubMed]
225. Chang, Z.; Liu, S.; Xiong, X.; Cai, Z.; Tu, G. A Survey of Recent Advances in Edge-Computing-Powered Artificial Intelligence of Things. *IEEE Internet Things J.* **2021**, *8*, 13849–13875. [CrossRef]
226. Seetohul, V.; Jahankhani, H.; Kendzierskyj, S.; Will Arachchige, I.S. Quantum Reinforcement Learning: Advancing AI Agents through Quantum Computing. In *Space Law and Policy*; Springer Nature Switzerland: Cham, Switzerland, 2024; pp. 55–73, ISBN 9783031640445.
227. Schmidt, B.; Hildebrandt, A. From GPUs to AI and Quantum: Three Waves of Acceleration in Bioinformatics. *Drug Discov. Today* **2024**, *29*, 103990. [CrossRef] [PubMed]
228. Stavropoulos, P.; Papacharalampopoulos, A.; Christopoulos, D. Use of Artificial Intelligence at the Level of Manufacturing Processes. In *Lecture Notes in Mechanical Engineering*; Springer Nature Switzerland: Cham, Switzerland, 2024; pp. 157–166, ISBN 9783031574955.
229. Massaro, A. Electronic Digital Twin in Industry 5.0: AI-based electronic Industry 5.0 models. In *AI-Driven Smart Industrial Technologies*; IGI Global: Hershey, PA, USA, 2025; Chapter 5, pp. 87–126. [CrossRef]

230. Taj, I.; Jhanjhi, N.Z. Towards Industrial Revolution 5.0 and Explainable Artificial Intelligence: Challenges and Opportunities. *Int. J. Comput. Digit. Syst.* **2022**, *12*, 285–310. [CrossRef]
231. Papacharalampopoulos, A.; Foteinopoulos, P.; Stavropoulos, P. Integration of Industry 5.0 requirements in digital twin-supported manufacturing process selection: A framework. *Procedia CIRP* **2023**, *119*, 545–551. [CrossRef]
232. Yin, X.; Liang, S.; Yu, J.; Zhou, F.; Liu, J. CNN-Based Digital Twin Model for Ultra-High Voltage Direct Current System Loss Measurement. *IEEE Access* **2024**, *12*, 70480–70488. [CrossRef]
233. Chen, G.; Yin, H. Health condition prediction method of the computer numerical control machine tool parts by ensembling digital twins and improved LSTM networks. *Nonlinear Eng.* **2025**, *14*, 20250095. [CrossRef]
234. Tong, S.; Yang, X.; Tong, Z.; Jiang, Y. Digital Twin-Driven Intelligent Monitoring of a Marine Gearbox Based on CNN-LSTM Network. In *Advances in Mechanical Design*; Tan, J., Liu, Y., Huang, H.Z., Yu, J., Wang, Z., Eds.; ICMD 2023, Mechanisms and Machine Science; Springer: Singapore, 2024; Volume 155. [CrossRef]

MDPI AG
Grosspeteranlage 5
4052 Basel
Switzerland
Tel.: +41 61 683 77 34

Machines Editorial Office
E-mail: machines@mdpi.com
www.mdpi.com/journal/machines

www.ingramcontent.com/pod-product-compliance
Lightning Source LLC
Chambersburg PA
CBHW061459210926
51850CB00008B/1768
* 9 7 8 3 7 2 5 8 6 4 2 8 7 *